Interpersonal Coordination and Performance in Social Systems

Interpersonal coordination is an important feature of all social systems. From everyday activities to playing sport and participating in the performing arts, our behaviour is constrained by the need to continually interact with others. This book examines how interpersonal coordination tendencies in social systems emerge, across a range of contexts and at different scales, with the aim of helping practitioners create learning environments to improve performance.

Showcasing the latest research from scientists and academics, this collection of studies examines how and why interpersonal coordination is crucial for success in sport and the performing arts. It explains the complex science of interpersonal coordination in relation to a variety of activities including competitive team sports, outdoor sports, racket sports and martial arts, as well as music and dance. Divided into four sections, these studies offer insight into:

- the nature, history and key concepts of interpersonal coordination
- factors that influence interpersonal coordination within social systems
- interpersonal coordination in competitive and cooperative performance contexts
- methods, tools and devices for improving performance through interpersonal coordination.

This book will be fascinating reading for students, researchers and educators interested in movement science, performance analysis, sport science and psychology, as well as for those working in the performing arts.

Pedro Passos is Assistant Professor at the Faculty of Human Kinetics at the University of Lisbon in Portugal. His research involves the study of the dynamics of interpersonal coordination in team sports. He has written numerous journal articles and book chapters, and is the author or editor of four books. His current research is on interpersonal coordination in social systems and team sports, extending the paradigm of analysis to video games and cooperative tasks, and searching for new methods of analysis in collaboration with researchers in Portugal, across Europe, Singapore, Australia and New Zealand.

Keith Davids is Professor of Motor Learning at the Centre for Sports Engineering Research at Sheffield Hallam University in the UK. His major research interest involves the study of movement coordination and skill acquisition in sport. He is particularly focused on understanding how to design representative learning and performance evaluation environments in sport.

Jia Yi Chow is an Associate Professor at the Physical Education and Sports Science Academic Group, and also Assistant Dean in the Office of Teacher Education at the National Institute of Education (NIE), Nanyang Technological University, Singapore. His area of expertise is in examining multi-articular coordination and a pedagogical approach underpinned by principles from a dynamical systems theory (Nonlinear Pedagogy).

Interpersonal Coordination and Performance in Social Systems

Edited by
Pedro Passos, Keith Davids
and Jia Yi Chow

Routledge
Taylor & Francis Group

LONDON AND NEW YORK

First published 2016 by Routledge

2 Park Square, Milton Park, Abingdon, Oxfordshire OX14 4RN
711 Third Avenue, New York, NY 10017

Routledge is an imprint of the Taylor & Francis Group, an informa business

First issued in paperback 2017

British Library Cataloguing-in-Publication Data
A catalogue record for this book is available from the British Library

Library of Congress Cataloging in Publication Data
Names: Passos, Pedro, editor. | Davids, K. (Keith), 1953- editor. | Chow, Jia
Yi, editor.
Title: Interpersonal coordination and performance in social systems / edited
by Pedro Passos, Keith Davids and Jia Yi Chow.
Description: Abingdon, Oxon ; New York, NY : Routledge, 2016. | Includes
bibliographical references and index.
Identifiers: LCCN 2015047121| ISBN 9781138901087 (hardback) | ISBN
9781315700304 (ebook)
Subjects: LCSH: Sports--Psychological aspects. | Performing
arts--Psychological aspects. | Athletes--Psychology. |
Entertainers--Psychology.
Classification: LCC GV706.4 .I5924 2016| DDC 796.01/9--dc23
LC record available at http://lccn.loc.gov/2015047121

ISBN: 978-1-138-90108-7 (hbk)
ISBN: 978-0-8153-7622-4 (pbk)

Typeset in Times New Roman
by Saxon Graphics Ltd, Derby

I dedicate this book to my wife Ana and my daughter Matilde. Their emotional support was crucial to keep my balance while working.

Pedro

I would like to dedicate this book to the memory of my parents Eric Joseph Davids and Pearl Jeannette Davids.

Keith

This book is dedicated to my wife Jessie and my lovely daughters Kirsten and Kyla.

Jia Yi

Contents

Contributors

David Adé, Université de Rouen, France

Duarte Araújo, Universidade de Lisboa, Portugal

Natalia Balagué, INEFC, University of Barcelona, Spain

Benoît G. Bardy, University of Montpellier and Institut Universitaire de France, France

Eric Berton, Aix-Marseille Université, France

Jean Philippe Boulenger, University of Montpellier and INSERM, France

Jerôme Bourbousson, University of Nantes, France

Delphine Capdevielle, University of Montpellier and INSERM, France

Jia Yi Chow, National Institute of Education, Singapore

Javier Cóteron, Technical University of Madrid, Spain

Laura S. Cuijpers, University of Groningen, the Netherlands

Rick Dale, University of California, USA

Keith Davids, Sheffield Hallam University, UK

Tehran Davis, University of Connecticut, USA

Anouk J. de Brouwer, University of Groningen and VU University Amsterdam, the Netherlands

Harjo J. de Poel, University of Groningen, the Netherlands

Jonathan Del-Monte, University of Montpellier and St. Etienne and University of Montpellier, France

Ana Diniz, Universidade de Lisboa, Portugal

Paula Fitzpatrick, Assumption College, USA

Marie Christine Gély-Nargeot, University of Montpellier and St. Etienne, France

Steven J. Harrison, University of Nebraska, USA

David Higham, Sheffield Hallam University, UK

Robert Hristovski, Sts. Cyril and Methodius University, Macedonia

Rachel W. Kallen, University of Cincinnati, USA

John Kelley, Sheffield Hallam University, UK

J. A. Scott Kelso, Florida Atlantic University, USA, and the University of Ulster, Northern Ireland

Akifumi Kijima, University of Yamanashi, Japan

Michel Laurent, Aix-Marseille Université, France

Rui Jorge Lopes, ISCTE, Instituto Universitário de Lisboa and Instituto de Telecomunicações, Portugal

Tim McGarry, University of New Brunswick, Canada

Ludovic Marin, University of Montpellier, France

Kerry L. Marsh, University of Connecticut, USA

Benjamin R. Meagher, Franklin and Marshall College, USA

Caroline Nicol, Aix-Marseille Université, France

Craig A. Nordham, Florida Atlantic University, USA

Motoki Okumura, Tokyo Gakugei University, Japan

Pedro Passos, Technical University of Lisbon, Portugal

Alexandra Paxton, University of California, USA

Stéphane Raffard, University of Montpellier and St. Etienne and University of Montpellier, France

João Paulo Ramos, Lusófona University and ISCTE, Instituto Universitário de Lisboa, Portugal

Robert Rein, German Sport University, Cologne

Ángel Ric, INEFC, University of Lleida, Spain

Daniel C. Richardson, University College London, UK

Mike Richardson, University of Cincinnati, USA

Kevin W. Rio, Brown University, USA

Robin N. Salesse, University of Montpellier, France

Jacques Saury, University of Nantes, France

Richard C. Schmidt, College of the Holy Cross, USA

Ludovic Seifert, Université de Rouen, France

Rita Sleimen-Malkoun, Aix-Marseille Université, France

Jean-Jacques Temprado, Aix-Marseille Université, France

Régis Thouvarecq, Université de Rouen, France

Carlota Torrents, INEFC, University of Lleida, Spain

Manuel Varlet, University of Montpellier, France

Sylvain Viry, Aix-Marseille Université, France

William H. Warren, Brown University, USA

Auriel Washburn, University of Cincinnati, USA

Jon Wheat, Sheffield Hallam University, UK

Yuji Yamamoto, Nagoya University, Japan

Keiko Yokoyama, Nagoya University, Japan

Preface

Aim of the book

Interpersonal coordination tendencies are an important general feature of social systems, since peoples' behaviours are constrained, among other things, by the need to continually interact with others. These interactive phenomena occur in everyday actions even if people are not aware of them, such as when entering a crowded bus or train, walking on a street while talking to a friend, or driving in a traffic jam. On the other hand, there are contexts where people are intentionally constrained to compete or cooperate with others. Here, interpersonal coordination tendencies are paramount, such as when performing in team games like rugby union or basketball, or when performing in a ballet where two dancers may coordinate with each other to form a creative, dyadic synergy. Competitive and cooperative contexts demand coordination of two or more independent individuals, grounded on variables that only emerge during interactive behaviours. The reciprocal and mutual influence that emerges among performing individuals within these contexts creates a behavioural dependency that is a general feature of interactive behaviours: nonlinearity. This means that it is not possible to predict with total accuracy what other individuals in a complex social system will do next. This feature leads to dyadic and collective system behaviours emerging in performance contexts where variability rules.

Thus, our purpose in this book is to invite numerous experts in this field of work to explain how interpersonal coordination tendencies in social systems emerge, not only based on an individualized analysis, but rather through the continuous interactions between individuals performing in complex social collectives. This rationale has become prominent from research over the last two decades, where a committed group of researchers, academics, teachers and coaches have intensively studied interpersonal coordination tendencies in complex social systems, aiming to enhance the knowledge that helps them to understand and interpret these complex behaviours. The authors in this edited text have accepted our challenge to write an informative set of chapters that constitute this book, written for those who work with, study and research on social systems where interpersonal coordination tendencies are crucial for successful performance. The readership includes undergraduate to graduate level university students, researchers in the areas of movement science, psychology, education and sport science, performance analysts, and professionals involved in the study of human movement that relates to interpersonal coordination (e.g. sports and dance).

The scope of the book

Being intentional or unintentional, desired or undesired, due to competition or cooperation, emergent interpersonal coordination tendencies are strongly constrained by the information that is locally created due to the interactive behaviours of individuals in a complex system. This information can be translated to simple rules that support dyadic (1v1) and collective system behaviours. Finding out what these rules are has been a major challenge for researchers in this field of study. The question is, what are the main reasons to use an interaction-based approach rather than an individual-based approach to study and describe interpersonal coordination tendencies in groups?

Different performance contexts such as team sports, dance, playing in an orchestra, walking side by side with someone or driving in a traffic jam have different demands, task and environmental constraints. Some tasks require competition whereas others require cooperation. To perform some tasks people need to use tools (e.g. rackets, violins, a steering wheel), whereas in other tasks people just need to maintain a 'functional' interpersonal distance to the next individual (e.g. avoid a defender and shoot at a basket in basketball). Despite differences in task and environmental constraints, there are general features that have characterized interpersonal coordination tendencies. Knowing the general features that led someone to coordinate with an(other) individual(s) will help practitioners to create learning environments that will enhance performance.

This book seeks to describe in detail the theoretical basis to understand and explain how interpersonal coordination tendencies emerge in different social contexts at different scales (e.g. local, global). In this field of research, a major issue is to identify variables that bound the way individuals relate to each other. A second issue is to understand how interpersonal coordination tendencies can be measured, not only from an individual point of view but also from an interactive perspective. Additionally, it is important to understand what tools can be used to measure and quantify coordinative behaviours among people. Here we move from the simplest to the most complex tools involving intricate statistical and mathematical procedures to describe and explain interactive behaviours. There is a progression from measuring kinematic variables, such as relative angles and velocities, to relative phase analysis, principal component analysis (PCA), the uncontrolled manifold hypothesis (UCM) and neural networks. We also aim to seek some depth by examining mathematical modelling that might create the future grounding for behavioural predictions.

The philosophy behind this book is to discuss what interpersonal coordination is, how it has been studied, and elucidate the main results and applications from current research programmes in this area of work.

How and why the book is organized as it is

The book is organized to drive the reader from the most general concepts to contextually specific performance situations where interpersonal coordination tendencies emerge. From there, we move on to a section that describes how interpersonal coordination tendencies are measured, and end with a chapter dedicated to discussing future trends in the study of interpersonal coordination.

Therefore, the first part of the book (i.e. Chapters 1 to 5) is dedicated to providing a historical perspective and conceptual background to interpersonal coordination tendencies. The second part of the book (from Chapter 6 to Chapter 16) is context-specific, where the chapter contents are related to how interpersonal coordination tendencies emerge in competitive performance contexts such as team ball games, but also in cooperative contexts like dance and sports where cooperation is an important task constraint required in team performance. This second part also includes research on interpersonal coordination tendencies in special populations, and between humans and other animals such as horse-rider interactions. Then we move on to the third part of the text dedicated to examining factors that influence interpersonal coordination, such as affordances and interpersonal coordination, and verbal communication (Chapters 17 and 18). In the fourth part we have commissioned a set of chapters dedicated to methods, tools and devices that are being used to study interpersonal coordination tendencies (Chapters 19 to 21). As previously stated, the book closes with a chapter dedicated to examining future trends for studies on interpersonal coordination (Chapter 22).

Why you should read this book

Interpersonal coordination has become a growing area of research in human movement sciences over the past decades, building on some excellent work in the biological sciences examining animal behaviours in complex systems. However, for practitioners who are interested in results and applications, the most relevant insights are typically scattered in a significant amount of scientific papers or in detached book chapters. This book seeks to bring together the work of experienced researchers on interpersonal coordination tendencies, capturing in a single book the main issues and applications regarding interpersonal coordination as a social system phenomenon. A second benefit for the reader relates to how the book contents are organized; our aim is to guide the reader from the theoretical origins of interest in interpersonal coordination, to context-specific performance situations, discussing issues in measurement and finishing with an analysis of future trends and applications. A third benefit is in each chapter's structure, which is focused on three main points: theoretical background, research methods and main results, applications.

Unique features of the book

This book is focused on the ability of individuals to relate with others based on the interaction that emerges when two or more individuals need to coordinate their actions, rather than rely on individual ability. That is why we have adopted an interaction-based approach to the study of interpersonal coordination.

The book structure covers different social contexts where interpersonal coordination is paramount, to cover a wide area of research and application. The authors of the individual chapters are from different areas of research and come from all over the world. Due to the development of writing skills in the past they are all experts in maintaining conceptual depth, while seeking methodological rigour in their work and at the same time communicating to readers using a suitably 'light' touch. Without their valued input this addition to knowledge would not occur. As co-editors we contributed four chapters to this book, and we would like to thank the following academics for their most valued and insightful contributions to this text, presented in order of appearance of the chapters:

Kevin W. Rio and William H. Warren from Brown University, USA; Richard C. Schmidt from Holy Cross College and Paula Fitzpatrick from Assumption College, USA; Craig A. Nordham and J. A. Scott Kelso from Florida Atlantic University, USA and Ulster University, Ireland; Tehran Davis from the University of Connecticut, USA; Mike Richardson, Auriel Washburn and Rachel W. Kallen from the University of Cincinnati, USA; Steven J. Harrison from the University of Nebraska, USA; Carlota Torrents from INEFC, University of Lleida, Spain; Natalia Balagué from INEFC, University of Barcelona, Spain; Robert Hristovski from Sts. Cyril and Methodius University, Macedonia; Javier Cóteron from the Technical University of Madrid, Spain; Ángel Ric from INEFC, University of Lleida, Spain; Ludovic Seifert, David Adé and Régis Thouvarecq from the Université de Rouen, France; Jacques Saury and Jerôme Bourbousson from the University of Nantes, France; Duarte Araújo from the Universidade de Lisboa, Portugal; Harjo J. de Poel, Anouk J. De Brouwer and Laura S. Cuijpers from the University of Groningen and VU University Amsterdam, The Netherlands; João Paulo Ramos from Lusófona University and ISCTE, the Instituto Universitário de Lisboa, Portugal; Rui Jorge Lopes from ISCTE, the Instituto Universitário de Lisboa and the Instituto de Telecomunicações, Portugal; Yuji Yamamoto and Keiko Yokoyama from Nagoya University, Japan; Motoki Okumura from Tokyo Gakugei University, Japan; Akifumi Kijima from the University of Yamanashi, Japan; Tim McGarry from the University of New Brunswick, Canada; Jonathan Del-Monte, Stéphane Raffard, Delphine Capdevielle, Manuel Varlet, Robin N. Salesse, Benoît G. Bardy, Jean Philippe Boulenger, Marie Christine Gély-Nargeot and Ludovic Marin from the University of Montpellier and St-Etienne, the University of Montpellier and the Institut Universitaire de France; Rita Sleimen-Malkoun, Jean-Jacques Temprado, Sylvain Viry, Eric Berton, Michel Laurent and Caroline Nicol from Aix-Marseille Université, France; Kerry L. Marsh from the University of Connecticut, USA; Benjamin R. Meagher from Franklin and

Marshall College, USA; Alexandra Paxton and Rick Dale from the University of California, USA; Daniel C. Richardson from University College London, UK; Robert Rein from the German Sport University in Cologne; Ana Diniz from the Universidade de Lisboa, Portugal; and John Kelley, David Higham and Jon Wheat from Sheffield Hallam University, UK.

Pedro Passos, Keith Davids and Jia-Yi Chow

Part I

The nature, historical perspective and conceptual background of interpersonal coordination tendencies

1 Interpersonal coordination in biological systems

The emergence of collective locomotion

Kevin W. Rio and William H. Warren

Collective locomotion is a ubiquitous feature of the natural world, found in living systems great and small – from migrating skin cells to flocks of starlings. In humans, collective locomotion is one form of *interpersonal coordination*, a term that encompasses many of the social and cultural activities that define our species. Whereas studies of interpersonal coordination often focus on the synchronization of rhythmic movement, the domain also includes non-rhythmic coordination, which calls for other tools of analysis. In this introductory chapter our goal is to examine collective locomotion in bird flocks, fish schools and human crowds as a case study for understanding this broader range of interpersonal coordination. Despite vast differences across species – including morphology, neurophysiology, perception and cognition – these seemingly diverse phenomena obey common principles of self-organization and may share similar local mechanisms.

Introduction

Collective locomotion as self-organization

Herring swim together in schools ranging well into the millions of individuals (Misund, 1993), forming characteristic shapes (Partridge, Pitcher, Cullen, & Wilson, 1980) and responding effectively when attacked by predators (Nottestad & Axelsen, 1999). How can this be? How can such large-scale order emerge out of the behaviour of relatively simple animals, with limited perceptual and cognitive capabilities?

The most complete and compelling answer comes from the study of *self-organization* (Couzin & Krause, 2003; Haken, 2006). Camazine et al. (2001) describe self-organization as "a process in which the pattern at the global level of a system emerges solely from numerous interactions among the lower-level components of a system. Moreover, the rules specifying interactions among the system's components are executed using only local information, without reference to the global pattern". Individual fish are perceptually coupled to their nearby neighbours and coordinate swimming with them; through a process of self-organization, these local interactions propagate and give rise to the global patterns of collective motion that characterize the school as a whole. The degree of order in the global pattern is characterized by Haken (2006) as the *order parameter*. Variables that take the ensemble between ordered and disordered states are called *control parameters*.

Levels of analysis

Local interactions give rise to global patterns – this is the central claim of the self-organization approach to collective behaviour. Analyses can be conducted at the local or global level, but must ultimately characterize the links between them. Sumpter, Mann and Pernea (2012) outline a cogent framework for such a research programme. They characterize studies at each level and the links in both directions, from local to global and from global to local. Making sense of these distinctions is crucial to formulating research strategies to understand collective behaviour, so it will be useful to review them in depth.

At the local or 'microscopic' level of analysis, researchers focus on the behaviour of individual agents – be they birds, fish, particles or people. The goal is to understand and ultimately predict how an individual moves in response to its immediate environment, including steering to goals, avoiding obstacles and interacting with other nearby agents. Studies at this level can range from deciphering the perceptual information that guides key behaviours, such as optic flow in walking and flying (Srinivasan, Zhang, Lehrer & Collett, 1996; Warren, Kay, Zosh, Duchon & Sahuc, 2001) or pressure waves in swimming (Partridge & Pitcher, 1980), to the control of steering and obstacle avoidance (Warren & Fajen, 2008), or the social factors that bind two people together while having a conversation (Shockley, Richardson & Dale, 2009). To draw an analogy with physical systems, analysis at the local level is comparable to the study of classical mechanics, where the aim is to work out the kinematic equations of motion for bodies acted upon by a system of forces. Thus, individual agents in a collective occupy the same role as, say, particles in a gas.

The global or 'macroscopic' level, on the other hand, is concerned with the ensemble properties of the collective as a whole. Analysis at the global level is comparable to the study of classical thermodynamics, which deals with large-scale quantities such as temperature, pressure, entropy and energy that are defined over an entire system. The goal is to understand and ultimately predict how such collective variables change over time and react to changes in the environment. The local properties of individual agents are not considered. Global analyses often consist of analysing the overall collective motion pattern, which can range from disordered chaos in swarms of insects (Kelley & Ouellette, 2013) to organized translational and rotational flows in schools of fish (Couzin, 2009).

The key to self-organization lies in understanding how these local and global scales are related. The most common approach is local-to-global analysis, which seeks to understand how simple interactions between agents at the local level give rise to ordered collective phenomena at the global level. Continuing the physical analogy, local-to-global analyses resemble statistical mechanics, which relates macroscopic thermodynamic properties (e.g. heat) to microscopic properties of interacting particles (e.g. velocity). This local-to-global approach often deploys multi-agent simulations of collective behaviour, in which the goal is to reproduce characteristic global patterns by modelling the behaviour of simple agents and their local interactions. By manipulating the "rules" or laws governing these interactions, such simulations can provide insight into the self-organization of collective locomotion, and help explain how complex behaviour emerges from seemingly simple agents. Ultimately, we seek general principles linking the local and global levels, analogous to physical equations that predict the velocity distribution of an ensemble from the equation of motion for individual particles.

A less common, but equally important, approach is global-to-local analysis. Here, the goal is to observe patterns at the global level and use them to infer properties of agents and their interactions at the local level. Regularities in the global patterns or their dynamics place constraints on the rules and models that characterize the local interactions. For example, a collective variable that indexes the degree of coordination between birds in a flock can provide clues about the local coupling, such as how many neighbours each bird responds to (Ballerini et al., 2008), how information about a predator propagates throughout the flock (Cavagna et al., 2010), or whether the coupling yields self-organized criticality in a flock (Bialek et al., 2013). However, due to the 'degeneracy' of large systems there are limitations to this approach: different local rules can give rise to identical global patterns (Vicsek & Zafeiris, 2012; Weitz et al, 2012). Specifying the rules thus requires experimental manipulation of individuals at the local level (Gautrais et al., 2012; Sumpter, Mann & Pernea, 2012).

Models of collective locomotion

The challenge of unraveling the complexity of collective locomotion has attracted an interdisciplinary community of scientists in fields as wide-ranging as biology, physics, mathematics, cognitive science, computer science, robotics, sociology, geography, architecture and evacuation planning. Since the 1970s, computational modelling has served as a common platform for these efforts. Whether studying aggregations of particles, schools of fish, crowds of people or swarms of robots, there is a familiar arc: researchers propose a set of local rules governing individual locomotion, simulate interactions between individuals, and observe the resulting patterns of collective motion. However, the connections between the local rules or global patterns on the one hand and the observations of actual human or animal behaviour on the other are often tenuous. In this section, several landmark models will be described.

One of the most influential models of collective locomotion was introduced in computer animation by Craig Reynolds (1987), drawing from earlier models of fish schooling (Aoki, 1982; Breder, 1954). Reynolds' Boids simulation is an example of agent-based modelling, in which a set of explicit rules defines how each "boid" (agent) behaves and interacts with other agents. His rules included: (1) *repulsion* – boids avoid collisions by moving away from nearby neighbours; (2) *alignment* – boids attempt to match the velocity (speed and heading direction) of nearby neighbours; and (3) *attraction* – boids move toward the centroid of nearby neighbours. Along with some additional assumptions, these three simple rules produce realistic-looking animations of what Reynolds called "happy aimless flocking". The boids form cohesive flocks, maintain a safe distance from each other, and avoid obstacles by splitting up and rejoining.

What accounts for the model's behaviour? First, two of the rules are position-based (repulsion and attraction) and yield a preferred interpersonal distance between agents. This accounts for compact flocks that avoid collisions and reform after splitting around an obstacle. The third rule is velocity-based (alignment), which yields common motion among neighbouring agents. Crucially, collective phenomena emerge from purely local interactions: each boid responds only to neighbours within a fixed radius. Agent-based models thus exhibit self-organization: they demonstrate that global patterns characteristic of collective animal locomotion can, in principle, be reproduced by many locally-interacting agents.

Many subsequent models share these basic components, yielding what Schellink and White (2011) call the *attraction-repulsion framework*. Couzin, Krause, Ruxton and Franks (2002) showed that a model with three similar rules – repulsion from neighbours in a near zone, attraction to neighbours in a far zone and alignment with neighbours in an intermediate zone (see Huth & Wissel, 1992) – could generate qualitatively different global patterns (Figure 1.1). Specifically, varying the radii of the attraction and repulsion zones produces four distinct forms of aggregation: swarm (high cohesion, low alignment), torus (rotational motion about an empty centre), dynamic parallel (loosely aligned translational motion), and highly parallel (highly aligned translational motion).

This finding illustrates that different large-scale behaviours can result from relatively small changes in the parameters of local rules governing individual agents. Such parameter changes might account for discontinuous transitions observed in animal behaviour, such as a sudden rearrangement in response to the detection of a predator. The model also exhibits hysteresis effects; that is, the threshold of the parameter value for changing from one mode to another depends on the current mode. This is a characteristic property of nonlinear systems (Haken, 2006; Kelso, 1995).

From a physical perspective, Vicsek (1995; Czirok, Stanley, & Vicsek, 1997) proposed a stripped-down model of collective motion, the *self-propelled particle* (SPP) model, which only includes a velocity-based alignment rule. All particles are assumed to move at the same speed, and on every time step each particle adopts the mean direction of all neighbours within a fixed radius. Noise is introduced into the coupling by adding a random angle to this mean direction at each time step. Remarkably, this minimal heading-matching model is sufficient to generate a noise-induced phase transition from disordered to translational motion as the noise parameter is decreased. *Canonical-dissipative* models (Ebeling & Schimansky-Geier, 2008; Erdmann, Ebeling & Mikhailov, 2005) add dissipative terms, such as forcing and damping, to the SPP model. This enables

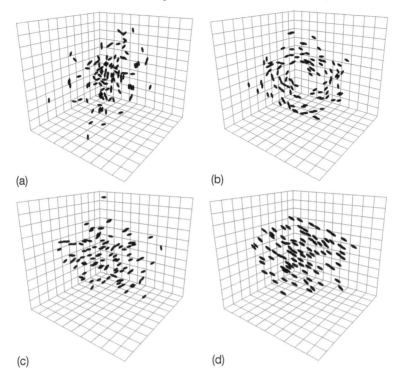

(a)   (b)

(c)   (d)

Figure 1.1 Four different types of aggregation, formed by different sets of parameter values in a single model.

Notes: Clockwise from top right: (A) swarm, (B) torus, (C) dynamic parallel group, and (D) highly parallel group. From Couzin et al. (2002).

them to account for transients to attractor states, and to exhibit noise-induced phase transitions from a translational to a rotational mode, with hysteresis.

In contrast to these local or microscopic models, the earliest models of crowd behaviour focused at the global or macroscopic level, based on analogies with fluid dynamics (Henderson, 1974). A crowd of individuals is treated as a fluid, their interactions approximated by ensemble parameters such as viscosity, pressure and temperature, and collective behaviour is measured by aggregate quantities like density and velocity distributions. Crowd behaviour is characterized using continuum equations, such as variations on the Navier-Stokes equations (Hughes, 2003). However, these *continuum models* generally failed to capture the behaviour of individuals, and made little effort to link the global and local levels.

These results demonstrate the power of using simplified, idealized models to capture patterns of collective locomotion. However, more realistic biological models aim to explain the behaviour of particular classes or species of animals. At issue is the generality of a basic set of local rules, and whether small variations in the rules or their parameter values can account for the variety of species-typical behaviour.

Helbing and Molnár (1995) introduced the most influential agent-based model of human crowd behaviour, the *social force model*. A social force is not a physical force exerted on the pedestrian, like gravity or electromagnetism, but the *"motivation to act"*. Typical forces include attraction to goals, repulsion from obstacles and a preferred walking velocity. The net force on each person in a crowd is thus the sum of these forces. In addition, there is a noise term to account for random variation in pedestrian behaviour, and a maximum possible acceleration.

Helbing and Molnar (1995) used this framework to simulate a variety of qualitative and quantitative crowd phenomena. For example, lanes form in a corridor when pedestrian density exceeds a critical value, and the number of lanes scales linearly with the width of the corridor. Counterflow at bottlenecks arises when two groups of pedestrians try to pass through a constriction in opposite directions. Because of variability in preferred velocity and goal attraction, pedestrians are more or less motivated to move through the constriction. Several pedestrians from one side pass through at a time, until more highly-motivated pedestrians from the other side begin moving in the opposite direction, and the cycle continues until all pedestrians have cleared the bottleneck.

To model high-density crowds with tightly-packed pedestrians, Helbing, Farkas and Vicsek (2000) added physical forces to the social force model. These include a body force that resists compression from neighbours, and a sliding friction force that reduces speed due to physical contact with neighbours. Social force models have been criticized because, while they seem to produce characteristic patterns of crowd behaviour, "they tend to create simulations that look more like particle animation than human movement" (Pechano, Allbeck & Badler, 2007). In other words, behaviour may appear realistic at the global level, but individual motion is unrealistic at the local level.

Although the social force model continues to dominate the field of pedestrian and crowd modeling, alternative approaches have been proposed. Moussaid, Helbing and Theraulaz (2011) described a model based on what they call *cognitive heuristics*, featuring a "synthetic vision" component that uses information in each agent's field of view. For all visual directions within the field of view, the model computes the distance to the first collision in that direction, taking into account the speed and motion direction of other objects. Two heuristics then drive behaviour. First, the agent steers in the direction that provides the most direct path to the goal without colliding with an obstacle. Second, if the time-to-collision in the current travel direction drops below some minimum value, the agent slows down or stops to prevent a collision.

Even though it is simpler than the social force model, the cognitive heuristics model can reproduce many of the same patterns of behaviour, such as lane formation and stop-and-go waves. By adopting a first-person viewpoint, it also simulates situations in which obstacles or other pedestrians are out of view, without additional components or assumptions. However, to deem the model "vision-based" or "cognitive" is problematic, for it makes a number of ungrounded assumptions about human vision. For example, the heuristics assume that the current 3D positions and velocities of neighbours are accurately perceived, and that steering is based on their predicted future positions. These are empirical claims that must be tested before the model can properly be called cognitive.

Grounding models in real behaviour

Agent-based models and continuum approaches have yielded insights into collective locomotion, and provided an existence proof that global patterns can emerge from local interactions through a process of self-organization. However, nearly all such models are predicated on rather ad hoc rules and assumptions, and the resulting behaviour is seldom tested against empirical evidence. Researchers have long acknowledged that theoretical models of collective locomotion must ultimately be grounded in real behaviour. But what kind of data? And for what purposes?

A local-to-global approach begins with experimental data on individual locomotor behaviour, models the control laws that govern local interactions, and then simulates many interacting agents to predict emergent patterns. Reciprocally, a global-to-local approach collects observational data on real collective motion, analyses this data to derive properties of the local interactions, and uses it to test the simulations. We are taking precisely this dual approach to build a *pedestrian model* that can account for human locomotion and crowd behaviour. Fajen and Warren (2003; Warren, 2006) introduced the behavioural dynamics framework, which synthesizes Gibson's (1979) approach to perception and the dynamical systems approach to action (Kelso, 1995; Kugler & Turvey, 1987). The aim is to understand how information about the environment modulates the dynamics of action to yield emergent, adaptive behaviour. At the individual level, Fajen and Warren (2003, 2007) decomposed the problem of locomotor control into four basic components – steering to a stationary or moving target, and avoiding a stationary or moving obstacle – modelling each as a second-order dynamical system. Combining these components can account for locomotor trajectories in more complex environments without the need for explicit path planning (Warren & Fajen, 2008).

The next step on the road from local to global is to consider interpersonal interactions between pairs of pedestrians (dyads). Cohen (2009) found that the steering dynamics model, originally developed for inanimate objects, could also account for pursuit and evasion of other pedestrians. The bridge to collective locomotion is to investigate interpersonal coordination: whether there are control laws governing pedestrian interactions analogous to the "rules" assumed in previous models. Do pedestrians adopt a common walking speed and heading direction (velocity-based alignment)? Is there a preferred interpersonal distance (position-based attraction/repulsion)? We are conducting a series of experiments to address these questions, building up from dyads (Figure 2) to pedestrian groups. Rio, Rhea and Warren (2014) found that a follower matches a leader's speed in accordance with a simple dynamical model, and does so by nulling the leader's optical expansion. Dachner and Warren (2014) found that a follower

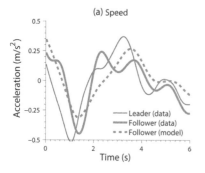
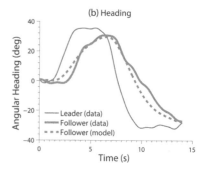

Figure 1.2 Sample time series of two participants (solid) and a dynamical model (dashed) in a following task.

Notes: (a) Speed control: follower acceleration is highly correlated with leader acceleration (mean of 462 trials, r: 0.68), with a slight visual-motor delay (M = 420 ms, SD = 373 ms). Speed-matching model captures follower acceleration (mean r: 0.67 m/s2). (b) Heading control: follower heading is highly correlated with leader heading (mean of 32 trials, r: 0.92), with a long visual-motor delay (M = 985 ms, SD = 786 ms). Heading-matching model reproduces follower heading (mean r: 0.71 deg). From Rio, Rhea, & Warren (2014) and Dachner & Warren (2014), respectively.

matches a leader's walking direction in accordance with a very similar model. Thus there appears to be a human analogue of the alignment rule. On the other hand, we have seen little evidence of a preferred interpersonal distance, casting doubt on position-based attraction/repulsion rules. These findings provide basic control laws for multi-agent simulations of crowd dynamics (Bonneaud & Warren, 2014).

To generate collective behaviour, all models assume that an individual agent interacts with multiple neighbours within some *neighbourhood*. However, the structure of a neighbourhood is unknown. To which neighbours is an agent visually coupled? Does coupling strength depend on distance and direction? How are the influences of multiple neighbours combined? Answering these questions requires experimental manipulation of neighbours and measurement of an agent's response. One approach is the use of biomimetic robots to probe the behaviour of animal collectives, such as Faria et al.'s (2010) "Robofish" (see also Marras & Porfiri, 2012). A complementary approach, which can be used to probe an individual's neighbourhood, is to manipulate multiple neighbours using virtual reality techniques.

We are investigating the neighbourhood of human pedestrians by immersing a walking participant in a virtual crowd. By perturbing the walking speed or direction of a subset of virtual neighbours, and recording the participant's speed and heading adjustments, we can infer the local coupling to multiple neighbours (Rio & Warren, 2014). We find that the participant's response increases linearly with the proportion of perturbed neighbours, but decreases with neighbour distance. The coupling can be modeled as a weighted average over the neighbourhood: a pedestrian is coupled to multiple neighbours, their influence is averaged, but coupling strength decays with distance.

Reciprocally, a global-to-local approach begins with observational data on collective locomotion. One of the most thorough observational studies to date used stereoscopic videography to record the 3D positions of thousands of starlings in a flock (Ballerini et al., 2008), allowing the authors to estimate the structure of a starling's neighbourhood. Nearly all models assume a *metric neighbourhood*, such that an agent is coupled to all neighbours within a fixed radius (e.g. in metres), and coupling strength decreases with metric distance. In a *topological neighbourhood*, by contrast, an agent is coupled to a fixed number of nearest neighbours regardless of how far away they are, and the coupling strength decreases with ordinal distance. Ballerini et al. (2008) found that the "anisotropy factor", a measure of the degree to which neighbours affect a given bird's behaviour, asymptoted at the sixth or seventh nearest neighbour. Crucially, this value remained constant despite changes in the flock's density: a bird was coupled to its six or seven nearest neighbours regardless of how far away they were, consistent with the topological hypothesis. Young et al. (2013) subsequently showed that this number of neighbours was optimal for robust consensus within the flock. Cavagna et al. (2010; Bialek et al., 2013) found that starling flocks exhibit scale-free correlations; that is, the response of a given bird is affected by all other birds via propagation, regardless of flock size.

While a topological neighbourhood may be adaptive for bird flocks and fish schools that must maintain cohesion despite large variations in density, terrestrial species may have a different neighbourhood structure. Our preliminary results for a pedestrian in a virtual crowd suggest that the human neighbourhood is not topologial (Rio & Warren, 2014). A participant's response to a subset of perturbed neighbours decays to zero by a distance of 4m in a crowd, and depends significantly on crowd density, contrary to the topological hypothesis. However, the radius also appears somewhat flexible, since the participant responds to a ring of neighbours up to 8m away. We are currently pursuing this issue.

Quantifying global coordination

The signature of collective behaviour is the emergence of global order among individuals in the collective, which Haken (2006) characterized as the order parameter. It is thus important to develop a measure of the degree to which individuals are coordinated with one another. Changes in this measure can reveal the influence of control parameters that take the collective through a phase transition between ordered and disordered states, as well as the response to environmental conditions such as attacks from predators. Several measures of global coordination have been introduced.

Group *polarization* and group *angular momentum* purportedly capture the overall linear and circular alignment of a swarm's heading and direction, respectively (Couzin et al., 2002). Both measures range from 0 to 1, with 0 representing no coordination and 1 representing perfect coordination. Group polarization (P) is essentially the vector sum of a unit velocity vector for each of N individuals (i.e. the heading direction, with a magnitude of 1), divided by N. If everyone's heading is perfectly aligned, the result will be a vector in that direction with a magnitude of 1, the maximum polarization. However, the minimum value of 0 can be produced in multiple ways. If individual headings are randomly distributed, they will cancel each other out, on average. Alternatively, if there are two streams of pedestrians moving in opposite directions, they will also cancel; thus, the overall group polarization will be 0, even though there is a high degree of organization.

Group angular momentum (M) measures the degree to which pedestrians orbit about the centre of mass of the group. Essentially, for each of N individuals, it computes a quantity proportional to the angle between the unit velocity vector and a unit vector pointing toward the crowd's centre of mass; taking its sum and dividing by N yields a magnitude between 0 and 1. If all pedestrians are moving on a circle about the centre of mass, their velocity vectors will be tangent to the circle and perpendicular to their centre vectors, and the angular momentum will be the maximum value of 1. However, once again the minimum value of 0 can be produced in multiple ways, including random motion, rotation about the centre in opposite directions, or linear flow. In addition, this measure is somewhat unreliable because it is overly dependent on a closed circular motion.

The combination of polarization and angular momentum can describe a wide range of motion patterns. Couzin et al. (2002) used them to characterize the four types of motion shown in Figure 1. The 'swarm' pattern (Figure 1a), largely random motion, is characterized by low polarization and low angular momentum. The torus (Figure 1b), in which agents circle around an empty centre, is characterized by low polarization and high angular momentum. The dynamic parallel (Figure 1c) and highly parallel (Figure 1d) patterns are both characterized by relatively high polarization and low angular momentum, with polarization close to 1 for the latter. These measures have been applied to cases ranging from tissue cell migration (Szabo, Szöllösi, Gönci, Jurányi, Selmeczi & Vicsek, 2006) to human crowds (Rio & Warren, 2013), indicating their general applicability. However, other measures based on the mean difference in heading between pairs of pedestrians may be more sensitive to the degree of organization within a group.

Conclusion

The purpose of this chapter was to introduce some of the key issues and approaches to research on collective behaviour, using collective locomotion as a case study. At this point, let us take a step back and consider some general conclusions that may apply to the study of interpersonal coordination more broadly.

First, the distinction between local and global levels of analysis is one that all researchers interested in interpersonal coordination must face, and this is especially true for interactions between many people. Consider the example of rowing. At the local level, it is important to understand the properties of individual rowers, as well as the local interactions between them. What is the preferred frequency of each rower, the informational coupling between them, and how does each rower coordinate their behaviour with others? But it is also important to consider the global level. What are appropriate measures of the team's collective behaviour? How does this global behaviour depend on the properties of the local coupling? Most importantly, can we formulate general principles that link the local and global levels? Similar questions can be asked for any coordinated activity between people.

A second general point is that understanding collective behaviour and interpersonal coordination requires a delicate balance between theoretical and empirical approaches. Theoretical work, such as the formal modelling described above, is essential, because it gives the field a framework within which empirical results can be situated. Theory drives us to ask the relevant questions, and it guides the choice of which experiments to run and which variables to manipulate and measure. But analysing empirical data, both from experiments and observations, is equally critical. If our ultimate goal is to understand how people and other animals coordinate with one another, it is not enough to propose ways in which this *could* be done; eventually, we must determine how it is actually done in real systems.

A third general consideration is the importance of developing appropriate measures and analyses to quantify coordination between individuals. Interpersonal coordination involves both rhythmic and non-rhythmic behaviour, and applying the relevant analytic tools may reveal common principles underlying seemingly disparate phenomena. For example, relative phase may be the relevant order parameter and coupled-oscillator dynamics the relevant model for rhythmic behaviour, whereas polarization and statistical mechanics may be more relevant to collective locomotion. In each case, identifying the appropriate order parameter and analytic tools make it possible to determine how potential control parameters affect the degree of coordination, and may lead us to the fundamental mechanisms of coordination governing a system.

As the other chapters in this book demonstrate, interpersonal coordination is a complex, multifaceted and compelling problem. It is our hope that the diversity of phenomena can ultimately be understood within a framework of self-organized collective behaviour, by bringing the relevant concepts, analytic tools and theoretical models to bear on how people coordinate with one another.

References

Aoki, I. (1982) A simulation study on the schooling mechanism in fish. *Bulletin of the Japanese Society of Scientific Fisheries*, 8, 1081–1088.

Ballerini, M., Cabibbo, N., Candelier, R., Cavagna, A., Cisbani, E., Giardina, I., Lecomte, V., Orlandi, A., Parisi, G., Procaccini, A., Viale, M. & Zdravkovic, V. (2008) Interaction ruling animal collective behavior depends on topological rather than metric distance: Evidence from a field study. *Proceedings of the National Academy of Sciences*, 105(4), 1232–1237.

Bialek, W., Cavagna, A., Giardina, I., Mora, T., Pohl, O., Silvestri, E., Viale, M. & Walczak, A. (2014) Social interactions dominate speed control in poising natural flocks near criticality. *Proceedings of the National Academy of Sciences*, *111*(20), 7212–7217.

Bonneaud, S. & Warren, W.H. (2014) An empirically-grounded emergent approach to modeling pedestrian behavior. In U. Weidmann, U. Kirsch and M. Schreckenberg (Eds), *Pedestrian and Evacuation Dynamics 2012*, Springer International: New York, p. 625–637.

Breder, C.M. (1954) Equations descriptive of fish schools and other animal aggregations. *Ecology*, *35*, 361–370.

Camazine, S., Deneubourg, J.-L., Franks, N.R., Sneyd, J., Theraulaz, G. & Bonabeau, E. (2001) *Self-Organization in Biological Systems*. Princeton University Press: Princeton, NJ.

Cavagna, A., Cimarelli, A., Giardina, I., Parisi, G., Santagati, R., Stefanini, F. & Viale, M. (2010) Scale-free correlations in starling flocks. *Proceedings of the National Academy of Sciences*, *107*(26), 11865–11870.

Cohen, J.A. (2009) *Perception of pursuit and evasion by pedestrians*. Unpublished PhD thesis Brown University, Providence, RI.

Couzin, I.D. (2009) Collective cognition in animal groups.*Trends in Cognitive Sciences*, *13*(1), 36–43.

Couzin, I.D. & Krause, J. (2003) Self-organization and collective behavior in vertebrates. *Advances in the Study of Behavior*, *32*, 1–75.

Couzin, I.D., Krause, J., James, R., Ruxton, G.D. & Franks, N.R. (2002) Collective memory and spatial sorting in animal groups. *Journal of Theoretical Biology*, *218*, 1–11.

Czirok, A., Stanley, H.E. & Vicsek, T. (1997) Spontaneously ordered motion of self-propelled particles. *Journal of Physics A: Mathematical and Theoretical*, *30*, 1375–1385.

Dachner, G. & Warren, W.H. (2014) Behavioral dynamics of heading alignment in pedestrian following. *Transportation Research Procedia*, *2*, 69–76.

Ebeling, W. & Schimansky-Geier, L. (2008) Swarm dynamics, attractors and bifurcations of active Brownian motion. *The European Physical Journal Special Topics*, *157*(1), 17–31.

Erdmann, U., Ebeling, W. & Mikhailov, A. (2005) Noise-induced transition from translational to rotational motion of swarms *Physical Review E*, *71*, 051904-051901–051904-051907.

Fajen, B.R. & Warren, W.H. (2003) Behavioral dynamics of steering, obstable avoidance, and route selection. *Journal of Experimental Psychology: Human Perception and Performance*, *29*(2), 343–362.

Fajen, B.R. & Warren, W.H. (2007) Behavioral dynamics of intercepting a moving target. *Experimental Brain Research*, *180*(2), 303–19.

Faria, J.J., Dyer, J.R., Clément, R.O., Couzin, I.D., Holt, N., Ward, A.J., ... & Krause, J. (2010) A novel method for investigating the collective behaviour of fish: Introducing 'Robofish'. *Behavioral Ecology and Sociobiology*, *64*(8), 1211–1218.

Gautrais, J., Ginelli, F., Fournier, R., Blanco, S., Soria, M., Chaté, H. & Theraulaz, G. (2012) Deciphering interactions in moving animal groups. *PLoS Comput Biology*, *8*(9).

Gibson, J.J. (1979) *The ecological approach to visual perception*. Houghton Mifflin: Boston.

Haken, H. (2006) *Information and Self-organization: A Macroscopic Approach to Complex Systems*. Springer: New York, NY.

Helbing, D. & Molnár, P. (1995) Social force model for pedestrian dynamics. *Physical Review E*, *51*(5), 4282–4286.

Helbing, D., Farkas, I. & Vicsek, T. (2000) Simulating dynamical features of escape panic. *Nature*, *407*(6803), 487–90.

Henderson, L.F. (1974) On the fluid mechanics of human crowd motion. *Transportation Research*, *8*(6), 509–515.

Hughes, R.L. (2003) The flow of human crowds. *Annual Review of Fluid Mechanics*, *35*, 169–182.

Huth, A. & Wissel, C. (1992) The simulation of the movement of fish schools. *Journal of Theoretical Biology*, *156*, 365–385.

Kelley, D.H. & Ouellette, N.T. (2013) Emergent dynamics of laboratory insect swarms. *Nature Scientific Reports*, *3*, 1073.

Kelso, J.A.S. (1995) *Dynamic Patterns: The Self-Organization of Brain and Behavior.* MIT Press: Cambridge, MA.

Kugler, P.N. & Turvey, M.T. (1987). *Information, Natural Law, and the Self-Assembly of Rhythmic Movement.* Erlbaum Associates: Hillsdale, NJ.

Marras, S. & Porfiri, M. (2012) Fish and robots swimming together: attraction towards the robot demands biomimetic locomotion. *Journal of The Royal Society Interface*, *9*(73), 1856–1868.

Misund, O.A. (1993) Dynamics of moving masses: variability in packing density, shape, and size among herring, sprat, and saithe schools. *ICES Journal of Marine Science*, *50*(2), 145–160.

Moussaïd, M., Helbing, D. & Theraulaz, G. (2011) How simple rules determine pedestrian behavior and crowd disasters. *Proceedings of the National Academy of Sciences*, *108*(17), 6884–6888.

Partridge, B.L. & Pitcher, T.J. (1980) The sensory basis of fish schools: relative roles of lateral line and vision. *Journal of Comparative Physiology*, *135*, 315–325.

Partridge, B.L., Pitcher, T., Cullen, J.M. & Wilson, J. (1980) The three-dimensional structure of fish schools. *Behavioral Ecology and Sociobiology*, *6*(4), 277–288.

Pechano, N., Allbeck, J.M. & Badler, N.I. (2007) Controlling individual agents in high-density crowd simulation. *Proceedings of the 2007 ACM SIGGRAPH/Eurographics Symposium on Computer Animation*, 99–108.

Reynolds, C.W. (1987) Flocks, herds and schools: A distributed behavioral model. *Computer Graphics*, *21*(4), 25–34.

Rio, K. & Warren, W.H. (2013) Visually-guided collective behavior in human swarms. *Journal of Vision*, *13*(9), 481.

Rio, K.W., Rhea, C.K. & Warren, W.H. (2014) Follow the leader: Visual control of speed in pedestrian following. *Journal of Vision*, *14*(2), 4:1–16.

Rio, K.W. & Warren, W.H. (2014) The visual coupling between neighbors in real and virtual crowds. *Transportation Research Procedia*, *2*, 132–140.

Schellinck, J. & White, T. (2011) A review of attraction and repulsion models of aggregation: Methods, findings and a discussion of model validation. *Ecological Modeling*, *222*, 1897–1911.

Shockley, K., Richardson, D.C. & Dale, R. (2009) Conversation and coordinative structures. *Topics in Cognitive Science*, *1*(2), 305–319.

Srinivasan, M., Zhang, S.W., Lehrer, M. & Collett, T. (1996) Honeybee navigation en route to the goal: Visual flight control and odometry. *Journal of Experimental Biology*, *199*(1), 237–244.

Sumpter, D.J., Mann, R.P. & Perna, A. (2012) The modelling cycle for collective animal behaviour. *Interface Focus*, rsfs20120031.

Szabo, B., Szöllösi, G.J., Gönci, B., Jurányi, Z., Selmeczi, D. & Vicsek, T. (2006) Phase transition in the collective migration of tissue cells: experiment and model. *Physical Review E, 74*(6), 061908.

Vicsek, T. (1995) Novel type of phase transition in a system of self-driven particles. *Physical Review Letters, 75*(6), 4–7.

Vicsek, T. & Zafeiris, A. (2012) Collective motion. *Physics Reports, 517*(3), 71–140.

Warren, W.H. (2006) The dynamics of perception and action. *Psychological Review, 113,* 358–389.

Warren, W.H. & Fajen, B.R. (2008) Behavioral dynamics of visually-guided locomotion. In A. Fuchs & V. Jirsa (Eds), *Coordination: Neural, behavioral, and social dynamics.* Springer: Heidelberg.

Warren, W.H., Kay, B.A., Zosh, W.D., Duchon, A.P. & Sahuc, S. (2001) Optic flow is used to control human walking. *Nature Neuroscience, 4*(2), 213–216.

Weitz, S., Blanco, S., Fournier, R., Gautrais, J., Jost, C. & Theraulaz, G. (2012) Modeling collective animal behavior with a cognitive perspective: A methodological framework. *PLoS One, 7*(6).

Young, G.F., Scardovi, L., Cavagna, A., Giardina, I. & Leonard, N.E. (2013) Starling flock networks manage uncertainty in consensus at low cost. *PLoS Comput Biol, 9*(1), 1–7.

2 The origin of the ideas of interpersonal synchrony and synergies

Richard C. Schmidt and Paula Fitzpatrick

Acknowledgments

This research was supported by National Institutes of Health Grant R01GM105045.

Introduction

Investigations of social motor or interpersonal coordination have burgeoned during the first decades of the new millennium. Research is being performed from many perspectives to try to understand how we move together and what it means for behavioural domains as diverse as sport science and psychotherapy. In truth, though, research in this area has been growing continually since technological advances allowed human movement behaviour to be recorded in the 1960s, developing into what it is today: an interdisciplinary domain in which researchers in clinical psychology, social psychology, developmental psychology, human movement science, neuroscience and complexity theory are all working to understand both the processes involved in interpersonal coordination as well as the significance of these processes across multiple domains.

Interpersonal alignment: interactional synchrony and behavioral matching

Interpersonal coordination refers to how the behaviour of two or more individuals is brought into alignment. However, for all of the diversity in the kinds of interpersonal coordination being studied (which form the contents of the chapters of this book), all the research must find a way to operationalize the 'alignment' of two persons' behaviour. The early history of the study of interpersonal coordination is in some sense a history of how this 'capturing' of coordination was done.

The modern investigation of the structure of bodily coordination in interpersonal interactions began with two papers on interactional synchrony by William S. Condon in the late 1960s (Condon & Ogston, 1966; 1967). Although Condon was very interested in how interpersonal coordination was disturbed in psychological pathology (e.g. in schizophrenia and learning disabilities), he studied the structure of interpersonal coordination for the first time and attempted to provide a theory of it. Two key elements formed the foundation of his investigation. First, Condon was able to evaluate the kinematic structure of bodily movement unfolding in time by using film recordings or "cinematographic techniques" to capture the complexity of the behavioural stream underlying social interactions (Condon, 1970). Using a projector which allowed film to be moved back and forth by hand, Condon performed a frame-by-frame "microanalysis" of interaction at a fine-grained (24 or 48 frames/s) time scale. Second, following the work of Scheflen (1964) and Birdwhistell (1970) and later verified by Dittmann and Llewellyn (1969), Condon noted that hierarchically more primitive to the interpersonal coordination phenomenon of interactional synchrony was the *intra*personal coordination phenomenon of self synchrony, namely, that speakers move various parts of their body synchronously and in time with their own speech:

> There are sustained relationships of movements of the body motion bundles, which occur isomorphically with the articulatory units of speech.
>
> (Condon, 1982, pg. 55)

Moreover, a person's behavioural stream is composed of a sequencing of motion bundle "processes units or quantal forms of movement" at different time scales, defined by linguistic units such as phone, syllable, word, or phrase, forming a rhythm hierarchy of behavior (Condon, 1975, pg. 42; Figure 1).

The film analysis, using a frame-by-frame microanalysis of intrapersonal process units, allowed Condon to identify the rhythm hierarchies of behaviour – the sequencing of processing units at different time scales – of both the speaker and listener in an interaction. Of course, what Condon discovered in addition was that although the listener was found to be moving less than the speaker, there was alignment in time of each person's rhythm hierarchy – that is to say, there was *interactional synchrony*. The synchronization between speaker and listener was thus defined by the sequential co-occurrence of each person's process unit boundaries. Consequently, Condon provided a rich description of the time unfolding nature of the behaviour within a conversation interaction, revealing that intrapersonal bodily coordination was nested within interpersonal bodily coordination:

> Analysis revealed harmonious or synchronous organizations of change between body motion and speech in both intra-individual and interactional behavior. Thus the body of the speaker dances in time with his speech. Further, the body of the listener dances in rhythm with that of the speaker.
>
> (Condon & Ogston, 1966, pg. 338)

Interest in Condon's interactional synchrony work was founded upon studies that investigated how such interpersonal coordination broke down in pathology and had its roots in mother infant interactions. Although some early work looked at behavioural matching in psychotherapy interactions and found that rapport between therapist and client was related to the degree of congruence of behavioural postures (e.g. Charney, 1966), Condon and Ogston (1966) looked at interactional movements in people with schizophrenia, and the intrapersonal movement abnormalities such as lack of head movement, fixed gaze and "self-dissynchrony" that formed the basis for a breakdown in interactional synchrony between therapist and client. This work has provided a foundation for a number of contemporary investigations of a breakdown in social motor synchrony in schizophrenia (Ramseyer & Tschacher, 2011; Varlet et al., 2012). In later investigations, Condon proposed that such asynchrony was a key aspect of many other pathologies including dyslexia, learning disability and autism (Condon, 1982). One study (Condon, 1975) specifically argued that disabled children not only responded in an asynchronous fashion but also exhibited multiple responding – that is, responding to auditory inputs in a delayed and repetitive fashion as if the stimulus had occurred more than once. However, subsequent research (Oxman, Webster & Konstantareas, 1978) was unable to replicate the multiple responding effect, although some evidence was found for dyssynchronous responding. Contemporary investigations of movement asynchrony in child pathologies (Fitzpatrick, Diorio, Richardson & Schmidt, 2013; Isenhower et al., 2012; Marsh et al., 2013; Trevarthen & Delafield-Butt, 2013) seem to provide support for the general idea of Condon's movement asynchrony hypothesis.

Condon's work on interactional synchrony garnered much interest, influenced others at that time (e.g. Kendon, 1970) and has formed the foundation of considerable subsequent research. However, this initial work on interactional synchrony suffered from a methodological deficiency and was consequently (and justly) criticized early on. Condon's claims about both interactional synchrony and self-synchrony suggested a grand omnipresence: "interactional synchrony … seems to occur constantly during normal interaction" (Condon & Ogston, 1971, p. 159), and "[n]o random movements, no non-shared changes, were found to occur during interactional synchrony" (Condon & Ogston, 1967, p. 229). The concern, however, was with the operational definition, and consequently with the reliability (consistency) and construct validity (accuracy) of behavioural synchrony measurement (Cappella, 1981; Rosenfeld, 1981). Condon's technique of microanalysis was performed by coding frame-by-frame movement boundaries and noting how these boundaries corresponded within a person for self-synchrony and across two people for interactional synchrony. However, there were questions about whether people were reliably coding the boundaries of movements and whether spurious movement co-occurrences happened just by chance – whether the construct was valid. What was needed and not performed by Condon was both a check of rater reliability and a method to statistically control for a baseline of chance co-occurrences to parry these two concerns. For the latter, as Deese (1971) pointed out early on: "They should compare synchronous sound-visual records with non-synchronous records. They need to show us that the synchrony is actually there." Unfortunately, when such control studies were performed (McDowall, 1978a, 1978b) interactional synchrony as measured by the microanalysis technique seemed to be problematic in terms of both reliability and construct validity. Although some (Gatewood & Rosenwein, 1981) questioned whether these empirical tests used the same assumptions and methods as Condon, for quite some time these studies were effective in convincing people that behavioural (self and interactional) synchrony estimated using the microanalysis film technique was a pseudo phenomenon.

With the notion of fine-grained, interactional synchrony falling out of favour, the period from the late 1970s through to the end of the 1980s was marked by investigations of interpersonal coordination using grosser, more molar body units. During this period the work of LaFrance (1982) demonstrated the important linkage between interpersonal bodily coordination and psychological connectedness as evaluated by rapport. Using measures coding posture similarity and mirroring, LaFrance (LaFrance, 1979; LaFrance & Broadbent, 1976) investigated interpersonal coordination in a classroom setting and found a significant relationship between interpersonal coordination and self-reported rapport. This important work laid the foundation for more contemporary investigations of the psychological consequences of interpersonal bodily connectedness (Miles, Nind & Macrae, 2009; Tickle-Degnen, 2006).

The re-entry of interpersonal coordination as measured via synchrony that is independent of behavioural content (Newtson, 1994) occurred in the late 1980s with the work of Frank Bernieri. Berneiri (1988) used human perceivers as a means for estimating the degree of synchrony in an interaction by asking them to rate such properties as simultaneous movement, tempo similarity and smoothness. This approach assumes that temporal unity and harmony in social interactions have a "gestalt"-like quality (Newtson, Hairfield, Bloomingdale & Cutino, 1987) to which humans are perceptually attuned. Bernieri's synchrony ratings obviated the two methodological problems of earlier interactional synchrony research, namely reliability and construct validity. Because the synchrony ratings were easier to perform than earlier methods, Bernieri was able to use multiple raters and perform inter-rater reliability checks (Bernieri, Reznick, & Rosenthal, 1988; Bernieri, 1988). Moreover, given advances in video technology, Bernieri was able to employ a pseudo synchrony control methodology to evaluate the reality of the interactional synchrony being perceived. The ratings of these pseudo synchrony control videos by the judges provided a baseline level of chance synchrony to compare with the judge's synchrony ratings of real interactions. Using this paradigm, Bernieri (1988; Bernieri et al., 1988; 1994) found that people rated real social interactions as having more synchrony than the pseudo synchrony baseline conditions, and found that these ratings increased with psychological properties of the interactors' relationship.

Of course, in spite of their inter-rater reliability and ability to discriminate real from chance synchrony, as an operational definition of interactional synchrony, synchrony ratings are subjective rather than objective measurements and do not have much content (validity) as they are not fine-grained but holistic. Although this rating methodology is less laborious, it does not provide as detailed a measurement of the coordination, and hence it limits the degree to which researchers can understand the processes underlying synchronization. These limitations have been remedied using computer-driven motion capture technology using optical or magnetic sensors which proliferated with increasing computer power in the late 1980s and 1990s.

These limitations have also been addressed by video-based motion capture methodologies that have evolved from Condon's frame-by-frame microanalysis. With the enhancement of digital video technology a number of researchers (Kupper, Ramseyer, Hoffman, Kalbermatten & Tschacher, 2010; Paxton & Dale, 2013; Ramseyer & Tschacher, 2011; Schmidt, Morr, Fitzpatrick, & Richardson, 2012; Schmidt, Nie, Franco, & Richardson, 2104) are measuring the amount of frame-to-frame pixel change to assess the moment-to-moment degree of bodily activity. The amount of pixel change between adjacent video frames corresponds to the amount of bodily activity of a participant if they are the only source of movement in that part of the frame. Although this measure is a measure of the amount of bodily activity not the movement of the body in space, the resulting time series are much more representative of the objective structure of behaviour across time than behavioural coding of punctate properties such as head nods or gestalt-like qualities such as interaction smoothness. Perhaps most importantly, the quantitative nature of the time series allows the whole battery of techniques from time series analysis and nonlinear dynamics to be brought to bear on the analysis of interactional synchrony. Such analyses were awaiting researchers in the 1990s.

Dynamics of synchrony: interpersonal synergies

The story of the research on interpersonal coordination in the 1970s and 1980s was all about getting the methodology correct – how to go about 'capturing' the interpersonal coordination, and how one could reliably and validly measure it. Condon's early methodological missteps in looking at the alignment of behaviour in time (interactional synchrony) led to a focus on the alignment of behaviour in terms of kind (behavioural matching). However, eventually technological advances made the fine-grained measurement of the temporal unfolding of behaviour possible and unbiased, and allowed researchers to investigate interpersonal coordination in terms of a synchronization of bodily activity independent of behavioural content (Grammer, Kruck & Magnusson, 1998; Newtson, 1994).

From the 1980s into the 1990s, the focus of the research transitioned to how to understand the processes underlying interpersonal coordination. Researchers began to interpret social synchronization in terms of the growing field of biological rhythms and their entrainment (Glass & Mackey, 1988). The text entitled *Interaction Rhythms* (Davis, 1982), which stemmed from an interdisciplinary conference that took place in 1980 at Columbia University, contains papers that are illustrative of the effort at this time. However, it was not until the end of the decade that the theory used to understand and model the entrainment of biological rhythms, namely the dynamics of synchronization, was proposed to understand the synchrony seen in social interactions. Newtson (Newtson et al., 1987), who was a social psychologist with leanings towards ecological psychology, proposed that the entrainment in social interactions could be understood as being governed by the same organizational processes that govern the rhythmic coordination of limbs in a single person, namely the dynamical processes of oscillatory entrainment:

> If action consists of coupled oscillator units, interaction between persons may consist of the coupling of such units across two persons. That is, since these are open systems, two persons may configure to the same set of environmental constraints and coordinate their behavior in the same manner as they coordinate any other multiple-configuration behavior, such as walking... If the two persons are acting jointly (e.g., carrying common objects), we would also expect their behavior to be coupled at 0° phase. Similar coupling, but at 180° phase, would produce alternating behavior on the part of the two persons, more like the taking of turns often observed in interaction.
>
> (Newtson et al., 1987, pg. 219)

With this statement the hypothesis that interpersonal coordination was governed by dynamical processes operating across perceptual information – the notion of a dynamical interpersonal synergy – was born.

An understanding of Newtson's initial hypothesis and the methodologies used to investigate it in subsequent studies requires a brief review of the dynamical perspective on rhythmic interlimb coordination that was being developed in the 1980s. The starting point for this perspective is that limbs are assembled into oscillators and that rhythmic interlimb coordination may be understood in terms of the self-organizing entrainment processes of oscillators (Kelso, Holt, Rubin, & Kugler, 1981). Haken, Kelso, and Bunz (1985) provided an explicit mathematical model (the HKB model) of this entrainment process. Kelso and colleagues initially developed this equation to understand the behavioural transitions that naturally occur in interlimb coordination (such as in quadruped gait transitions) and which can be investigated in laboratory bimanual tasks (Kelso, 1984). In particular, the most striking phenomenon was a bimanual phase mode transition: anti-phase coordination of wrist or index fingers becomes increasingly less stable as the frequency of oscillation is increased, eventually breaking down and leading to a transition to in-phase coordination. The HKB model allowed these behavioural transitions to be understood using the language of synergetics (Haken, 1977) as dynamical phase transitions (i.e. a general nonlinear way that a dynamical system reorganizes after destabilization), in which the qualitative state of an order parameter (in this case, relative phase) changes its state *catastrophically* as a consequence of the scaling of a control parameter – in this case, the frequency of oscillation.

However, the HKB model also made predictions about the coordination patterns that occur when the two oscillators are not identical in their frequencies through the scaling of another dynamical control parameter, namely frequency detuning or the difference in the inherent frequencies (eigenfrequencies) of the oscillators (Kelso, DelColle & Schöner, 1990). Experimentally, frequency detuning was manipulated by varying the inertial loadings of the limbs to be coordinated (Turvey, Rosenblum, Schmidt & Kugler, 1986; Kelso & Jeka, 1992; Fitzpatrick, Schmidt & Carello, 1996). One methodology used to employ this manipulation is a paradigm developed by Kugler and Turvey (1987) in which hand-held pendulums are swung at the wrist joint in the sagittal plane where the inherent frequencies of the movements can be manipulated by varying the inertial loadings of the pendulums. The HKB model predicts that although the pendulums are being swung isochronously, the inherently slower oscillator (i.e. the wrist swinging the larger pendulum) will lag in its cycle and that the increased frequency difference between the oscillators will increase this lag as well as the variability (and hence, decrease the stability) of their coordination. A number of studies (Rosenblum & Turvey, 1988; Schmidt, Shaw & Turvey, 1993; Schmidt & Turvey, 1995) have substantiated this increase in the relative phase lag and standard deviation of relative phase with increases of frequency detuning for bimanual coordination, and helped to substantiate the claim that intrapersonal rhythmic coordination is underwritten by a motor synergy governed by synchronization dynamics (Kelso et al., 1981).

It was in the context of this line of research that Newtson (Newtson et al., 1987) proposed his hypothesis of similarity of action and (social) interaction and the idea of a dynamical interpersonal synergy. The empirical validation of this intuition came quickly with a series of experiments performed by one of the authors of this chapter (Schmidt) for his dissertation in 1988. The goal of these studies at the time was less to provide a mechanism to understand the interpersonal coordination in social interactions, and more to demonstrate the generality of a dynamical theory of behaviour – that the same organizing processes can coordinate limbs within a person with one CNS or between two people with two CNSs. Turvey, Kelso and Kugler in their writings (e.g. Kugler, Kelso & Turvey, 1980) stressed that the dynamical organizing principles that were guiding rhythmic interlimb coordination were seen at all scales of nature and were universal and incorporeal. Indeed, similar equations to the HKB model had up to this time been used to model central pattern generators at the neural scale (Stein, 1974) and cockroach locomotion (Foth & Graham, 1983), as well as the entrainment process in firefly flashing (Hanson, 1978). Schmidt's dissertation investigated the universality of these behavioural dynamics by seeing whether they organized interpersonal rhythmic coordination. If, indeed, in-phase and anti-phase were canonical steady states that arose as a consequence of the dynamics of oscillators and their interactions, shouldn't they also be differentially stable steady states in rhythmic movements coordinated across two people, across two neurally-based oscillators linked by perceptual information (Schmidt, 1988)?

The studies in question used laboratory tasks that were adapted from single-person interlimb coordination research. The methodology of these studies was to record rhythmic limb movements intentionally coordinated between two people and evaluate the relative phase predictions of the HKB model that had already been verified for intrapersonal bimanual coordination. The first study published,

Schmidt, Carello, and Turvey (1990), investigated whether an anti- to in-phase transition occurred interpersonally when the frequency of oscillation of two people visually coordinating rhythmic movements of their lower legs was scaled. Interestingly, and much to everyone's surprise, increases in the frequency of oscillation of the legs resulted in a breakdown of interpersonal anti-phase (as well as other hallmark properties of a dynamical change such as critical fluctuations), replicating the "phase" transition phenomena found by Kelso in single-person bimanual finger coordination. The next study in the series (Schmidt & Turvey, 1994) used the interpersonal coordination of wrist-pendulums to scale eigenfrequency differences to affirm the HKB predictions for manipulations of frequency detuning (Kelso et al., 1990).[1] The findings demonstrated a weaker dynamic for interpersonal coordination resulting in greater phase lags and higher relative phase SDs as frequency detuning increased (i.e. as the oscillators had greater and greater differences in their inherent frequencies).

These two studies (Schmidt et al., 1990; Schmidt & Turvey, 1994) were the first to demonstrate that dynamical principles that govern coupled oscillator processes constrain the emergence of the coordination of movements between two people who are linked only by perceptual information. Their results have been replicated many times subsequently both during the 1990s (Amazeen, Schmidt & Turvey, 1995; Fitzpatrick et al., 1996; Schmidt, Bienvenue, Fitzpatrick & Amazeen, 1998; Schmidt, Christianson, Carello, & Baron, 1994) and more recently (de Rugy, Salesse, Oullier & Temprado, 2006; Fine, Likens, Amazeen & Amazeen, 2015; Richardson, Marsh, Isenhower, Goodman & Schmidt, 2007; Schmidt, Fitzpatrick, Caron & Mergeche, 2011; Temprado, Swinnen, Carson, Tourment & Laurent, 2003; Temprado & Laurent, 2004; Varlet et al., 2012; 2014). Although the original studies were devised to specifically test the generality of the dynamical theory of motor coordination, they were also the first empirical tests of Newtson's dynamical interpersonal synergy hypothesis – that bodily movements of socially interacting individuals can be entrained through synchronization dynamics. Indeed, if the coupled oscillatory processes were the basis of the functional synergies or coordinative structures underlying bimanual interlimb coordination, these studies demonstrated that the same processes can be the basis for similar dynamical synergies written across two individuals to form interpersonal interactions.

However, could such processes provide a basis for understanding the phenomenon of natural interactional synchrony? Deliberately coordinating wrist-pendulums interpersonally is far from a natural social interaction. It was necessary to demonstrate that these dynamical organizing principles also underlie interpersonal interactions where bodily synchrony arises spontaneously. Although the first studies described above demonstrated that a synchronization dynamic can constrain dyadic motor coordination, these studies did so only for intentional coordination, i.e. when the social goal is the coordination itself. However, such a conscious goal to coordinate bodies is not present in many natural social interactions where interactional synchrony has been studied, such as in conversations. In this case, the explicit goal of the dyad is to communicate information, not to coordinate their movements. The question is whether a synchronization dynamic constrains the bodily coordination in these more natural instances where the social synchrony arises spontaneously within the functional context of a social interaction.

To determine this, Schmidt and O'Brien (1997, 1998) devised a method to investigate whether such *spontaneous* entrainment of rhythmic movements would occur if the participants had a social goal other than the coordination itself. In this study, two people sitting side-by-side swinging pendulums were told first to not look at one another and then told to look at each other's swinging pendulums, but to try to 'keep their original tempo' while doing so. The participants thought that seeing the pendulum was a distractor to the real task goal of keeping their tempo. The results indicated that when the individuals were looking at each other's movements, they could not help but become entrained with the other person's movements. The movements were more correlated and had a greater in-phase and anti-phase patterning in the "looking" part of the trials than in the "not looking" part. Rather than being phase-locked coordination, the entrainment pattern observed was meta-stable and intermittent (Kelso & Ding, 1993) – what von Holst (1973) called relative coordination. Importantly, more spontaneous entrainment was found when the participants swung pendulums that were the same size (i.e. no frequency detuning) than when they swung pendulums that were different sizes. Both the meta-stable and intermittent phase locking as well the effect of frequency detuning are predicted by a weakly parameterized synchronization dynamic model such as the HKB. Like the intentional coordination studies described above, the results of these studies have been replicated many times using various tasks and alternative measures of coordination (Issartel, Marin & Cadopi, 2007; Oullier, De Guzman, Jantzen, Lagarde & Kelso, 2008; Richardson, Marsh & Schmidt, 2005; Richardson et al., 2007; Tognoli, Lagarde, de Guzman & Kelso, 2007; Varlet et al., 2012; 2014). All these studies suggest that the dynamics of synchronization underlies the interactional synchrony, originally noticed by Condon, that arises spontaneously without awareness. Indeed, they suggest that dynamical interpersonal synergies are formed not only in laboratory tasks but also in everyday social interactions.

Conclusion

These early ideas about interactional synchrony and interpersonal synergies planted in the late decades of the twentieth century have been continuing to yield fruit in the new millennium. Much of the information from this 'second round' of research on interpersonal coordination is summarized in the chapters that follow in the present volume. A few areas of this more recent research are more explicit extensions of the themes from the heyday of interactional synchrony and original studies on interpersonal synergies.

For example, recent research has been echoing Condon's work in investigating how social/interactional synchrony breaks down in pathology in children (Fitzpatrick et al., 2013; Marsh et al., 2013) and adults (Lavelle, Healey &

McCabe, 2012; Ramseyer & Tschacher, 2014; Varlet et al., 2012, 2014). This research suggests that social motor coordination may be a biomarker for these pathologies and could be used in early detection as well as have implications for treatment. Other recent investigations have updated and broadened the interpersonal synergy concept. Riley, Richardson, Shockley and Ramenzoni (2011) have attempted to provide support for their more formal properties as motor synergies. Using the techniques of principal components analysis and the uncontrolled manifold method, they demonstrated the motor synergy properties of both dimensional compression (degrees of freedom are linked to yield fewer controllable aspects) and reciprocal compensation (ability of one component of a synergy to react to changes in others) in a different interpersonal coordination tasks. Fusaroli, Raczaskzek-Leonardi and Tylen (2013) have argued that the interpersonal synergy concept can be used to understand the mechanisms of conversational dialogue (Pickering & Garrod, 2004). In order to understand the progressive entrainment seen in interlocutors' linguistic behaviour during conversation, Fusaroli and colleagues argue that one needs to employ the concept of a dynamically-based interpersonal synergy with the formal characteristics described by Riley et al. (2011). These studies provide additional evidence to support Newtson's original insight that an interpersonal synergy is governing interpersonal coordination in natural interactions. Indeed, this second round of research on interpersonal coordination at the beginning of the twenty-first century holds great promise in illuminating how minds are combined, embodied and embedded in the environment on the basis of dynamical organizing principles.

Note

1   In spite of the fact that a frequency detuning term was added to the HKB equation in this 1990 book chapter to explain the *metastability* of finger oscillations with a metronome, the effect of frequency detuning on the lag-lead relationship in *stable* phase locking was less than clear. The six years between the dissertation (1988) and the publication of the paper (1994) was a consequence of trying to understand how the steady-state relative phase lag seen in the bimanual (and interpersonal) wrist-pendulum studies could be explained in terms of the dynamics of coupled oscillators. Investigations of frequency detuning in invertebrate locomotion (e.g. Rand, Cohen & Holmes, 1982) explicitly addressed this steady-state relative phase lag using coupled oscillator models similar to the HKB model and were incorporated in the first paper (Schmidt et al., 1993) that addressed the effect of frequency detuning control parameter on stable state wrist-pendulum swinging. Subsequent dialogue on the relevant merits of these models versus the HKB model can be found in Fuchs and Kelso (1994) and Schmidt and Turvey (1995).

References

Amazeen, P.G., Schmidt, R.C. & Turvey, M.T. (1995). Frequency detuning of the phase entrainment dynamics of visually coupled rhythmic movements. *Biological Cybernetics*, 72, 511–518.

Bernieri, F.J. (1988). Coordinated movement and rapport in teacher-student interactions. *Journal of Nonverbal Behavior, 12,* 120–138.

Bernieri, F.J., Reznick, J.S. & Rosenthal, R. (1988). Synchrony, pseudosynchrony, and dissynchrony: Measuring the entrainment process in mother-infant interactions. *Journal of Personality and Social Psychology, 54,* 243–253.

Bernieri, F.J., Davis, J.M., Rosenthal, R. & Knee, C.R. (1994). Interactional synchrony and rapport: Measuring synchrony in displays devoid of sound and facial affect. *Personality and Social Psychology Bulletin, 20,* 303–311.

Birdwhistell, R.L. (1970). *Kinesics and context: Essays on body motion communication.* University of Pennsylvania Press: Philadelphia.

Cappella, J.N. (1981). Mutual influence in expressive behavior: Adult-adult and infant-adult dyadic interaction. *Psychological Bulletin, 89,* 101–132.

Charney, J.E. (1966). Psychosomatic manifestations of rapport in psychotherapy. *Psychosomatic Medicine, 28,* 305–315.

Condon, W.S. (1970). Method of micro-analysis of sound films of behavior. *Behavior Research Methods and Instrumentation, 2,* 51–54.

Condon, W.S. (1975). Multiple response to sound in dysfunctional children. *Journal of Autism and Childhood Schizophrenia, 5,* 37–56.

Condon, W.S. (1982). Cultural micro-rhythms. In M.D. Davis (Ed.), *Interaction rhythms* (pp. 53–57), Human Sciences Press: New York.

Condon, W.S. & Ogston, W.D. (1966). Sound film analysis of normal and pathological behavior patterns. *Journal of Nervous Mental Disorders, 143,* 338–347.

Condon, W.S. & Ogston, W.D. (1967). A segmentation of behavior. *Journal of Psychiatric Research,* 5, 221–235.

Condon, W.S. & Ogston, W.D. (1971). Speech and body motion synchrony of the speaker-hearer. In D.L. Horton & J.J. Jenkins (Eds.), *Perception of language.* Charles E. Merrill: Columbus, Ohio.

Davis, M.E. (Ed.) (1982). *Interaction rhythms: Periodicity in communicative behavior.* Human Science Press: New York.

Deese, J. (1971). General discussion of the conference on the perception of language. In D.L. Horton & J.J. Jenkins (Eds.), *Perception of language.* Charles E. Merrill: Columbus, Ohio.

de Rugy A., Salesse R., Oullier O. & Temprado, J.J. (2006). A neuro-mechanical model for interpersonal coordination. *Biological Cybernetics, 94,* 427–443.

Dittmann, A.T. & Llewellyn, L.G. (1969). Body movement and speech rhythm in social conversation. *Journal of Personality and Social Psychology, 11,* 98–106.

Fine, J.M., Likens, A.D., Amazeen, E.L. & Amazeen, P.G. (2015). Emergent complexity matching in interpersonal coordination: Local dynamics and global variability. *Journal of Experimental Psychology: Human Perception and Performance, Mar 23,* no pagination.

Fitzpatrick, P., Diorio, R., Richardson, M.J. & Schmidt, R.C. (2013). Dynamical methods for evaluating the time-dependent unfolding of social coordination in children with autism. *Frontiers in Integrative Neuroscience, 7,* 1–13.

Fitzpatrick, P.A., Schmidt, R.C. & Carello, C. (1996). Dynamical patterns in clapping behavior. *Journal of Experimental Psychology: Human Perception and Performance, 22,* 707–724.

Foth, E. & Graham, D. (1983). Influence of loading parallel to the body axis on the walking coordination of an insect. *Biological Cybernetics, 48,* 149–157.

Fuchs, A. & Kelso, J.A.S. (1994). A theoretical note on models of interlimb coordination. *Journal of Experimental Psychology: Human Perception and Performance, 20,* 1088–1097.

Fusaroli, R., Raczaszek-Leonardi, J. & Tylen, K. (2013). Dialogue as interpersonal synergy. *New Ideas in Psychology, 32,* 147–157.

Gatewood, J.B. & Rosenwein, R. (1981). Interactional synchrony: Genuine or spurious? A critique of recent research. *Journal of Nonverbal Behavior, 6,* 12–29.

Glass, L. & Mackey, M.C. (1988). *From clocks to chaos: The rhythms of life.* Princeton University Press: Princeton, NJ.

Grammer, K., Kruck, K.B. & Magnusson, M.S. (1998). The courtship dance: Patterns of nonverbal synchronization in opposite-sex encounters. *Journal of Nonverbal Behavior, 22,* 3–29.

Haken, H. (1977). *Synergetics: An introduction.* Springer: Berlin.

Haken, H., Kelso, J.A.S. & Bunz, H. (1985). A theoretical model of phase transitions in human hand movements. *Biological Cybernetics, 51,* 347–356.

Hanson, F.E. (1978). Comparative studies of firefly pacemakers. *Federation Proceedings, 37,* 2158–2164.

Issartel, J., Marin, L. & Cadopi, M. (2007). Unintended interpersonal co-ordination: "Can we march to the beat of our own drum?" *Neuroscience Letters, 411,* 174–179.

Isenhower, R.W., Marsh, K.L., Richardson, M.J., Helt, M., Schmidt, R.C. & Fein, D. (2012). Rhythmic bimanual coordination is impaired in children with autism spectrum disorder. *Research in Autism Spectrum Disorders, 6,* 25–31.

Kelso, J.A.S. (1984). Phase transitions and critical behavior in human bimanual coordination. *American Journal of Physiology: Regulatory, Integrative and Comparative Physiology, 15,* R1000–R1004.

Kelso, J.A.S., DelColle, J. & Schöner, G. (1990). Action-perception as a pattern formation process. In M. Jeannerod (Ed.), *Attention and Performance XIII* (pp. 139–169). Hillsdale, NJ: Erlbaum.

Kelso, J.A.S. & Ding, M. (1993). Fluctuations, intermittency and controllable chaos in biological coordination. In K.M. Newell & D.M. Corcos (Eds.), *Variability and motor control.* Human Kinetics: Champaign, IL.

Kelso, J.A.S., Holt, K.G., Rubin, P. & Kugler, P.N. (1981). Patterns of human interlimb coordination emerge from the properties of non-linear oscillatory processes: Theory and data. *Journal of Motor Behavior, 13,* 226–261.

Kelso, J.A.S. & Jeka, J.J. (1992). Symmetry breaking dynamics of human multilimb coordination. *Journal of Experimental Psychology: Human Perception and Performance, 18,* 645–668.

Kendon, A. (1970). Movement coordination in social interaction: Some examples described. *Acta Psychologica, 32,* 100–125.

Kugler, P.N., Kelso, J.A.S. & Turvey, M.T. (1980). On the concept, of coordinative structures as dissipative structures: I. Theoretical lines of convergence. In G.E. Stelmach & J. Requin (Eds.), *Tutorials in motor behavior* (pp. 3–47). North-Holland: Amsterdam.

Kugler, P.N. & Turvey, M.T. (1987). *Information, natural law, and the self-assembly of rhythmic movement.* Erlbaum: Hillsdale, NJ.

Kupper, Z., Ramseyer, F., Hoffmann, H., Kalbermatten, S. & Tschacher, W. (2010). Video-based quantification of body movement during social interaction indicates the severity of negative symptoms in patients with schizophrenia. *Schizophrenia Research, 121,* 90–100.

LaFrance, M. (1979). Nonverbal synchrony and rapport: Analysis by the cross-lag panel technique. *Social Psychology Quarterly, 42*, 66–70.

LaFrance, M. (1982). Posture mirroring and rapport. In M.D. Davis (Ed.), *Interaction rhythms* (pp. 279–297), New York: Human Sciences Press.

LaFrance, M. & Broadbent, M. (1976). Group rapport: Posture sharing as nonverbal indicator. *Group and Organization Studies, 1*, 328–333.

Lavelle, M., Healey, P.G. & McCabe, R. (2012). Is nonverbal communication disrupted in interactions involving patients with schizophrenia? *Schizophrenia Bulletin, 39*, 1150–1158.

Marsh, K.L., Isenhower, R.W., Richardson, M.J., Helt, M., Schmidt, R.C. & Fein, D. (2013). Autism and social disconnection in interpersonal rocking. *Frontiers in Integrative Neuroscience, 7*(4). doi: 10.3389/fnint.2013.00004

McDowall, J.J. (1978a). Microanalysis of filmed movement: The reliability of boundary detection by observers. *Environmental Psychology and Nonverbal Behaviour, 3*, 77–88.

McDowall, J.J. (1978b). Interactional synchrony: A reappraisal. *Journal of Personality and Social Psychology, 36*, 963–975.

Miles, L.K., Nind, L.K. & Macrae, C.N. (2009). The rhythm of rapport: Interpersonal synchrony and social perception. *Journal of Experimental Social Psychology, 45*, 585–589.

Newtson, D. (1994). The perception and coupling of behavior waves. In R.R. Vallacher & A. Nowak (Eds.), *Dynamical systems in social psychology* (pp. 139–167). Academic Press: San Diego.

Newtson, D., Hairfield, J., Bloomingdale, J. & Cutino, S. (1987). The structure of action and interaction. *Social Cognition, 5*, 191–237.

Oullier, O., De Guzman, G.C., Jantzen, K.J., Lagarde, J. & Kelso, J.A.S. (2008). Social coordination dynamics: Measuring human bonding. *Social Neuroscience, 3*, 178–192.

Oxman, J., Webster, C.D. & Konstantareas, M.M. (1978). Condon's multiple-response phenomenon in severely dysfunctional children: An attempt at replication. *Journal of Autism Childhood Schizophrenia, 8*(4), 395–402.

Paxton, A. & Dale, R. (2013). Frame-differencing methods for measuring bodily synchrony in conversation. *Behavior Research Methods, 45*(20), 329–343.

Pickering, M.J. & Garrod, S. (2004). Toward a mechanistic psychology of dialogue. *Behavioral and Brain Sciences, 27*, 169–190.

Ramseyer, F. & Tschacher, W. (2011). Nonverbal synchrony in psychotherapy: Coordinated body movement reflects relationship quality and outcome. *Journal of Consulting and Clinical Psychology, 79*, 284–295.

Ramseyer, F. & Tschacher, W. (2014). Nonverbal synchrony of head- and body-movement in psychotherapy: Different signals have different associations with outcome. *Frontiers in Psycholology, 5*, 979.

Rand, R.H., Cohen, A.H. & Holmes, P.J. (1988). Systems of coupled oscillators as models of central pattern generators. In A.H. Cohen, S. Rossignol & S. Grillner (Eds.), *Neural control of rhythmic movements in vertebrates* (pp. 333–367). Wiley: New York.

Richardson, M.J., Marsh, K.L., Isenhower, R., Goodman, J. & Schmidt, R.C. (2007). Rocking together: Dynamics of intentional and unintentional interpersonal coordination. *Human Movement Science, 26*, 867–891.

Richardson, M.J., Marsh, K.L. & Schmidt, R.C. (2005). Effects of visual and verbal interaction on unintentional interpersonal coordination. *Journal of Experimental Psychology: Human Perception & Performance, 31*, 62–79.

Riley, M.A., Richardson, M.J., Shockley, K. & Ramenzoni, V.C. (2011). Interpersonal synergies. *Frontiers in Psychology*, *2*, 1–7.

Rosenblum, L.D. & Turvey, M.T. (1988). Maintenance tendency in coordinated rhythmic movements: Relative fluctuations and phase. *Neuroscience*, *27*, 289–300.

Rosenfeld, H.M. (1981). Whither interactional synchrony? In K. Bloom (Ed.), *Prospective issues in infancy research* (pp. 71–97). Lawrence Erlbaum Associates: New York .

Scheflen, A.E. (1964). The significance of posture in communication systems. *Psychiatry*, *27*, 316–331.

Schmidt, R.C. (1988). *Dynamical constraints on the coordination of rhythmic limb movements between two people.* Unpublished PhD Thesis, University of Connecticut, Storrs, CT.

Schmidt, R.C., Bienvenu, M., Fitzpatrick, P.A. & Amazeen, P.G. (1998). A comparison of within- and between-person coordination: Coordination breakdowns and coupling strength. *Journal of Experimental Psychology: Human Perception and Performance*, *24*, 884–900.

Schmidt, R.C., Carello, C. & Turvey, M.T. (1990). Phase transitions and critical fluctuations in the visual coordination of rhythmic movements between people. *Journal of Experimental Psychology: Human Perception and Performance*, *16*, 227–247.

Schmidt, R.C., Christianson, N., Carello, C. & Baron, R. (1994). Effects of social and physical variables on between-person visual coordination. *Ecological Psychology*, *6*, 159–183.

Schmidt, R.C., Fitzpatrick, P., Caron, R. & Mergeche, J. (2011). Understanding social motor coordination. *Human Movement Science*, *30*, 834–845.

Schmidt, R.C., Morr, S., Fitzpatrick, P.A. & Richardson, M.J. (2012). Measuring the dynamics of interactional synchrony. *Journal of Nonverbal Behavior*, *36*, 263–279.

Schmidt, R.C., Nie, L., Franco, A. & Richardson, M.J. (2014). Bodily synchronization underlying joke telling. *Frontiers in Human Neuroscience*, *8*, 633. doi: 10.3389/fnhum.2014.00633

Schmidt, R.C. & O'Brien, B. (1997). Evaluating the dynamics of unintended interpersonal coordination. *Ecological Psychology*, *9*, 189–206.

Schmidt, R.C. & O'Brien, B. (1998). Modeling interpersonal coordination dynamics: Implications for a dynamical theory of developing systems. In P.C. Molenaar & K. Newell (Eds.), *Dynamics Systems and Development: Beyond the metaphor* (pp. 221–240). Erlbaum: Hillsdale, NJ.

Schmidt, R.C., Shaw, B.K. & Turvey, M.T. (1993). Coupling dynamics in interlimb coordination. *Journal of Experimental Psychology: Human Perception and Performance*, *19*, 397–415.

Schmidt, R.C. & Turvey, M.T. (1994). Phase-entrainment dynamics of visually coupled rhythmic movements. *Biological Cybernetics*, *70*, 369–376.

Schmidt, R.C. & Turvey, M.T. (1995). Models of interlimb coordination: Equilibria, local analyses, and spectral patterning. *Journal of Experimental Psychology: Human Perception and Performance*, *21*, 432–443.

Stein, P.S.G. (1974). The neural control of interappendage phase during locomotion. *American Zoologist*, *14*, 1003–1016.

Temprado, J.J., Swinnen, S.P., Carson, R.G., Tourment, A. & Laurent, M. (2003). Interaction of directional, neuromuscular and egocentric constraints on the stability of preferred bimanual coordination patterns. *Human Movement Science*, *22*, 339–363.

Temprado, J.J. & Laurent, M. (2004). Attentional load associated with performing and stabilizing a between-persons coordination of rhythmic limb movements. *Acta Psychologica, 115*, 1–16.

Tickle-Degnen, L. (2006). Nonverbal behavior and its functions in the ecosystem of rapport. In V. Manusov & M. Patterson (Eds.), *The SAGE handbook of nonverbal communication*. Sage: Thousand Oaks, CA.

Tognoli, E., Lagarde, J., de Guzman, G.C. & Kelso, J.A.S. (2007). The phi complex as a neuromarker of human social coordination. *Proceedings of the National Academy of Science of the United States of America, 104*, 8190–8195.

Trevarthen, C. & Delafield-Butt, J.T. (2013). Autism as a developmental disorder in intentional movement and affective engagement. *Frontiers in Integrative Neuroscience, 7*, 49. doi: 10.3389/fnint.2013.00049

Turvey, M.T., Rosenblum, L.D., Schmidt, R.C. & Kugler, P.N. (1986). Fluctuations and phase symmetry in coordinated rhythmic movement. *Journal of Experimental Psychology: Human Perception and Performance, 12*, 564–583.

Varlet, M., Marin, L., Raffard, S., Schmidt, R.C., Capdevielle, D., Boulenger, J.P., Del-Monte, J. & Bardy, B.G. (2012). Impairments of social motor coordination in schizophrenia. *PLOS ONE, 7*, e29772.

Varlet, M., Marin, L., Capdevielle, D., Del-Monte J., Schmidt, R.C., Salesse, R., Boulenger, J. P., Bardy, B.G. & Raffard, S. (2014). Difficulty leading interpersonal coordination: Towards an embodied signature of social anxiety disorder. *Frontiers in Behavioral Neuroscience, 8*, 29.

von Holst, E. (1973). *The behavioral physiology of animal and man*. University of Miami Press: Coral Gables, FL. (Original work published 1939.)

3 The nature of interpersonal coordination

Why do people coordinate with others?

Craig A. Nordham and J. A. Scott Kelso

Introduction

Social interaction is a pervasive aspect of human existence. To behave with others is to coordinate with others, which may happen intentionally, e.g. as a shared goal, or spontaneously, merely due to human contact and information exchange. Examples include mother-infant interactions, jazz improvisation, sexual intercourse and team sports. It is hard to imagine behaving socially without movement – glancing, speaking, shrugging, kissing or nodding – leading many researchers to investigate the essential role of movement in social behaviour.

With technological advances, studies of social interaction have sought to approximate real-life situations – for example, face-to-face interactions recorded with high-resolution video on sub-second timescales. Researchers increasingly employ so-called hyperscanning, a new experimental paradigm to simultaneously record brain activity and behaviour from interacting persons (Babiloni & Astolfi, 2014; Dumas et al., 2011; Hasson et al., 2012; Konvalinka & Roepstorff, 2012; Montague et al., 2002; Sänger et al., 2011; Schilbach et al., 2013; Tognoli et al., 2007). Hyperscanning studies of social interaction have investigated live, delayed, pre-recorded, virtual or imagined partners to manipulate social variables. By hyperscanning experimental participants, researchers address both theoretical and empirical questions about the dynamics of live interaction on several levels of description.

Along with hyperscanning, various minimalist social interaction paradigms have cropped up, such as the mirror game (Noy et al., 2011) and the tactile perceptual crossing task (Auvray et al., 2009; see also Froese et al., 2014). These studies measure simple social behaviours with the goal of understanding issues such as leader-follower roles and the detection of agency. Recording two people simultaneously provides a means to measure and analyse the dyad as a whole, thereby yielding insights beyond the individual level. Past efforts to investigate social phenomena were limited to single subjects presented with social stimuli such as pictures or descriptions of another person. Present work investigates real-time social interaction, which requires at least two people to be present and engaged.

In this chapter, we review similarities and differences between individual and social coordination. Humans belong to nature, yet purpose and goal-directedness seem to set them apart from inanimate things. Since the late 1970s our approach has been grounded in the concepts, methods and tools of self-organizing dynamical systems (Synergetics, Dissipative Structures). From such beginnings arose a theoretical and empirical framework called Coordination Dynamics (CD), tailored specifically to the activities of living things such as moving, perceiving, feeling, deciding, learning and remembering (see Kelso, 2009a for review; Section 1). In CD, patterns of coordination, say between limbs of individual people, may be used to predict patterns of coordination between people, such as differential stability, phase transitions, symmetry breaking and remote compensation (Section 2). CD recognizes that the obvious separation between people in social settings may result in novel behaviours such as the way individuals enter and exit different coordination patterns with each other, and the consequences of shared information (Section 3). Finally, we explore how the dynamics of social coordination at the behavioural level serve as a useful foundation for understanding how coordination works within and between brains (Section 4).

Pattern formation in nature informs the study of human movement

Although our bodies may move in ways described by the laws of motion, classical mechanics falls short of an appropriate description of human behaviour (Gibson, 1979; Kelso, 1995). As Schrödinger (1944) remarked, 'new laws are to be expected in the organism' (Kelso & Haken, 1995). Two hallmarks of life include the "order-from-order" principle (Schrödinger, 1944) and "synergies of meaningful movement" (Kelso et al., 1984; Sheets-Johnstone, 2011). "Order-from-order" reminds us that living things adaptively switch from one ordered state to another, and that life resists disorder, to which all matter inevitably draws. "Meaningful movement" suggests that biological motion does not occur randomly, but rather is goal-directed and reflects selection among possible patterns. Were an organism's body parts to move completely separately of one another, the number of degrees of freedom involved would be astronomical (Beek & Turvey, 1992; Bernstein, 1967; Kelso, 1982; Kelso & Tuller, 1984; Kugler et al., 1980; Latash, 2008; Turvey, 1990), yet in practice, seldom if ever do effectors move completely independently (for an early example, see Kelso et al., 1979). Instead, elements develop functionally ordered relations – coordinative structures or synergies. Coordination Dynamics studies how such structures or functional synergies form and adapt to circumstances, their selection, persistence, and capacity to change (Kelso, 1995).

Animal locomotion marks a prime example of pattern formation and change in natural systems. For instance, take horse gait, in which increasing movement speed elicits transitions, i.e. from walk to trot and again from trot to gallop (Schöner et al., 1990). The order of footfalls undergoes qualitative change at certain critical frequencies (Hoyt & Taylor, 1981). Furthermore, these transitions are metabolically efficient; they accompany optimal consumption of oxygen. Similarly, in human movement – likely for the same reasons – qualitative transitions from walk to run occur (Diedrich & Warren, 1995).

Transitions offer a mechanism for pattern change, alerting us to which variables change during reorganization (to form new patterns and functions) and which parameters promote change. Selecting among patterns constitutes a simple decision-making process and provides an entry point into how complex systems work. Following Haken's (1977) synergetics, a measure of spatiotemporal order is the order parameter, also termed a collective variable. So-called control parameters lead the system through changes of patterns described by the order parameter(s)/collective variable.

Hundreds of laboratory experiments and 'real-life' settings have identified order parameters and often their dynamics at many levels of analysis: cellular/ neural, muscle-joint, EMG, kinematic, biomechanical, brain (e.g. EEG, MEG, fMRI), between-brain and so forth (Kelso et al., 2013). Collective variables and their dynamics have been used to understand a wide variety of processes such as perception, attention, pattern recognition, decision-making, learning, memory, development, motor control, posture, respiration, locomotion, social coordination, human–machine interaction, and so on (Kelso, 2009a for review). Here we focus mainly on coordination dynamics as it pertains to intra- and interpersonal coordination.

The same coordination dynamics occur whether within or between people

All human behaviour requires movement, whether of limbs, vocal cords, tongue, eyes or fingers. Living things have moving parts that work together seemingly effortlessly to attain goals. Coordination Dynamics (CD) aims to capture the details of the patterns formed by interactions among parts and processes, as well as the elements themselves. The prototypical task in CD – bimanual flexion–extension (rhythmically moving both index fingers) – has proved to be a key model system because it represents some of the essential properties of animate movement (Kelso, 1984). Many experiments have verified that bimanual coordination captures many of the same coordination features as other systems. Notably, an extended version of the Haken-Kelso-Bunz model (Haken, Kelso & Bunz, 1985), which describes basic coordinated limb movement patterns within a person, also describes how people coordinate with external stimuli (Kelso et al., 1990) and with each other (Schmidt et al., 1990; Schmidt & O'Brien, 1997).

Stability

A crucial aspect of coordination is stability. The metaphorical language of attractors is often used. One may ask, however, the stability of what? In CD, stability refers to collective patterns into which a behaving system tends to settle regardless of initial conditions. Recall that changing a control parameter can induce transitions (bifurcations, instabilities), resulting in changes in stability. Two moving limbs tend to mirror their movements either in-phase or anti-phase both inside (Kelso, 1984) and outside laboratory settings (Howard et al., 2009). In-phase has 0 radian phase difference or relative phase (0 degrees), whereas anti-phase has π radian phase difference (180 degrees). Remarkably, the same relative phase patterns seen in normal populations also occur in so-called 'split-brain' patients, people with a severed corpus callosum (the structure bridging brain hemispheres; Tuller & Kelso, 1989). Despite no direct pathway linking the hands, the two most stable patterns (in-phase and anti-phase) still take place.

Laws of coordination are materially independent: they require only a medium of interaction to manifest themselves (Kelso, 1994). For example, the same in-phase and anti-phase patterns occur in social settings, despite lacking a physical connection between the limbs. One experiment instructed two people to produce rhythmic movements at their self-chosen pace. Researchers examined the effects of visual coupling in social (dyadic) interaction using an electronic liquid crystal screen that controlled the visibility partners had of each other's hand movements (see Figure 3.1; Tognoli et al., 2007). Before visual contact, individual hand movements occurred independently from one another (Figure 3.1, top). After the screen opened, participants tended to coordinate spontaneously, despite no explicit instructions. Notice in the first half of the trial, the relative phase visited each value of ϕ approximately equally. After visual contact, relative phase stabilized near 0 (Figure 3.1, bottom). Thus, stable coordination can occur spontaneously with visual coupling.

Coordinative stability occurs across different social situations and experimental instructions. People coordinate intentionally with their legs (Schmidt et al., 1990), unintentionally while moving handheld pendulums (Schmidt & O'Brien, 1997), or while instructed to resist coordination while moving their arms (Issartel et al., 2007). Spontaneous coordination also occurs between two people rocking

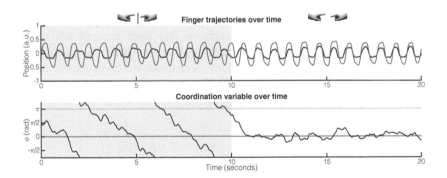

Figure 3.1 Example of individual to collective behaviour.

Notes: At top are finger movement trajectories of two participants (Tognoli et al., 2007). From the start, they move without sight of their partner (shaded region). At ten seconds an opaque screen becomes transparent, allowing partners to see each other. Notice that in the no vision stage, trajectories are disordered and all values of ϕ are visited nearly equally, showing no coordination. In the visual contact stage, trajectories become ordered and ϕ concentrates around zero, signifying coordination.

in rocking chairs (Richardson et al., 2007), while moving fingers (Oullier et al., 2008; Tognoli et al., 2007) or during standing postural sway (Varlet et al., 2011). The variety of cases suggests that stability of relative phase depends neither on particular body parts, nor on whether they belong to one or more people. In this sense, Coordination Dynamics is universal, capturing basic forms of biological coordination.

Phase transitions

Opposite stability is instability, which accompanies phase transitions. Kelso (1984) drove the bimanual system to instability, resulting in a phase transition to a more stable dynamic state. By increasing movement frequency, the anti-phase pattern destabilized and the system switched to in-phase, the last remaining stable steady state. Thus, a phase transition occurred (verified by predicted fluctuation enhancement and critical slowing down, see below) when the control parameter (frequency) reached a critical value and changed the order of the system.

An interesting dynamical property of the bimanual system also occurs when the direction of the control parameter is changed, namely *hysteresis*. Again, when starting anti-phase in this system, *increasing* pacing frequency led to a transition from π to 0 rad. However, *decreasing* frequency resulted in no subsequent switch to π despite that state being possible. Furthermore, starting at 0 relative phase and increasing frequency did not cause a transition (Kelso, 1984).

Haken, Kelso and Bunz's (1985) dynamical model explained all these effects at both collective (relative phase) and component levels. Here we restrict discussion to the former (see e.g. Fuchs & Kelso, 2009 for component level details). The HKB equation consists of a superposition of two sine functions with coefficients a and b, whose ratio ($b/a = k$) directly relates to frequency in the Kelso experiments and effectively weighs the relative dominance of each mode of coordination. The critical value of this ratio corresponds to the experimental control parameter, provoking the transition. Symmetries – invariant patterns under transformation – constrain the formulation of the model. For instance, the system has spatial symmetry (left and right hand are considered equivalent, i.e. $\phi = -\phi$) and periodic symmetry (i.e. each cycle begins anew, 0 = plus or minus 2π).

$$\dot{\phi} = -a \sin - 2b \sin 2\phi, \text{ where } a, b \geq 0 \quad \text{(Eq. 1)}$$

A phase transition, corresponding to a subcritical pitchfork bifurcation, occurs when $b/a = 0.25$. Note "phase" in "phase transition" is different from in "relative phase". The term "phase transition" means a qualitative change of state – as in water changing to steam. Bimanual experiments also display a change in coordination state, hence the appropriateness of "phase transition". It just so happens that the coordination state is quantifiable as relative phase undergoes a phase transition! As the system nears criticality, the HKB model predicts that variability in the order parameter should occur, a phenomenon called critical fluctuations. Critical fluctuations signify anticipation of change: the attractor deforms (cf. Figure 3.2) and variability of ϕ increases as k decreases. This prediction was confirmed by Kelso, Scholz and Schöner (1986): fluctuations increased upon reaching the critical k and decreased after switching to the more stable mode.

Environmental coupling provides a conceptual bridge between intra- and interpersonal coordination. Participants asked to syncopate (anti-phase coordination) with an auditory metronome exhibited the characteristic transition to synchronization (in-phase coordination) when the driving frequency reached the critical value (Kelso et al., 1990). Further demonstrations showed phase transitions in coordination between people: Schmidt et al. (1990) asked partners to swing their legs anti-phase, paced by a metronome increasing in frequency. With this increase, participants who watched each other's movement displayed a transition from anti-phase to in-phase, just as individual people do in bimanual and sensorimotor coordination. Leading up to the transition, standard deviation

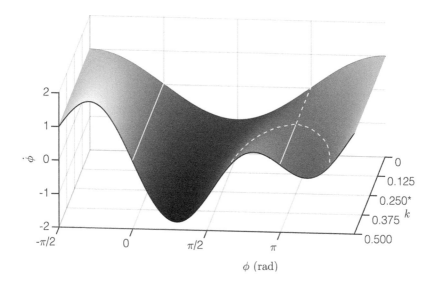

Figure 3.2 Phase space and bifurcation diagram of the HKB model.

Notes: This surface depicts the intrinsic dynamics of bimanual coordination. The bimanual system forms a trajectory in this space, which consists of stable fixed points (attractors), where the slope of the function is negative and $\dot{\phi}$ is zero, denoted by solid, white lines. Unstable fixed points (repellers) sit where the slope is positive and $\dot{\phi}$ is zero, denoted by dashed, white lines. Note that ϕ, a circular variable, repeats along the x-axis. At low rates of movement (high values of $k = b/a$ in Eq. 1), the system is bistable, which means both 0 and π are possible. When the driving frequency is increased, (k decreases) a bifurcation occurs creating a repeller at $\phi = \pi$, and the system transitions to monostable with only $\phi = 0$ remaining stable.

of ϕ increased and then decreased after the transition. Taken together, these observations of (obviously nonequilibrium) phase transitions provide specific evidence that self-organizing processes underlie coordination within and between people. Component behaviour may appear programmed, but collective order is emergent.

Symmetry breaking

Previously, we considered coordination between components that have approximately equal intrinsic frequencies. When differences in individual frequencies exist, the spatiotemporal symmetry of the HKB model is broken. Studies of non-homologous limb movements such as between arms and legs have revealed characteristic changes in spatiotemporal patterns of relative phase (Kelso & Jeka, 1992; Jeka & Kelso, 1995) as have basic studies of sensorimotor interaction (Kelso et al., 1990).

To accommodate changes in symmetries (and the empirical consequences thereof) the HKB model had to be extended (Kelso et al., 1990):

$$\dot{\phi} = \delta\omega - \sin\phi - 2k\sin 2\phi \quad \text{(Eq. 2)}$$
$$V(\phi) = -\delta\omega\phi - \cos\phi - k\cos 2\phi \quad \text{(Eq. 3)}$$

$$\text{where } k = \tfrac{b}{a} \geq 0$$

The introduction of $\delta\omega$ breaks spatiotemporal symmetry and changes the dynamics of the original HKB model: the form of the bifurcation changes from a pitchfork (Figure 3.2) to a saddle-node. Many paths to system change exist: both deterministic (e.g. changing k, changing $\delta\omega$) and stochastic (e.g. critical fluctuations, cf. Schöner, Haken & Kelso, 1986). Decrease k and the potential at π becomes convex and the "ball" rolls out. Increase the magnitude of $\delta\omega$, and the potential function tilts. Attractors shift until only tendencies to dwell remain. The behavioural trajectory continues to fall without settling into any attractor, but slowing at their 'ghosts' – the so-called metastable regime (see the portion of the potential where coupling is fixed and $\delta\omega$ is large, Figure 3.3).

For an example of broken symmetry in social situations, consider a parent and child walking together on the beach (Kelso, 1995). They have the opportunity to couple haptically and visually. The walking pattern is subject to both collective and individual tendencies (k and $\delta\omega$). When $\delta\omega$ is small, the two components generally couple at close to the same frequency. For a few cycles, they walk together but lose the pattern when they walk at their preferred speed, and then catch up. If $\delta\omega$ is too large, then little to no relative phase coordination remains. However, different frequency-lockings (e.g. 2:1 or 3:2) may be possible (e.g. Assisi et al., 2005).

Weighted props have allowed the manipulation of the intrinsic frequency of movement. For instance, in a study of social coordination, using a within-subjects design, experimenters used different weights during one experimental session,

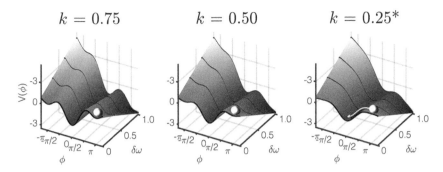

Figure 3.3 Extended HKB potential function with symmetry breaking term (Eq. 3).

Notes: We can visualize the stability of the system as a ball travelling on a plane. $V(\phi)$ varies according to changes in $\delta\omega$ and k parameters. The system settles into local minima (attractors), represented here as valleys. When driving frequency increases, the value of k decreases and when the system reaches the critical ratio (*), the valley at π rad becomes convex and the system falls into the remaining well at 0. Note that ϕ, a circular variable, repeats as $-\pi = \pi$. When $\delta\omega$ increases, the valleys (attractors) shift positively along the ϕ-axis until eventually they become smooth lines with no minima. In this metastable regime, trajectories slow – but do not stop – near 'ghosts' where valleys used to exist.

thereby affording many combinations of $\delta\omega$. In experiments using hand-held pendulums or rocking chairs, experimenters asked pairs of participants to coordinate their movements with each other (Amazeen et al., 1995; Richardson et al., 2007; Schmidt & Turvey, 1994). In all cases, the expected attractor shift scaled with the frequency difference of the props. Thus, broken symmetry, a key and quantifiable aspect of coordination dynamics (Eqs. 2, 3), describes both individual and social behaviour.

Remote compensation

Degeneracy – a key property of all biological systems – is the ability of combinations of different components to perform the same function (Edelman & Gally, 2001; Kelso et al., 1980; Prinz et al., 2004). Consider the seemingly infinite number of ways to accomplish a goal; many combinations of movements are possible. For instance, take walking and then recovering from tripping or from being nudged: recovery of walking posture occurs rapidly and automatically. Recruitment of remotely linked components may occur, all in the service of a goal. Remote compensation, related to reciprocal compensation, occurs system-wide in response to natural or experimentally induced perturbations (Fusaroli, Rączaszek-Leonardi & Tylén, 2014; Kelso, 2009b; Kelso et al., 1984; Latash, 2008; Riley, Richardson, Shockley & Ramenzoni, 2011; Turvey & Carello, 1996). It reveals the extent to which a system will reorganize itself to preserve functionally stable patterns of coordination (coordinative structures); perturbation of components reveals coupling and degree of involvement (Zaal & Bongers, 2014).

To test this phenomenon, experimenters applied a mechanical perturbation to the jaw during different phases of sound production (Kelso et al., 1984; see also Shaiman, 1989). Participants asked to articulate /bæb/ had their jaw briefly perturbed during the final stage of the speech sound. As a result, the upper and lower lips compensated to create closure for the consonant. In another context, when participants pronounced /bæz/, the lips did not respond, but the tongue compensated. Both compensatory actions occurred quickly (20–30 ms.) and without noticeable interference to sound production. The highly degenerate speech apparatus reflects a complex system's adaptive capacity: seemingly passive, non-involved elements are available to spring into action. In cases with just two components in a bimanual paradigm, subjects moved their arms in-phase or anti-phase (Post et al., 2000). When experimenters perturbed a single arm, both arms compensated to fulfill task requirements (see also Kelso et al., 1981).

In parallel with studies of individual coordination, compensation has also been investigated in social settings. Peper and colleagues (2013) asked two people to maintain in-phase coordination and quickly restore it when interrupted. To perturb movement, the experimenters halted one person's limb movement at peak extension. Assuming no coupling, i.e. two people moving simultaneously but not with each other, one would expect adjustment only in the perturbed arm. However, perturbing one arm resulted in both partners compensating equally with each other. The authors recommended using this method to probe the symmetry of partners' engagement in social interaction. Characteristic social deficiencies in autism spectrum disorders may affect responses to a partner's perturbation – by over- or under-compensating, for instance.

Uniquely social aspects of coordination

Individuals acting alone or together share previously described features of coordination. Still, how individuals act cannot predict all of social coordination. Take rule-based dancing. Traditional couples' dancing illustrates social behaviour where one person leads and the other follows. The combination of asymmetrical goals and independence of components permits some uniquely interpersonal aspects of coordination: competition, patterns of entering and exiting, and strategies to recruit and enhance informational flow.

Competitive goals and novel individual behaviour

One framework for studying social coordination is the human dynamic clamp paradigm (HDC). The HDC uses a computer-controlled visualization of a moving finger (e.g. via the HKB model of a virtual partner, VP), whose movement reciprocally couples with a real-life partner's finger movements (Kelso et al., 2009; Dumas et al., 2014). By manipulating the VP's parameters, we can see their effects on emergent patterns of reciprocal interaction with a human being. Given that the VP uses empirically verified models of coordination dynamics with controlled parameters, "the unknown", as insightfully noted by Dr. Gonzalo de Guzman, "is the human" (cf. Dumas et al., 2014). Indeed, the HDC paradigm allows us to parametrically explore a wider scope of human behaviour than previously possible.

By creating competition between the human and VP, one experiment yielded an unexpected outcome: When experimenters programmed the VP to coordinate anti-phase and instructed experimental participants to coordinate in-phase, the VP exhibited a drastic decay in amplitude, to the point where participants perceived no motion. To elicit change in the inanimate partner and regain the desired in-phase pattern, participants spontaneously employed three novel parametric strategies: amplitude-scaling, baseline shifting, and temporarily switching to anti-phase (Figure 3.4). The effects of these "strategies" were to increase or maintain the VP's amplitude and stabilize the in-phase pattern. Thus, people successfully altered the VP's behaviour to achieve their goal via unique and uninstructed movement strategies.

An analogous example comes from a classic study, which demonstrated disruption of naturalistic coordination between parent and child. In this "still-face" paradigm a mother and infant played together, after which the mother was instructed to look at the child but keep her face still (Tronick et al., 1979). Infants attempted to greet or get their mother's attention but became upset or withdrawn until the mother resumed normal behaviour. However, one of the older infants (five months old) laughed heartily, which sufficiently interrupted the mother's stone-face into joined laughter. Thus, the mature infant successfully achieved the (presumed) preferred outcome, the return to normal play, by a novel intervention.

Note, however, in the case of human and VP interaction, that we know which parameters to change and by how much in a well-defined task with known dynamics. The HDC paradigm allows us to characterize quantitatively the manner in which individuals attempt to exert control or change a partner's behaviour. Likewise, we might better understand parent-infant interactions if we knew which variables were relevant during play and attentive communication.

Entering and exiting social coordination

Christiaan Huygens, a Dutch polymath who lived in the seventeenth century, invented the pendulum clock and discovered its "odd sympathy": the pendulums of two clocks synchronize despite apparent separation between them (Hugenii, 1673; Pikovsky et al., 2001; Winfree, 2001). He discovered that small vibrations in the material connecting the clocks (a coupling medium) caused the synchronization. Given nonlinear coupling, simple oscillators such as pendulum clocks should enter and exit coordination in understandable ways.

So how do people enter and exit social coordination? Are they like Huygens's clocks? To manipulate information exchange, Oullier et al. (2008) instructed pairs of participants to flex and extend their fingers and open/close/open their eyes for twenty-second periods over the course of each sixty-second trial. They measured the power spectrum overlap of behaviour, finding more in common

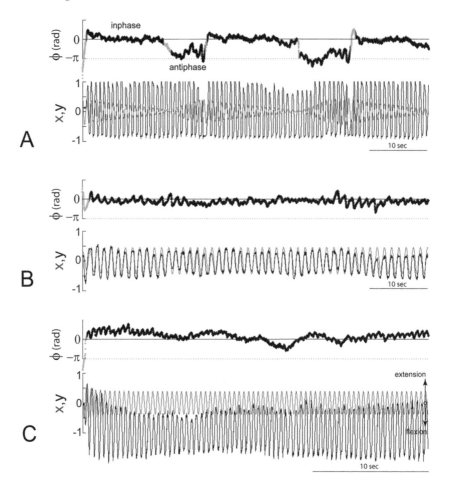

Figure 3.4 Spontaneous novel behaviours produced by humans in a competitive interaction
with a virtual partner (VP).

Notes: Participants employed three novel strategies to achieve the goal of moving in-phase with the
coupled VP programmed to coordinate anti-phase. The upper plots show the relative phase time series
showing the collective behaviour of the pair and the lower plots show the component trajectories of
individual performances (VP is x, grey line; human is y, black line); **A**: Temporarily switching to anti-
phase. **B**: Amplitude reduction. **C**: Shifting baseline of the oscillation toward flexion (adapted from
Kelso et al., 2009).

after the partners had interacted than earlier, referring to this effect as social
memory. Individual initial conditions (peak of frequency spectrum before visual
contact) dictated future behaviour, specifically the direction of change of frequency
after visual contact. With vision, the initially slower partner increased frequency
to match the other. After removing sight of the partner, they tended to return to
their initial frequency. However, the faster-moving partner who reduced frequency
to match his/her partner continued at that rate even after removal of vision.

In a similarly structured experiment, we examined the evolution of individual participants' instantaneous frequency and found that after visual contact had ended, persistence at the socially adopted frequency either ceased or continued (Nordham et al., 2012). That is, a participant's frequency either quickly returned to their intrinsic frequency or persisted throughout the remainder of the recording. One person's persistence did not necessarily depend on the persistence of their partner. Furthermore, we found that some pairs spontaneously coordinated at a slowly increasing frequency, across trials over the entire experiment. These results suggest social coordination tunes individual components, lasting even after the interaction ends and over timescales longer than the duration of visual contact or even single trials. Such persistence calls for revision to the ancient saying of Heraclitus: you may never step into the same river twice, but you are affected by the previous occasion!

Enhancement of information flow in interpersonal joint action

Previous examples considered coordination within and between people using hand movements. However, coordination also occurs between hands acting on an object. Joint actions require two hands – for typing, kneading dough, crocheting, reeling a fishing rod or playing bass guitar. In the social case of joint action, two people collaborate to work with the same object, such as rowing a boat, moving furniture, playing jump rope or sawing a tree down.

It is of interest to compare a joint action performed by a single person with the same joint action performed by two people. With this in mind, Foo et al. (2002) introduced the novel joint action coordination paradigm of pole balancing (see also Foo et al., 2000; van der Wel et al., 2011). The apparatus consisted of an inverted pendulum (pole) with two opposing strings attached near its base (Figure 3.5 A, D). In experiments, individual participants and pairs of participants were asked to balance the pole by pulling on an attached string. Thus, the goal was to stabilize the pole upright, which required the application of force from both ends of the string. Notice that individuals use both hands (one brain); in dyads, each member uses one hand (two brains!).

Foo and colleagues (2002) found that amplitude and period were the same for individuals and dyads, but the amount of force applied was not. Dyads used more overlapping force than individuals, especially when the pole was nearly vertical and therefore least stable (Figure 3.5 B, E). Participants applied a steady force with an oscillating strategy to stabilize the pole. Furthermore, individuals moved anti-phase, whereas partners tended to use in-phase (Figure 3.5 C, F; Foo et al., 2002).

Such differences between intra- and interpersonal performance show how joint tasks constrain component behaviour. Successful pole balancing requires prospective, anticipatory control – perceiving and reacting faster than the characteristic fall time of the pole (Foo et al., 2000). Within one person, neural communication between hands allows the individual to catch the pole. Recall that participants accomplished this with opposing, alternating tugs on the

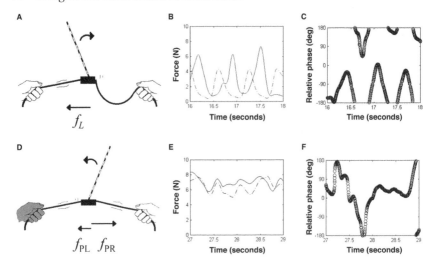

Figure 3.5 Intra- and interpersonal stabilization of an unstable dynamical system.

Notes: The task is to balance a pole by applying force to strings. *Top row:* Intrapersonal case. *Bottom row:* Interpersonal case. The schematics show participants' coordination strategies (A & D), representative time series of hand forces (B & E), and corresponding relative phases (C & F). Intrapersonal anti-phase strategy. A: One person alternates pulling on each string. B: Magnitudes of alternating forces exerted by the left (f_L: dashed line) and right (f_R: solid line) hands. C: Corresponding continuous relative phase values centre around ± 180 degrees ($\pm \pi$ rad). Interpersonal in-phase strategy. D: Two individuals pull simultaneously on the strings. E: Simultaneous forces exerted by the left (f_{PL}: dashed line) and right (f_{PR}: solid line) participants. F: The corresponding continuous relative phase values between participants centre around 0 degrees (0 rad). Figure comes from Foo et al. (2002).

strings. Whereas an individual has both visual information and proprioceptive coupling via the nervous system, the dyad has to rely on vision alone, which is insufficient. In this case, successful performance requires both people to enhance haptic information by applying simultaneous, continuous tension on the string (Foo et al., 2002; van der Wel et al., 2011). Only when the string is taut can forces propagate via a haptic information channel, allowing both partners to sense the pole and partner movement. The two joint action strategies demonstrate yet another way between- and within-person coordination can differ.

The nature of interpersonal coordination in behaviour and brain

Past studies established that rhythmic movements of different muscle groups or limbs follow the laws of Coordination Dynamics (cf. Kelso, 1995). Studies of movement and environmental signals demonstrated the informational basis of coupling (Kelso et al., 1990) and extended to social behavior (Schmidt et al., 1990). Remarkably, pattern stability, pattern switching, symmetry breaking and remote compensation occur regardless of whether the component effectors belong to one person or two. Although the collective dynamics of *intra*personal coordination have much in common with those of *inter*personal coordination, examples of novel behaviour, individuals entering and exiting coordination, and joint action illustrate some of the unique aspects of interpersonal coordination. We can understand such apparently diverse phenomena in terms of the causal loops of coordination dynamics.

The causal loops of coordination dynamics are pathways by which self-organization can occur (Figure 3.6, left side). The first loop (1, 2, 3) refers to *reciprocal* or *circular* causality, where an outside control parameter acts as a driving force on the components (1). Order emerges from nonlinear interactions among the parts; the direction of flow is bottom-up, from local to global (2). Subsequently, the parts conform to the emerging order via a top-down, global to local flow (3) as in Haken's (1977) famous slaving principle. Bimanual, sensorimotor and social coordination paradigms in which the coupling (k) increases with driving frequency and produces bifurcations can all be viewed in the light of circular or reciprocal causality (see Kelso, 1995). The second causal loop concerns novel *self-referential* effects on the control parameter, where the order parameter acts back on the control parameter closing the loop (4) (Kelso & Engstrøm, 2006; Tschacher & Haken, 2007).

For instance, the causal loops of coordination dynamics can help us to understand the human~virtual partner interaction experiment (Figure 3.6, right side; Kelso et al., 2009). Competitive goals (human's goal in-phase, virtual partner programmed anti-phase) set the boundary constraints (Figure 3.6, outline). In response to the in-phase behaviour stabilized by the human, the virtual partner decreases its amplitude (Pathways 2 & 3). Behaviour becomes uncertain because of the competing goals, which have the effect of diminishing visual information to the human. Only when the human finds a successful strategy will the virtual partner's amplitude grow, increasing the visual information necessary for the human to accomplish their goal. In these instances, the order parameter (relative phase) and perception of it (Bingham, 2004; Haas, et al., 1995; Haken, et al., 1990; Wilson & Bingham, 2008) changes the control parameter (Pathway 4).

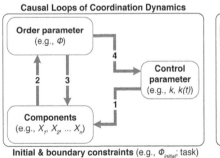 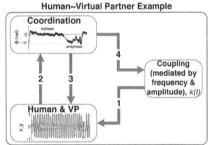

Figure 3.6 The causal loops of coordination dynamics (left panel) exemplified by the Human Dynamic Clamp (Virtual Partner) Paradigm (right panel, see text for details).

More broadly in neuroscience, the discovery of mirror neurons, originally recorded from area F5 in the brains of Macaque monkeys, has catalyzed human social neuroscience. These single neurons selectively show activity during observation of goal-directed movements regardless of whether they are performed by the monkey or the experimenter (di Pellegrino et al., 1992; Gallese et al., 1996). Macaque monkeys, like humans, spontaneously coordinate movements with one another (Nagasaka et al., 2013). Vittorio Gallese (2001, 2003), a co-discoverer of mirror neurons, has proposed that a conceptual "shared manifold" or "shared multidimensional interpersonal space" underlies social functions such as action understanding, which "motor resonances" between people may fulfil (for a possibility, see Dumas et al., 2010). However, a quantitative and experimentally tested manifold exists already – the extended HKB model – which has been demonstrated to work both within and between people.

It is desirable to know how CD forms at the brain level. While brain activity organizes movement, movement also organizes the brain's activities. To circumvent this apparent dilemma, we can simply examine the simultaneous activity of behaviour and brain. Furthermore, if behaviour and brain use the same dynamical language, then a common currency for cross-scale descriptions exists (Kelso, 1995). Identifying the appropriate spatiotemporal variables allows observation of the direct relationship between movement and neural patterns (e.g. Kelso et al., 1998; Fuchs et al., 2000).

The stability of social coordination has led to the discovery of a ~10Hz neuromarker with a right-centroparietal scalp topography called the phi complex, which appears to distinguish individual from social movement (Tognoli et al., 2007). Other alpha-mu band rhythms, originating in the motor cortex are suppressed (desynchronized) during self-movement, subtle self-movement, imagining movement, and a partner's observed movement (Pineda et al., 2000; Nikulin et al., 2008). In a dual-EEG experiment recording simultaneous movement and EEG, individuals were instructed to ignore their partner or move together in-phase or anti-phase (Naeem et al., 2012a, b). For all conditions, lower mu band (8–10 Hz) suppression was equal between hemispheres. However, for upper mu (10–12 Hz), suppression was particularly right-lateralized for the anti-phase case but left-lateralized for the in-phase case. Anti-phase coordination, being the least stably coupled pattern and requiring a high level of perceptuomotor engagement, showed the most suppression of upper mu. Though these are early days, the preceding work shows how interpretations of brain activity rely on real-time, simultaneous measures of social interaction and their stability.

Application of hyperscanning methods has provided evidence for hyper-brain theories – interpersonal coordinative structures at the inter-brain level (Dumas et al., 2010). During a task in which one person mirrored their partner's hand movements, Dumas and colleagues (2010) found a symmetrical pattern of synchrony between brains over centroparietal scalp areas in the alpha-mu band, likely to involve the phi complex. Much like relative phase for behaviour, relevant measures of brain activity during interpersonal coordination pertain to between- rather than within-brains (Dodel et al., 2013). Such quasisynchrony between brains represents the first stage of characterizing interpersonal, between-brain coordinative structures, which are expected to include phase transitions, symmetry breaking, remote compensation and more. The CD approach and insights about causal loops can guide brain researchers to find and describe essential variables – the control and order parameters of intra- and interbrain activity – and their dynamics.

The editors of this book charged us to answer the question: why do people coordinate socially? The answer is that we do not really know. The evidence presented here suggests that how people coordinate interpersonally is a result of self-organizing processes including both intrinsic factors and bi-directional coupling. Elementary forms of social coordination are a realization of generic biological coordination laws such as the extended HKB model (op. cit.; Kelso, 2012). When information is available, organisms couple to it and use it to couple to each other. When information between organisms is mutually available, they couple reciprocally. It is crucial to acknowledge that both coupling strength (which promotes togetherness) and intrinsic frequency differences (which represent individual autonomy) are both essential for flexible and robust social interaction. CD says that if individual components are too separate, no new information about their coordination is created. On the other hand, if individual components couple too strongly, people get stuck in a pattern and behaviour is too rigid. The subtle interplay of individual and collective properties is what matters. Rather than an either-or situation, both tendencies for individual expression and tendencies to work together co-exist. This metastable regime best characterizes societies where individuals can live freely and independently yet work together.

Acknowledgments

The work described herein was supported by the National Institute of Mental Health (MH080838), the National Science Foundation (BCS0826897), the US Office of Naval Research (N000140910527), and the Chaire d'Excellence Pierre de Fermat.

References

Amazeen, P.G., Schmidt, R.C. & Turvey, M.T. (1995) Frequency detuning of the phase entrainment dynamics of visually coupled rhythmic movements. *Biological Cybernetics*, *72*(6), 511–518.

Assisi, C.G., Jirsa, V.K. & Kelso, J.A.S. (2005) Dynamics of multifrequency coordination using parametric driving: theory and experiment. *Biological Cybernetics*, *93*(1), 6–21.

Auvray, M., Lenay, C. & Stewart, J. (2009) Perceptual interactions in a minimalist virtual environment. *New Ideas in Psychology*, *27*(1), 32–47.

Babiloni, F. & Astolfi, L. (2014) Social neuroscience and hyperscanning techniques: Past, present and future. *Neuroscience & Biobehavioral Reviews*, *44*, 76–93.

Beek, P.J. & Turvey, M.T. (1992) Temporal patterning in cascade juggling. *Journal of Experimental Psychology: Human Perception and Performance*, *18*(4), 934–947.

Bernstein, N. (1967) *The coordination and regulation of movement*. Pergamon: London.

Bingham, G. (2004) A perceptually-driven dynamical model of bimanual rhythmic movement (and phase perception). *Ecological Psychology, 16*, 45–53.

Diedrich, F.J. & Warren, W.H. (1995) Why change gaits? Dynamics of the walk-run transition. *Journal of Experimental Psychology: Human Perception and Performance, 21*(1), 183–202.

di Pellegrino, G., Fadiga, L., Fogassi, L., Gallese, V. & Rizzolatti, G. (1992) Understanding motor events: a neurophysiological study. *Experimental Brain Research, 91*(1), 176–180.

Dodel, S., Tognoli, E. & Kelso, J.A.S. (2013) The geometry of behavioral and brain dynamics in team coordination. In D.D. Schmorrow & C.M. Fidopiastis (Eds.), *Foundations of Augmented Cognition* (pp. 133–142). Berlin, Heidelberg: Springer.

Dumas, G., de Guzman, G.C., Tognoli, E. & Kelso, J.A.S. (2014) The human dynamic clamp as a paradigm for social interaction. *Proceedings of the National Academy of Sciences, 111*(35), E3726–E3734.

Dumas, G., Lachat, F., Martinerie, J., Nadel, J. & George, N. (2011) From social behaviour to brain synchronization: Review and perspectives in hyperscanning. *IRBM, 32*(1), 48–53.

Dumas, G., Nadel, J., Soussignan, R., Martinerie, J. & Garnero, L. (2010) Inter-brain synchronization during social interaction. *PLoS ONE, 5*(8), e12166.

Edelman, G.M. & Gally, J.A. (2001) Degeneracy and complexity in biological systems. *Proceedings of the National Academy of Sciences, 98*(24), 13763–13768.

Foo, P., de Guzman, G.C. & Kelso, J.A.S. (2002) Intermanual and interpersonal stabilization of unstable systems. Unpublished ms., available on request.

Foo, P., Kelso, J.A.S. & de Guzman, G.C. (2000) Functional stabilization of unstable fixed points: Human pole balancing using time-to-balance information. *Journal of Experimental Psychology: Human Perception and Performance, 26*(4), 1281–1297.

Froese, T., Iizuka, H. & Ikegami, T. (2014) Using minimal human-computer interfaces for studying the interactive development of social awareness. *Frontiers in Psychology, 5*, 1061.

Fuchs, A., Jirsa, V.K. & Kelso, J.A.S. (2000) Theory of the Relation between Human Brain Activity (MEG) and Hand Movements. *NeuroImage, 11*(5), 359–369.

Fuchs, A. & Kelso, J.A.S. (2009) Movement Coordination. In A.R. Meyers (Ed.), *Encyclopedia of Complexity and Systems Science* (pp. 5718–5736). New York: Springer.

Fusaroli, R., Rączaszek-Leonardi, J. & Tylén, K. (2014) Dialog as interpersonal synergy. *New Ideas in Psychology, 32*, 147–157.

Gallese, V. (2001) The "shared manifold" hypothesis. From mirror neurons to empathy. *Journal of Consciousness Studies, 8*(5–6), 33–50.

Gallese, V. (2003) The manifold nature of interpersonal relations: the quest for a common mechanism. *Philosophical Transactions of the Royal Society B: Biological Sciences, 358*(1431), 517–528.

Gallese, V., Fadiga, L., Fogassi, L. & Rizzolatti, G. (1996) Action recognition in the premotor cortex. *Brain, 119*(2), 593–609.

Gibson, J.J. (1979) *The ecological approach to visual perception.* Houghton Mifflin: Boston.

Haas, R., Fuchs, A., Haken, H., Horvath, E., Pandya, A.S. & Kelso, J.A.S. (1995) Pattern recognition of Johansson point light displays by synergetic computer. In P. Kruse & M. Stadler (Eds.), *Ambiguity in Mind and Nature* (pp. 139–155), Springer-Verlag: Heidelberg.

Haken, H. (1977) *Synergetics: An introduction.* Springer-Verlag: Berlin.

Haken, H., Kelso, J.A.S. & Bunz, H. (1985) A theoretical model of phase transitions in human hand movements. *Biological Cybernetics, 51*(5), 347–356.

Haken, H., Kelso, J.A.S., Fuchs, A. & Pandya, A. (1990) Dynamic pattern recognition of coordinated biological motion. *Neural Networks, 3,* 395401.

Hasson, U., Ghazanfar, A.A., Galantucci, B., Garrod, S. & Keysers, C. (2012) Brain-to-brain coupling: A mechanism for creating and sharing a social world. *Trends in Cognitive Sciences, 16*(2), 114–121.

Howard, I.S., Ingram, J.N., Körding, K.P. & Wolpert, D.M. (2009) Statistics of natural movements are reflected in motor errors. *Journal of Neurophysiology, 102*(3), 1902–1910.

Hoyt, D.F. & Taylor, C.R. (1981) Gait and the energetic of locomotion in horses. *Nature, 229*(5820), 239–240.

Hugenii, C. (1673) *Horologium oscillatorium.* Apud F. Muguet: Paris.

Issartel, J., Marin, L. & Cadopi, M. (2007) Unintended interpersonal co-ordination: "Can we march to the beat of our own drum?" *Neuroscience Letters, 411*(3), 174–179.

Jeka, J.J. & Kelso, J.A.S. (1995) Manipulating symmetry in the coordination dynamics of human movement. *Journal of Experimental Psychology: Human Perception and Performance, 21*(2), 360–374.

Kelso, J.A.S. (1982) *Human motor behavior: An introduction.* Erlbaum: Hillsdale, NJ.

Kelso, J.A.S. (1984) Phase transitions and critical behavior in human bimanual coordination. *The American Journal of Physiology, 246*(6 Pt 2), R1000–R1004.

Kelso, J.A.S. (1994) The informational character of self-organized coordination dynamics. *Human Movement Science, 13*(3–4), 393–413.

Kelso, J.A.S. (1995) *Dynamic patterns: The self-organization of brain and behavior.* MIT Press: Cambriddge, MA.

Kelso, J.A.S. (2009a) Coordination Dynamics. In R.A. Meyers (Ed.), *Encyclopedia of Complexity and System Science,* Heidelberg: Springer (pp. 1537–1564).

Kelso, J.A.S. (2009b) Synergies: Atoms of brain and behavior. *Advances in Experimental Medicine and Biology, 629,* 83–91. [Also in D. Sternad (Ed.), *A multidisciplinary approach to motor control.* Heidelberg: Springer.]

Kelso, J.A.S. (2012) Multistability and metastability: Understanding dynamic coordination in the brain. *Philosophical Transactions of the Royal Society of London B: Biological Sciences, 367*(1591, 906–918.

Kelso, J.A.S. & Tuller, B. (1984) A dynamical basis for action systems. In M.S. Gazzaniga (Ed.), *Handbook of Cognitive Neuroscience* (pp. 321–356), Plenum: New York.

Kelso, J.A.S., Southard, D.L. & Goodman, D. (1979) On the nature of human interlimb coordination. *Science, 203*(4384), 1029–1031.

Kelso, J.A.S., de Guzman, G.C., Reveley, C. & Tognoli, E. (2009) Virtual partner interaction (VPI): Exploring novel behaviors via coordination dynamics. *PLoS ONE, 4*(6), e5749.

Kelso, J.A.S., DelColle, J.D. & Schöner, G. (1990) Action-perception as a pattern formation process. In M. Jeannerod (Ed.), *Attention and performance 13: Motor representation and control* (pp. 139–169). Lawrence Erlbaum Associates, Inc.: Hillsdale, NJ.

Kelso, J.A.S., Dumas, G. & Tognoli, E. (2013) Outline of a general theory of behavior and brain coordination. *Neural Networks, 37,* 120–131.

Kelso, J.A.S. & Engstrøm, D.A. (2006) *The complementary nature.* MIT Press: Cambridge, MA.

Kelso, J.A.S., Fuchs, A., Lancaster, R., Holroyd, T., Cheyne, D. & Weinberg, H. (1998) Dynamic cortical activity in the human brain reveals motor equivalence. *Nature, 392*(6678), 814–818.

Kelso, J.A.S. & Haken, H. (1995) New laws to be expected in the organism: synergetics of brain and behaviour. In M. Murphy & L.A.J. O'Neill (Eds.), *What is life?: The next fifty years: speculations on the future of biology* (pp. 137–160). Cambridge, UK: Cambridge University Press.

Kelso, J.A.S., Holt, K.G. & Flatt, A.E. (1980) The role of proprioception in the perception and control of human movement: Toward a theoretical reassessment. *Perception & Psychophysics, 28*(1), 45–52.

Kelso, J.A.S., Holt, K.G., Rubin, P. & Kugler, P.N. (1981) Patterns of human interlimb coordination emerge from the properties of non-linear, limit cycle oscillatory processes. *Journal of Motor Behavior, 13*(4), 226–261.

Kelso, J.A.S. & Jeka, J.J. (1992) Symmetry breaking dynamics of human multilimb coordination. *Journal of Experimental Psychology: Human Perception and Performance, 18*(3), 645–668.

Kelso, J.A.S., Scholz, J.P. & Schöner, G. (1986) Nonequilibrium phase transitions in coordinated biological motion: critical fluctuations. *Physics Letters A, 118*(6), 279–284.

Kelso, J.A.S., Tuller, B., Vatikiotis-Bateson, E. & Fowler, C.A. (1984) Functionally specific articulatory cooperation following jaw perturbations during speech: evidence for coordinative structures. *Journal of Experimental Psychology: Human Perception and Performance, 10*(6), 812–832.

Konvalinka, I. & Roepstorff, A. (2012) The two-brain approach: how can mutually interacting brains teach us something about social interaction? *Frontiers in Human Neuroscience, 6,* 215.

Kugler, P.N., Kelso, J.A.S. & Turvey, M.T. (1980) Coordinative structures as dissipative structures I. Theoretical lines of convergence. In G.E. Stelmach & J. Requin (Eds.), *Tutorials in motor behavior* (pp.1–40), North Holland: Amsterdam.

Latash, M.L. (2008) *Synergy.* Oxford University Press: New York.

Montague, P.R., Berns, G.S., Cohen, J.D., McClure, S.M., Pagnoni, G., Dhamala, M., Wiest, M.C., Karpov, I., King, R.D., Apple, N. & Fisher, R.E. (2002) Hyperscanning: simultaneous fMRI during linked social interactions. *NeuroImage, 16*(4), 1159–1164.

Naeem, M., Prasad, G., Watson, D.R. & Kelso, J.A.S. (2012a) Electrophysiological signatures of intentional social coordination in the 10–12 Hz range. *NeuroImage, 59*(2), 1795–1803.

Naeem, M., Prasad, G., Watson, D.R. & Kelso, J.A.S. (2012b) Functional dissociation of brain rhythms in social coordination. *Clinical Neurophysiology, 123*(9), 1789–1797.

Nagasaka, Y., Chao, Z.C., Hasegawa, N., Notoya, T. & Fujii, N. (2013) Spontaneous synchronization of arm motion between Japanese macaques. *Scientific Reports, 3,* 1151.

Nikulin, V.V., Hohlefeld, F.U., Jacobs, A.M. & Curio, G. (2008) Quasi-movements: A novel motor-cognitive phenomenon. *Neuropsychologia, 46*(2), 727–742.

Nordham, C., de Guzman, G.C., Kelso, J.A.S. & Tognoli, E. (2012) *Neural and behavioral evidence for 'social memory' in humans following real-time interaction.* Program No. 695.11. 2012 Neuroscience Meeting Planner. Society for Neuroscience: Washington, DC. http://www.abstractsonline.com/Plan/ViewAbstract.aspx?sKey=fe9cb251-1ec7-4891-9164-7d28c5126a8f&cKey=32f35096-11cc-47f0-a801-b972404e1439

Noy, L., Dekel, E. & Alon, U. (2011) The mirror game as a paradigm for studying the dynamics of two people improvising motion together. *Proceedings of the National Academy of Sciences, 108*(52), 20947–20952.

Oullier, O., de Guzman, G.C., Jantzen, K.J., Lagarde, J. & Kelso, J.A.S. (2008) Social coordination dynamics: Measuring human bonding. *Social Neuroscience, 3*(2), 178–192.

Peper, C. (Lieke) E., Stins, J.F. & de Poel, H.J. (2013) Individual contributions to (re-) stabilizing interpersonal movement coordination. *Neuroscience Letters, 557, Part B,* 143–147.

Pikovsky, A., Rosenblum, M. & Kurths, J. (2001) *Synchronization: A universal concept in nonlinear sciences.* Cambridge University Press: Cambridge.

Pineda, J.A., Allison, B.Z. & Vankov, A. (2000) The effects of self-movement, observation, and imagination on mu; rhythms and readiness potentials (RPs): Toward a brain-computer interface (BCI). *IEEE Transactions on Rehabilitation Engineering, 8*(2), 219–222.

Post, A.A., Peper, C.E. & Beek, P.J. (2000) Relative phase dynamics in perturbed interlimb coordination: the effects of frequency and amplitude. *Biological Cybernetics, 83*(6), 529–542.

Prinz, A.A., Bucher, D. & Marder, E. (2004) Similar network activity from disparate circuit parameters. *Nature Neuroscience, 7*(12), 1345–1352.

Richardson, M.J., Marsh, K.L., Isenhower, R.W., Goodman, J.R.L. & Schmidt, R.C. (2007) Rocking together: Dynamics of intentional and unintentional interpersonal coordination. *Human Movement Science, 26*(6), 867–891.

Riley, M.A., Richardson, M., Shockley, K. & Ramenzoni, V.C. (2011) Interpersonal synergies. *Frontiers in Psychology, 2,* 38.

Sänger, J., Lindenberger, U. & Müller, V. (2011) Interactive brains, social minds. *Communicative & Integrative Biology, 4*(6), 655–663.

Schilbach, L., Timmermans, B., Reddy, V., Costall, A., Bente, G., Schlicht, T. & Vogeley, K. (2013) Toward a second-person neuroscience. *Behavioral and Brain Sciences, 36*(4), 393–414.

Schmidt, R.C., Carello, C. & Turvey, M.T. (1990) Phase transitions and critical fluctuations in the visual coordination of rhythmic movements between people. *Journal of Experimental Psychology: Human Perception and Performance, 16*(2), 227–247.

Schmidt, R.C. & O'Brien, B. (1997) Evaluating the dynamics of unintended interpersonal coordination. *Ecological Psychology, 9*(3), 189–206.

Schmidt, R.C. & Turvey, M.T. (1994) Phase-entrainment dynamics of visually coupled rhythmic movements. *Biological Cybernetics, 70*(4), 369–376.

Schöner, G., Haken, H. & Kelso, J.A.S. (1986) A stochastic theory of phase transitions in human hand movement. *Biological Cybernetics, 53*(4), 247–257.

Schöner, G., Jiang, W.-Y. & Kelso, J.A.S. (1990) A synergetic theory of quadrupedal gaits and gait transitions. *Journal of Theoretical Biology, 142*(3), 359391.

Schrödinger, E. (1944) *What is life?* Cambridge University Press: Cambridge.

Shaiman, S. (1989) Kinematic and electromyographic responses to perturbation of the jaw. *The Journal of the Acoustical Society of America, 86,* 78–88.

Sheets-Johnstone, M. (2011) *The primacy of movement* (2nd ed.). John Benjamins Publishing Company: Amsterdam.

Tognoli, E., Lagarde, J., de Guzman, G.C. & Kelso, J.A.S. (2007) The phi complex as a neuromarker of human social coordination. *Proceedings of the National Academy of Sciences, 104*(19), 8190–8195.

Tronick, E., Als, H., Adamson, L., Wise, S. & Brazelton, T.B. (1979). The infant's response to entrapment between contradictory messages in face-to-face interaction. *Journal of the American Academy of Child Psychiatry, 17*(1), 1–13.

Tschacher, W. & Haken, H. (2007) Intentionality in non-equilibrium systems? The functional aspects of self-organized pattern formation. *New Ideas in Psychology*, *25*(1), 1–15.

Tuller, B. & Kelso, J.A.S. (1989) Environmentally-specified patterns of movement coordination in normal and split-brain subjects. *Experimental Brain Research*, *75*(2), 306–316.

Turvey, M.T. (1990) Coordination. *American Psychologist*, *45*(8), 938–953.

Turvey, M.T. & Carello, C. (1996) Dynamics of Bernstein's level of synergies. In M.L. Latash & M.T. Turvey (Eds.), *Dexterity and Its Development* (pp. 339–377). Erlbaum: Mahwah, NJ.

van der Wel, R.P.R.D., Knoblich, G. & Sebanz, N. (2011) Let the force be with us: Dyads exploit haptic coupling for coordination. *Journal of Experimental Psychology: Human Perception and Performance*, *37*(5), 1420–1431.

Varlet, M., Marin, L., Lagarde, J. & Bardy, B.G. (2011) Social postural coordination. *Journal of Experimental Psychology: Human Perception and Performance*, *37*(2), 473–483.

Wilson, A. & Bingham, G. (2008) Identifying the information for the perception of relative phase. *Perception & Psychophysics*, *70*, 465–476.

Winfree, A.T. (2001). *The geometry of biological time* (2nd ed.). Springer: Berlin.

Zaal, F.T.M. & Bongers, R.M. (2014) Movements of individual digits in bimanual prehension are coupled into a grasping component. *PLoS ONE, 9*(5): e97790.

4 The ties that bind

Unintentional spontaneous synchrony in social interactions

Tehran Davis

Introduction

A pair of strangers walking down the street pace lock-step with one another. During a curtain call, the individual claps of one thousand people settle, quite naturally, into the rhythmic applause of an audience. These common everyday experiences provide anecdotal evidence for what is becoming a well-studied phenomenon: people unconsciously, unintentionally and sometimes uncontrollably entrain their behaviours with one another.

The present chapter offers a brief entry into the scientific study of the behavioural entrainments and alignments that occur between co-actors. Here, we will focus our attention upon the unintended coordination of activity between agents. Indeed, many joint actions – such as dancing, passing a plate and engaging in conversation – demand that people actively control their behaviours in space and time to achieve some desired ends. In this case, the resulting coordination of dance steps, arm movements and verbal discourse are normally explicitly guided by the actors; that is, their coordination is *intentional*. In contrast, unintentional coordination occurs outside of the conscious control of actors, such as the subtle entrainment of head nods during a conversation (Hadar, Steiner & Rose, 1985). At first glance unintentional coordination may appear spurious or incidental to the primary social interaction at hand. However, contemporary theories of psychology and human behaviour treat the routine presence of unintentional interpersonal coordination in social interactions as anything but accidental, and propose instead that the unintentional coordination of movements and behaviours between actors is indicative of a larger, more fundamental phenomenon that lies at the foundations of what it means to be social.

Two general types of unintentional interpersonal coordination have been well studied in literature: *non-conscious mimicry* and *spontaneous synchrony*. Given the constraints of the current volume, I only briefly highlight *non-conscious mimicry*. Research concerning non-conscious mimicry dates back to the 1960s, when early empirical work by Condon and Ogston (1966) noted the presence of an "interactional synchrony" between partners engaged in a verbal interaction. In the time since, a large body of work has demonstrated that actors routinely align their posture (Bernieri, Reznick & Rosenthal, 1988; Kendon, 1970;

LaFrance & Broadbent, 1976; Shockley, Santana & Fowler, 2003), facial expressions (Dimberg, Thunberg & Elmehed, 2000; Hess & Blairy, 2001; Wallbott, 1991), mannerisms (Chartrand & Bargh, 1999) and limb movements (Schmidt & Turvey, 1994; Schmidt, Carello, & Turvey, 1990), while interlocutors often match each other's vocal intensities and inflections (Natale, 1975; Neumann & Strack, 2000), as well as their speaking rates and patterns (Cappella & Planalp, 1981; Giles & Powesland, 1975; Street, 1984) and move in synchrony with one another's rhythms of speech (Condon & Ogston, 1966; Hadar et al., 1985; Newtson, 1994). Each of these phenomena are an instance of unintended interpersonal coordination – the actors are neither instructed nor required to align these behaviours with one another and are often unaware that they are doing so. A more detailed account of non-conscious mimicry phenomena may be found in Chartrand and Lakin's (2013) excellent review. In what follows we will outline key discoveries as well as describe methodologies employed in the study of unintentional, spontaneous synchrony in interpersonal coordination.

An at-home demonstration of spontaneous synchrony

Research on *spontaneous synchrony* focuses on the emergence of temporal alignments and entrainments between the fine-grained movements of co-effectors that occur over time. Those investigating spontaneous synchrony in the interpersonal domain place special significance on the dynamic and reciprocal adaptation in the temporal structure of co-actors' behaviour (Schmidt & Richardson, 2008). Before continuing to interpersonal considerations, however, let us take a break to provide an example of spontaneous synchrony as it occurs within a single actor. Though the following example may seem a bit odd, the conclusions drawn from similar experiments have had a huge impact on interpersonal synchrony research for the last two decades. After establishing the basic phenomena here, we will see how these basic findings extend into broader domains.

First, a simple example: Place this book down somewhere so that you will still be able to read and follow along. Extend your arms with the palms of your hands facing outward. Abduct your right hand (rotating your wrist inward to the midline of your body) as far as is comfortable – this position we may describe as $\theta = 0°$. Then adduct your right hand (rotating your wrist away from your body) as far as possible – here $\theta = 180°$. Waving your right hand from full abduction to full adduction and back again may be characterized as a *rhythmic phase cycle* moving from 0° to 180° to 360°/0°; the same can be defined for the left hand (abduction: moving the hand toward 0°; adduction: moving toward 180°). The collective positions of both hands may be described in terms of their *relative phase* φ, where $\varphi = \theta_{RH} - \theta_{LH}$. Now, begin waving your right hand back and forth, and after a few moments wave your left in tandem. If you are like most people, your hands will almost immediately settle into one of two extremely reliable coordination patterns – either: 1) your hands were abducting and adducting in unison, with both approaching 0° and 180° at nearly the same time; or 2) as one hand was abducting the other was adducting, such that as one approached 0° the other approached 180°. In the former case, when both hands (or limbs, heads or other effectors) are continuously abducting and adducting simultaneously, their movements are said to be *in-phase* ($\varphi = 0°$) with one another. In contrast, *anti-phase* ($\varphi = 180°$) coordination describes instances where as one hand abducts, the other adducts. Without providing any explicit instruction, I would be willing to bet that either of these two patterns emerged almost immediately and remained quite stable.

Okay, now for an experiment: wave your hands in unison, but this time try to avoid either of these two patterns of coordination. For example, you might try to make it so that as your left hand approaches 0° your right hand is at the midpoint of its movement around 90°. You may note that it is more difficult but not impossible to maintain this pattern (it helps if you reduce the frequency of movement). Once you have settled into a stable solution, slowly start to speed up your hands until you cannot move them any faster. Resist as much as you like, but eventually you will settle into a relatively stable in-phase pattern. In fact, if you start your hands in any other phase relation, including anti-phase, they will eventually settle into an in-phase solution.

Congratulations – you have just replicated one of the seminal findings in our understanding of inter-limb coordination (Kelso, 1984). The study of oscillating effectors in organisms (e.g. arms, legs, antennae or fins) has for some time noted the existence of preferred modes of coordination (Holst, 1937). In their landmark paper Haken, Kelso, and Bunz (1985) developed a dynamical model for describing the "emergence of" and "transitions between" stable patterns of coordination for pairs of limbs moving in a rhythmic, sinusoidal fashion. This model, named after its authors (HKB), originally described the self-organization of coordination during bi-manual finger tapping in terms of a minimal dynamical model of coupled nonlinear oscillators (Haken, 1983). Importantly, the HKB model makes specific predictions about the emergence of the two aforementioned stable states (*in-phase, anti-phase*) in the inter limb coordination of an actor given several control parameters, including frequency of oscillation as demonstrated above.

Extensions to interpersonal synchrony

In the previous section we highlighted spontaneous interlimb synchrony within an actor. When only considering an individual actor, there is an obvious appeal to explain observed constraints on limb coordination as a function of inherent physiology, where observed coordination dynamics may reflect the characteristics of the connections between bone, ligature, muscle and the central nervous system. However, shortly after the development of the HKB model, research began to suggest that when pairs of individuals coordinate the movements of their limbs with one another, the resulting patterns of synchrony were commensurate in kind with the coordination dynamics predicted by HKB (Schmidt et al., 1990; Schmidt & Turvey, 1994). Here, actors were asked to swing pendulums in a sinusoidal fashion at their preferred speed, while facing toward or away from one another.

When actors faced one another (i.e., had visual information about the movements of their confederate) their swings became entrained. Even though the limbs of co-actors are not physiologically or mechanically bound, it appears that they may become informationally coupled via the perception of one another's activity. Building on this paradigm, many investigations of interpersonal synchrony have focused on the rhythmic movement of arms, legs and other manipulanda (Marsh, Richardson & Schmidt, 2009; Richardson, Garcia, Frank, Gergor & Marsh, 2012; Schmidt, Bienvenu, Fitzpatrick, & Amazeen, 1998). A large number of these studies have confirmed the predictions of the HKB model – stable patterns of in-phase and anti-phase coordination emerge when actors are *instructed* to entrain their movements with another actor or metronome, with in-phase being the stronger attractor than anti-phase.

But what of cases when participants were not asked to intentionally synchronize their movements? Although similar patterns of interpersonal synchrony occur spontaneously outside of the conscious control of co-actors (Coey, Varlet, Schmidt & Richardson, 2011; Issartel, Marin & Cadopi, 2007; Miles, Lumsden, Richardson & Neil Macrae, 2011; Oullier, de Guzman, Jantzen, Lagarde & Kelso, 2008; Schmidt & O'Brien, 1997), only intermittent periods of in-phase or anti-phase coordination are typically observed. That is, the rhythmic movements of actors tend to drift in and out of stable regions, in part due to the relative strengths of the coupling between actors and the tendency for individual movements to maintain their intrinsic dynamics. For example, Schmidt and O'Brien (1997) asked pairs of participants to swing pendulums with different natural frequencies at their preferred tempo. When co-actors did not face one another (i.e., when they were not visually linked), each tended to maintain their own preferred frequency. However, when actors were asked to look at one another's pendulums, a mutual influence was observed – each actor deviated from their preferred frequency. Of note, in-phase and anti-phase coordination, while not absolute, dominated the coordination regime, with in-phase coordination being more prevalent.

Richardson and colleagues (2005) investigated this phenomenon using a similar paradigm with rocking-chairs. Here, co-actors were given a puzzle to "solve" by discussing differences between two similar pictures. As each participant could only see one of the pictures, they had to converse with one another in order to complete the task. Unintended coordination of rocking was observed in trials where participants watched one another's movements while performing the task. Notably, verbal discussion by itself had no impact on the presence or degree of unintended synchrony in this task. However, it has been demonstrated that verbal discourse may lead to the spontaneous interpersonal coordination of postural sway (Shockley et al., 2003; Shockley, Baker, Richardson & Fowler, 2007), the low-amplitude fluctuations of the body's centre of mass that occur during upright stance. As might be expected, the unintended coordination of postural sway may be visually mediated as well (Varlet, Marin, Lagarde & Bardy, 2011).

Unintended coordination may also emerge nested within intentional coordination tasks. For example, many joint actions, such as passing a cup, demand a high level of spatiotemporal coordination of the limbs to be successful.

In this case, coordination at the hands may "trickle down" into other regions of actors' bodies resulting in unconscious coordination of effectors more removed from the immediate task, such as the torso (Ramenzoni, Davis, Shockley & Baker, 2011; Ramenzoni, Shockley & Baker, 2012). Moreover, Athreya and colleagues (Athreya, Riley & Davis, 2014) observed that torsos become spontaneously entrained during a super postural alignment task even during trials where participants had no visual information related to each other's actions other than the movements of a small target projected on a screen. This suggests that that the observed unintentional coordination of torsos may be mediated by inherit task constraints in addition to visual entrainment.

In some cases, spontaneous synchrony has been observed to coincide with higher-order socialization and cognitive performance in interpersonal settings. Spontaneous synchrony enhances altruism and empathy between co-actors (Valdesolo & DeSteno, 2011), affects (Miles, Nind & Macrae, 2009; Reddish, Bulbulia & Fischer, 2013) and is affected by (Cohen, Mundry, & Kirschner, 2013; Miles, Griffiths, Richardson & Macrae, 2010; Paxton & Dale, 2013) rapport or shared belief between actors, strengthens feelings of connectedness (Hove & Risen, 2009), is mediated by affiliation (Miles et al., 2011), and promotes pro-social behaviour in infants (Cirelli, Einarson, & Trainor, 2014a; Cirelli, Wan, & Trainor, 2014b; Trainor & Cirelli, 2015; Tunçgenç, Cohen, & Fawcett, 2015).

Many of the studies highlighted thus far are more broadly framed within a dynamical systems approach to social interaction (Coey, Varlet, & Richardson, 2012; Marsh, 2010; Marsh et al., 2009; Schmidt & Richardson, 2008). This approach argues that social interactions, particularly joint actions that involve the motor coordination of two or more actors, may best be understood as the formation of "softly-assembled" coordinative structures between people (Riley, Richardson, Shockley & Ramenzoni, 2011; Shockley, Richardson & Dale, 2009). Rather than direct appeals to physiological or cognitive mechanisms, dynamical systems approaches to motor coordination appeal to physical principles of self-organization for explaining how the large number of degrees of freedom (e.g. in neuromuscular groups) might become functionally coordinated without a centralized controller or mechanism (i.e. motor programmes) – a significant issue when "control" is spread out among different actors. In individuals, the "solving" of this problem involves the recruitment and reduction of degrees of freedom through gradual, emergent constraints among components of the system (e.g. parts of the body) during the execution of an action. In the case of joint action, it may be argued that the coordination of action between individuals is the result of similar processes – mutual constraint and synergistic processes across two or more people's bodily and cognitive states.

Studies highlighting lawful regularities in rhythmic synchrony perhaps offer the clearest articulation of this approach. A successful model of unintentional interpersonal inter-limb coordination can be produced by coupling the two oscillatory components (the tendency for limbs to become entrained) and introducing a parameter that accounts for the difference in the inherent tendencies of each limb along with a control parameter, such as the frequency of oscillation, that drives the system through its behavioural modes (Kelso, DelColle & Schoner, 1990). Thus interpersonal inter-limb coordination may be successfully described using a generic, dynamical model that makes no specific claims about the physiology (bones, muscles, neurons) of its components. The relative stabilities of in-phase and anti-phase coordination suggest the existence of two attractors at $0°$ and $180°$. The empirical observation that anti-phase coordination is weaker than in-phase coordination is predicted in this model – as the system is pushed (e.g. velocity is increased) the $180°$ attractor is annihilated, leaving only the in-phase solution. The predicted asymmetry between the strength of phase modes also raises an important theoretical question – not only may the presence of unintended synchrony be fundamental to a given social outcome, but also the *mode* of unintended synchrony may influence characteristics of "higher order" interactions. For example, the unintended *in-phase* synchrony of hand movements with a confederate enhances memory recall contrasted with both asynchrony and *anti-phase* synchrony (Macrae, Duffy, Miles & Lawrence, 2008).

More broadly, dynamical accounts have also been offered for observed non-conscious mimicry phenomena. For example, rhythmic motor interference (RMI) occurs when an actor is tasked with producing oscillatory limb movements while at the same time observing a co-actor producing movements in the orthogonal plane. This often leads to increased movement variability in the incongruent plane (Bouquet, Shipley, Capa & Marshall, 2011; Capa, Marshall, Shipley, Salesse & Bouquet, 2011; Kilner, Paulignan & Blakemore, 2003; Stanley, Gowen & Miall, 2007). A common interpretation of these results is that each actor co-represents the other's actions, and this in turn interferes with one's own action planning and execution. Interestingly, RMI has been demonstrated to be independent of co-actors' effectors – an actor moving his arm experiences interference when observing a co-actor move her leg in the orthogonal plain (Fine, Gibbons & Amazeen, 2013), which is inconsistent with embodied simulation interference. Work by Richardson and colleagues (Richardson, Campbell & Schmidt, 2009; Romero, Coey, Schmidt & Richardson, 2012) suggests that the observed patterns of RMI may be explained as a consequence of the task dynamics, reflecting the spontaneous recruitment of degrees of freedom in the non-instructed plane.

Experimental methods and data collection

Researchers have employed a wide variety of measures and methods in the quantitative analysis of unintended interpersonal coordination. Perhaps the simplest involves taking an aggregate measure of some activity over a window of time. Contemporary synchrony research has increasingly focused on measures that capture the spatial-temporal relationships of the interaction by obtaining or creating time series of activity. By "unfolding the temporal dimension" (Coco & Dale, 2014) of an interaction, researchers are able to investigate its underlying patterns of organization. In turn, these patterns may yield important insights into the predictably, complexity, stability and temporal relations inherent in the emerging social coordination. The use of time series analysis is especially prevalent in synchrony research, with its emphasis on the dynamics of the temporal alignments and entrainments between co-actors.

As mentioned before, the rhythmic swinging of limbs and other manipulanda is a common paradigm for investigating spontaneous interpersonal synchrony. As a relatively simple measure, the continuous rocking back and forth of limbs, pendulums and chairs may be expressed using the phase angle of its circular trajectory on a position-by-velocity phase space. In turn, the positions of two or more actors may be combined into a single measure of their relative positions in the oscillatory cycle, or their *relative phase*. While models of inter-limb coordination predict stable modes of in-phase and anti-phase modes coordination, in contrast to intentional rhythmic synchrony (Schmidt et al., 1998; Schmidt & Turvey, 1994), unintended interpersonal rhythmic coordination is not phase-locked (i.e. a constant relative phase) but intermittent – actors tend to periodically enter into and fall out of in-phase and anti-phase coordination. Entrainment may be assessed by both the distribution of relative phase angles and a measure of cross-spectral coherence (Coey et al., 2011; Lopresti-Goodman, Richardson, Silva & Schmidt, 2008; Richardson, Marsh, Isenhower, Goodman & Schmidt, 2007b; Schmidt & O'Brien, 1997). The former involves creating a histogram from the relative phase angles at each point in the time series where stable entrainment is indicated by a high frequency of relative phase angles in any one of the resulting bins (normally bins near 0° or 180°, indicating in-phase or anti-phase entrainment). The latter measure begins by performing a cross-spectral analysis of participants' position time series, expressing the original time series in the frequency domain. Cross-spectral coherence identifies the frequency-domain correlation between the two time series where a value of 0 indicates no entrainment and 1 indicates full entrainment.

Many natural forms of interpersonal coordination involve movements that are too complex to be captured by a relative phase measure. Investigating interactional synchrony of these sorts require more involved analysis techniques. Cross-recurrence quantification analysis (CRQA) (Coco & Dale, 2014; Marwan, Carmen Romano, Thiel & Kurths, 2007; Shockley, 2005) is a non-linear analysis technique that is able to capture the spatial and temporal structure of coordinated activity between confederates. Most notably, CRQA is able to deal with movement time series that are complex, irregular and non-stationary. CRQA begins with projecting participants' time series into a multidimensional phase space and proceeds by identifying moments in time where one time series is in a state that was previously occupied by the other. The observed patterns of co-visitation can yield important insights into the deterministic characteristics of interpersonal coordination, including the degree and temporal structure of shared activity, stability, complexity and lead-lag relationships (Athreya et al., 2014; Richardson & Dale, 2005; Richardson, Dale, & Kirkham, 2007a).

A more recent development is the use of cross-wavelet analysis (Maraun & Kurths, 2004; Prokoph & Bilali, 2008) in interpersonal coordination studies (Fine et al., 2013; Issartel et al., 2007; Schmidt, Nie, Franco & Richardson,

2014; Washburn et al., 2014). As with CRQA, cross-wavelet analysis is able to handle non-stationary signals, making it well equipped for the study of more natural forms of movement synchrony. Moreover, cross-wavelet analysis is also able to provide a measure for interactions that involve several nested temporal scales. This method builds upon spectral methods by expressing component frequencies as a function of time, thus allowing of the analyses of coherence and phase relation across multiple time scales (Issartel, Bardainne, Gaillot & Marin, 2015).

Conclusions

While our everyday experience of social interactions is dominated by the intentional coordination of goals, actions and ideas, beneath the surface we are constantly, automatically and unintentionally coupling ourselves to our social others. Converging evidence indicates that the covert coordination of gestures and rhythms that takes place outside our awareness is not only crucial to the fluidity of our interactions, but also has an important influence on the development of empathy, understanding and bonding with others. Research in this domain, while still relatively new, is yielding profound results, and it is becoming increasingly apparent that the principles that guide spontaneous social entrainments may be a reflection of some deeper, more fundamental laws. The development of new technologies and analysis techniques to capture and quantify coordination at multiple scales promises to provide new insights and raise more questions in the endeavour to understand our social nature.

References

Athreya, D.N., Riley, M.A. & Davis, T.J. (2014) Visual influences on postural and manual interpersonal coordination during a joint precision task. Experimental Brain Research, 232, 2741–2751.

Bernieri, F.J., Reznick, J.S. & Rosenthal, R. (1988) Synchrony, pseudosynchrony, and dissynchrony: Measuring the entrainment process in mother-infant interactions. Journal of Personality and Social Psychology, 54, 243–253.

Bouquet, C.A., Shipley, T.F., Capa, R.L. & Marshall, P.J. (2011) Motor contagion: goal-directed actions are more contagious than non-goal-directed actions. Experimental Psychology, 58, 71–78.

Capa, R.L., Marshall, P.J., Shipley, T.F., Salesse, R.N. & Bouquet, C.A. (2011) Does motor interference arise from mirror system activation? The effect of prior visuo-motor practice on automatic imitation. Psychological Research, 75, 152–157.

Cappella, J.N. & Planalp, S. (1981) Talk and Silence Sequences in Informal Conversations iii: Interspeaker Influence. Human Communication Research, 7, 117–132.

Chartrand, T.L. & Bargh, J.A. (1999) The chameleon effect: The perception-behavior link and social interaction. Journal of Personality and Social Psychology, 76, 893–910.

Chartrand, T.L. & Lakin, J.L. (2013) The antecedents and consequences of human behavioral mimicry. Annual Review of Psychology, 64, 285–308.

Cirelli, L.K., Einarson, K.M. & Trainor, L.J. (2014a) Interpersonal synchrony increases prosocial behavior in infants. Developmental Science, 17, 1003–1011.

Cirelli, L.K., Wan, S.J. & Trainor, L.J. (2014b) Fourteen-month-old infants use interpersonal synchrony as a cue to direct helpfulness. Philosophical Transactions of the Royal Society B: Biological Sciences, 369, 20130400–20130400.

Coco, M.I. & Dale, R. (2014) Cross-recurrence quantification analysis of categorical and continuous time series: an R package. Frontiers in Psychology, 5. doi: 10.3389/fpsyg.2014.00510

Coey, C.A., Varlet, M. & Richardson, M.J. (2012) Coordination dynamics in a socially situated nervous system. Frontiers in Human Neuroscience, 6. doi:10.3389/fnhum.2012.00164

Coey, C., Varlet, M., Schmidt, R.C. & Richardson, M.J. (2011) Effects of movement stability and congruency on the emergence of spontaneous interpersonal coordination. Experimental Brain Research, 211, 483–493.

Cohen, E., Mundry, R. & Kirschner, S. (2013) Religion, synchrony, and cooperation. Religion, Brain & Behavior, 4, 20–30.

Condon, W.S. & Ogston, W.D. (1966) Sound Film Analysis of Normal and Pathological Behavior Patterns. The Journal of Nervous and Mental Disease, 143, 338.

Dimberg, U., Thunberg, M. & Elmehed, K. (2000) Unconscious Facial Reactions to Emotional Facial Expressions. Psychological Science, 11, 86–89.

Fine, J.M., Gibbons, C.T. & Amazeen, E.L. (2013) Congruency Effects in Interpersonal Coordination. Journal of Experimental Psychology: Human Perception and Performance. doi:10.1037/a0031953

Giles, H. & Powesland, P.F. (1975) Speech style and social evaluation. Academic Press: Cambridge.

Hadar, U., Steiner, T.J. & Rose, F.C. (1985) Head movement during listening turns in conversation. Journal of Nonverbal Behavior, 9, 214–228.

Haken, H. (1983) Advanced Synergetics (Vol. 20). Springer: Berlin, Heidelberg.

Haken, H., Kelso, J.A.S. & Bunz, H. (1985) A theoretical model of phase transitions in human hand movements. Biological Cybernetics, 51, 347–356.

Hess, U. & Blairy, S. (2001) Facial mimicry and emotional contagion to dynamic emotional facial expressions and their influence on decoding accuracy. International Journal of Psychophysiology: Official Journal of the International Organization of Psychophysiology, 40, 129–141.

Holst, von, E. (1937) On the nature of order in the central nervous system. The Collected Papers of Erich Von Holst Vol. 1, the Behavioral Physiology of Animal and Man, 133–155.

Hove, M.J. & Risen, J.L. (2009) It's all in the timing: Interpersonal synchrony increases affiliation. Social Cognition, 27, 949–961.

Issartel, J., Bardainne, T., Gaillot, P. & Marin, L. (2015) The relevance of the cross-wavelet transform in the analysis of human interaction – a tutorial. Frontiers in Psychology, 5. doi:10.3389/fpsyg.2014.01566

Issartel, J., Marin, L. & Cadopi, M. (2007) Unintended interpersonal co-ordination: "can we march to the beat of our own drum?". Neuroscience Letters, 411, 174–179.

Kelso, J.A.S. (1984) Phase transitions and critical behavior in human bimanual coordination. The American Journal of Physiology, 246, R1000–4.

Kelso, J.A.S., DelColle, J.D. & Schoner, G. (1990) Action-perception as a pattern formation process. In M. Jeannerod, Attention and performance XIII: Motor Representations and Control (Vol. 5, pp. 139–169). Erlbaum Hillsdale, NJ.

Kendon, A. (1970) Movement coordination in social interaction: some examples described. Acta Psychologica, 32, 100–125.

Kilner, J.M., Paulignan, Y. & Blakemore, S.J. (2003) An interference effect of observed biological movement on action. Current Biology, 13, 522–525.

LaFrance, M. & Broadbent, M. (1976) Group Rapport: Posture Sharing as a Nonverbal Indicator. Group & Organization Management, 1, 328–333.

Lopresti-Goodman, S.M., Richardson, M.J., Silva, P. & Schmidt, R.C. (2008) Period Basin of Entrainment for Unintentional Visual Coordination. Journal of Motor Behavior, 40, 3–10.

Macrae, C., Duffy, O., Miles, L. & Lawrence, J. (2008) A case of hand waving: Action synchrony and person perception. Cognition, 109, 152–156.

Maraun, D. & Kurths, J. (2004) Cross wavelet analysis: significance testing and pitfalls. Nonlinear Processes in Geophysics, 11, 505–514.

Marsh, K.L. (2010) Sociality, from an ecological, dynamical perspective. In G.R. Semin & G. Echterhoff, Grounding sociality neurons, mind, and culture (pp. 43–71). Psychology Press: London.

Marsh, K.L., Richardson, M.J. & Schmidt, R.C. (2009) Social Connection Through Joint Action and Interpersonal Coordination. Topics in Cognitive Science, 1, 320–339.

Marwan, N., Carmen Romano, M., Thiel, M. & Kurths, J. (2007) Recurrence Plots for the Analysis of Complex Systems. Physics Reports, 438, 237–329.

Miles, L.K., Griffiths, J.L., Richardson, M.J. & Macrae, C.N. (2010) Too late to coordinate: Contextual influences on behavioral synchrony. European Journal of Social Psychology, 40, 52–60.

Miles, L.K., Lumsden, J., Richardson, M.J. & Neil Macrae, C. (2011) Do birds of a feather move together? Group membership and behavioral synchrony. Experimental Brain Research, 211, 495–503.

Miles, L.K., Nind, L.K. & Macrae, C.N. (2009) The rhythm of rapport: Interpersonal synchrony and social perception. Journal of Experimental Social Psychology, 45, 585–589.

Natale, M. (1975) Convergence of mean vocal intensity in dyadic communication as a function of social desirability. Journal of Personality and Social Psychology, 32, 790–804.

Neumann, R. & Strack, F. (2000) "Mood contagion": The automatic transfer of mood between persons. Journal of Personality and Social Psychology, 79, 211–223.

Newtson, D. (1994) The perception and coupling of behavior waves. In R.R. Vallacher & A. Nowak, Dynamical systems in social psychology (pp. 139–167). Academic Press: San Diego, CA.

Oullier, O., de Guzman, G.C., Jantzen, K.J., Lagarde, J. & Kelso, J.A.S. (2008) Social coordination dynamics: Measuring human bonding. Social Neuroscience, 3, 178–192.

Paxton, A. & Dale, R. (2013) Argument disrupts interpersonal synchrony. The Quarterly Journal of Experimental Psychology, 66, 2092–2102.

Prokoph, A. & Bilali, El, H. (2008) Cross-Wavelet Analysis: a Tool for Detection of Relationships between Paleoclimate Proxy Records. Mathematical Geosciences, 40, 575–586.

Ramenzoni, V.C., Davis, T., Shockley, K. & Baker, A.A. (2011) Joint action in a cooperative precision task: nested processes of intrapersonal and interpersonal coordination. Experimental Brain Research, 211, 447–457.

Ramenzoni, V.C., Shockley, K. & Baker, A.A. (2012). Interpersonal and intrapersonal coordinative modes for joint and single task performance. Human Movement Science, 31, 1253–1267.

Reddish, P., Bulbulia, J. & Fischer, R. (2013) Does synchrony promote generalized prosociality? Religion, Brain & Behavior, 4, 3–19.

Richardson, D.C. & Dale, R. (2005) Looking to understand: The coupling between speakers' and listeners' eye movements and its relationship to discourse comprehension. Cognitive Science, 29, 1045–1060.

Richardson, D.C., Dale, R. & Kirkham, N.Z. (2007a) The art of conversation is coordination: Common ground and the coupling of eye movements during dialogue. Psychological Science, 18, 407–413.

Richardson, M.J., Campbell, W.L. & Schmidt, R.C. (2009) Movement interference during action observation as emergent coordination. Neuroscience Letters, 449, 117–122.

Richardson, M.J., Garcia, R.L., Frank, T.D., Gergor, M. & Marsh, K.L. (2012) Measuring group synchrony: a cluster-phase method for analyzing multivariate movement time-series. Frontiers in Physiology, 3. doi:10.3389/fphys.2012.00405

Richardson, M.J., Marsh, K.L. & Schmidt, R.C. (2005) Effects of Visual and Verbal Interaction on Unintentional Interpersonal Coordination. Journal of Experimental Psychology: Human Perception and Performance, 31, 62–79.

Richardson, M.J., Marsh, K.L., Isenhower, R.W., Goodman, J.R.L. & Schmidt, R.C. (2007b) Rocking together: Dynamics of intentional and unintentional interpersonal coordination. Human Movement Science, 26, 87–891.

Riley, M.A., Richardson, M.J., Shockley, K. & Ramenzoni, V.C. (2011) Interpersonal Synergies. Frontiers in Psychology, 2. doi:10.3389/fpsyg.2011.00038

Romero, V., Coey, C., Schmidt, R.C. & Richardson, M.J. (2012) Movement Coordination or Movement Interference: Visual Tracking and Spontaneous Coordination Modulate Rhythmic Movement Interference. PLoS ONE, 7, e44761.

Schmidt, R.C. & O'Brien, B. (1997) Evaluating the Dynamics of Unintended Interpersonal Coordination. Ecological Psychology, 9, 189–206.

Schmidt, R.C. & Richardson, M.J. (2008) Dynamics of Interpersonal Coordination. In A. Fuchs & V.K. Jirsa, Understanding Complex Systems (pp. 281–308). Springer: Berlin Heidelberg.

Schmidt, R.C. & Turvey, M.T. (1994) Phase-entrainment dynamics of visually coupled rhythmic movements. Biological Cybernetics, 70, 369–376.

Schmidt, R.C., Bienvenu, M., Fitzpatrick, P.A. & Amazeen, P.G. (1998) A comparison of intra- and interpersonal interlimb coordination: Coordination breakdowns and coupling strength. Journal of Experimental Psychology: Human Perception and Performance, 24, 884–900.

Schmidt, R.C., Carello, C. & Turvey, M.T. (1990) Phase Transitions and Critical Fluctuations in the Visual Coordination of Rhythmic Movements Between People. Journal of Experimental Psychology, 16, 227–247.

Schmidt, R.C., Nie, L., Franco, A. & Richardson, M.J. (2014) Bodily synchronization underlying joke telling. Frontiers in Human Neuroscience, 8, 633.

Shockley, K. (2005) Cross recurrence quantification of interpersonal postural activity. In Tutorials in Contemporary Nonlinear Methods for the Behavioral Sciences: Proceedings of the National Science Foundation Workshop on Nonlinear Methods In Psychology (pp. 142–177).

Shockley, K., Baker, A.A., Richardson, M.J. & Fowler, C.A. (2007). Articulatory constraints on interpersonal postural coordination. Journal of Experimental Psychology: Human Perception and Performance, 33, 201–208.

Shockley, K., Richardson, D.C. & Dale, R. (2009) Conversation and coordinative structures. Topics in Cognitive Science, 1, 305–319.

Shockley, K., Santana, M.V. & Fowler, C.A. (2003) Mutual interpersonal postural constraints are involved in cooperative conversation. Journal of Experimental Psychology: Human Perception and Performance, 29, 326–332.

Stanley, J., Gowen, E. & Miall, R.C. (2007) Effects of agency on movement interference during observation of a moving dot stimulus. Journal of Experimental Psychology: Human Perception and Performance, 33, 915–926.

Street, R.L. (1984) Speech convergence and speech evaluation in fact-finding interviews. Human Communication Research, 11, 139–169.

Trainor, L.J. & Cirelli, L. (2015) Rhythm and interpersonal synchrony in early social development. PubMed NCBI. Annals of the New York Academy of Sciences, 1337, 45–52.

Tunçgenç, B., Cohen, E. & Fawcett, C. (2015) Rock With Me: The Role of Movement Synchrony in Infants' Social and Nonsocial Choices. Child Development, 86, 976–984.

Valdesolo, P. & DeSteno, D. (2011) Synchrony and the social tuning of compassion. Emotion, 11, 262–266.

Varlet, M., Marin, L., Lagarde, J. & Bardy, B.G. (2011) Social postural coordination. Journal of Experimental Psychology: Human Perception and Performance, 37, 473–483.

Wallbott, H.G. (1991) Recognition of emotion from facial expression via imitation? Some indirect evidence for an old theory. British Journal of Social Psychology, 30, 207–219.

Washburn, A., DeMarco, M., de Vries, S., Ariyabuddhiphongs, K., Schmidt, R.C. & Richardson, M.J. (2014) Dancers entrain more effectively than non-dancers to another actor's movements. Frontiers in Human Neuroscience, 8, 800.

5 Symmetry and the dynamics of interpersonal coordination

Michael J. Richardson, Auriel Washburn,
Steven J. Harrison and Rachel W. Kallen

Introduction

Imagine you are at a wedding, seated at one of those circular and immaculately set dinner tables, with all the glasses positioned symmetrically around the table. Although the arrangement is pleasing to the eye, you are often left with a question – which glass is mine? If you are unfamiliar with the social convention, the solution to the dilemma of "whose glass is whose" is ill-defined. Indeed, from the perspective of any one individual seated at the table, their glass could be the one to their right or the one to their left. Accordingly, the collective behaviour of the group of individuals seated at the table exists at the precipice of a right- or left-glass state. This dilemma is quickly solved, of course, as soon as any one individual has the courage to grasp a glass. For example, as soon as one individual chooses a glass to their right the collective order of the group immediately collapses to a right-glass state, with every other individual in the group choosing the glass to the right.

Although somewhat idealized, the example above highlights a deep and meaningful point: the order of behaviour emerges from symmetry-breaking events. Here the term order refers to the collective organization or patterning of system behaviour. Thus, within the context of the "whose glass is whose" example, the selection of an individual glass breaks the symmetrical arrangement of the glasses as a whole and thus defines the organization or pattern of the table's group behaviour.[1] With this point in mind, the current chapter is centred on the following claim: the dynamic stabilities and behavioural patterning of coordinated social activity is defined by the symmetries and symmetry-breaking events that characterize a joint-action task. As we will explain in more detail below, by symmetries we are referring to the set of task-relevant properties that are invariant with respect to some form of transformation. These properties could refer to the physical or informational aspects of the task environment, the biomechanical or perceptual-motor abilities of the co-acting agents, or the agents' intentional states, goals or psychological dispositions.

To demonstrate how the dynamics of social coordination and joint-action are defined by the symmetries, asymmetries and symmetry-breaking events of environmentally situated social action, we first briefly introduce the formal concepts of symmetry and symmetry-breaking. Then, using two empirical examples we illustrate how the principles of symmetry and symmetry-breaking can be employed to understand the behavioural order of social coordination. We feel that is important to acknowledge that the argument presented here has been recently articulated by the authors in Richardson and Kallen (2015), which provides an expansive tutorial on the formal and theoretical principles of symmetry and symmetry-breaking. Here, we provide a more concise discussion of the implications of symmetry and symmetry breaking for understanding human and social behavioural coordination, specifically focused on rhythmic interpersonal motor coordination.

Symmetry

Most people have an intuitive idea of what the term symmetry means, often defining it as the correspondence in configuration of an object or image. As such, when people hear the term symmetry they often think of the bilateral or mirror symmetry of the human face or a butterfly's wings. One can also readily think of numerous geometric shapes, such as a circle or star, as having symmetry, in that these shapes look exactly the same when rotated in certain ways. Although this everyday understanding of symmetry is by no means wrong, the principle of symmetry does not only apply to pictures or objects; it is a much more abstract concept that can be applied to understand the order of anything, from mathematical functions, chemical compounds and crystals to the fundamental laws of nature and the universe. In short, the term symmetry simply refers to an equivalence or invariance of some kind, given some form of transformation. The vagueness of this definition is deliberate, in that 'some kind' and 'some form' can stand for almost anything. Indeed, an equivalence or invariance could refer to the fact that an equilateral triangle looks exactly the same when rotated by 120°, or that a line of four basketballs stacked on a shelf looks the same if one rearranges the order in which they are lined up, or finding that the results of an experiment conducted on Monday November 14th 2014 at 10am at the University of Cincinnati replicate the results obtained using the same experimental method on Friday October 22nd 2014 at 3pm at the University of Aberdeen.

With this more abstract understanding of symmetry in hand, one can also start to see how the symmetry of an object, phenomena or 'thing' can also be quantified, in that some object, phenomena or thing may have more or less symmetry than some other object, phenomena or thing. To use a common example, consider the geometric shapes of a circle and an equilateral triangle. There are an infinite number of rotations and midpoint reflections that leave a circle looking the same, whereas there are only three rotations and three reflections that leave an equilateral triangle the same (see Figure 5.1). Therefore a circle has a larger set or group of symmetries than an equilateral triangle. A square has eight symmetries (four rotations and four reflections) and therefore also has a greater number, a larger group, of symmetries than an equilateral triangle, but still much less that the symmetries of a circle.

The formal quantification of symmetry is achieved using Group Theory, such that the symmetry group of an object or phenomena corresponds to a closed set of transformations that leave the system unchanged with respect to some defined property. For instance, the symmetry group of an equilateral triangle and a square are the dihedral groups D_3 and D_4, respectively. The symmetry group of a circle is the orthogonal group $O(2)$ (see Figure 5.1 for more details). The line of four basketballs described above has the permutation symmetry group S_4, which means that there are n-factorial (4! = 4×3×2×1) ways of indistinguishably rearranging the basketballs.

Unfortunately, a detailed discussion of Group Theory is beyond the scope of the current chapter and we refer the reader to Richardson and Kallen (2015) for a more detailed overview of Group Theory and how it can be employed to develop a formal explanation of how symmetry defines and constrains the dynamics and behavioural possibilities of joint-action and interpersonal coordination (also see e.g. Rosen, 1995; Weyl, 1952 for an introduction to Group Theory). We mention Group Theory here, however, as we will to a limited extent employ the symmetry group concept to explicate the role of symmetry in the examples employed later on in this chapter. Moreover, it is important to note that Group Theory plays an important role in understanding and defining symmetry, including how the symmetries of one thing relate to the symmetries of something else. Indeed, Group Theory cannot only be used to formally define the symmetries of an object or phenomena, but can also be employed to determine whether the symmetries of two or more different objects or phenomena are isomorphic. Accordingly, one can use symmetry groups to generalize a theoretical understanding of the ordered relations that exist across completely different objects or phenomena.

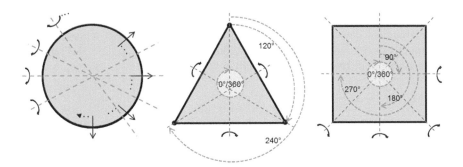

Figure 5.1 Symmetries of a circle, equilateral triangle, and square.

Notes: Left: a circle has an infinite (continuous) number of rotations and reflections that are captured by orthogonal group O(2). Middle: an equilateral triangle has six discrete symmetries, three rotations = {0°/360°, 120°, 240°} and three reflections, that are captured by the dihedral group D3. Right: a square has eight symmetries, four rotations = {0°/360°, 90°, 180°, 270°} and four reflections, that are captured by the dihedral group D4.

Broken symmetry

Symmetry implies asymmetry in that the symmetry of something can only be defined with respect to something that is not symmetric, i.e. it is a gauge that captures when a transformation has occurred (Kugler & Shaw, 1990; Shaw, Kugler & Kinsella-Shaw, 1990; Shaw, Mcintyre & Mace, 1974; Warren & Shaw, 1985). Moreover, complete and absolute symmetry corresponds to complete and absolute asymmetry, both of which equal an indistinguishable nothingness. In order to fully comprehend this later point, imagine if all the matter and energy in the universe was maximally and evenly distributed throughout the universe (i.e. a state of universal maximum entropy). In this state of perfect symmetry the universe would look the same to an observer no matter what point in space (or time) the observer was located. Every place in the universe would therefore be equivalent, leaving an observer simultaneously everywhere and nowhere.

The duality of symmetry and asymmetry and the knowledge that complete symmetry corresponds to the absence of differentiation (the lack of something) implies that the existence of something is dependent on symmetry being broken. To paraphrase the famous words of Pierre Carrie (1894), "dissymmetry [broken symmetry] is what creates phenomena" (see Brading & Castellani, 2003; Castellani, 2003). Symmetry-breaking is therefore essential and fundamental for the emergence and manifestation of structure, organization and behavioural order. Of particular significance for the current discussion is that the redistribution or breaking of symmetry can occur in two ways: *spontaneously* or *explicitly*.

Spontaneous symmetry-breaking

Spontaneous symmetry-breaking refers to the situation where the break in symmetry is entailed by the symmetry of the system itself. The implication is that *the symmetry of the symmetry-breaking effect is related to the symmetry of the causes that bring about that effect* (Stewart & Golubitsky, 1992). With respect to dynamical laws (equations of a time-evolving process), a spontaneous symmetry-break corresponds to solutions of a symmetric law that are not symmetric. Take the differential equation

$$\dot{x} = \alpha x - x^3 \qquad \text{(Eq. 1)}$$

This system has global symmetry with respect to inversion (i.e. $x \rightarrow -x$), which can be represented by the cyclic group, Z_2. As illustrated in Figure 5.2, this global (higher order) symmetry is easily discerned when one plots Eq. 1 as a potential function. When $\alpha > 0$, the higher order symmetry is locally preserved by the stable fixed-point solution at $x = 0$. However, as α is scaled from a positive real number to a negative real number the local symmetry of the system breaks (a bifurcation occurs), with the higher order symmetry of the system being spontaneously redistributed across two new stable states. More specifically, the stable solution at $x = 0$ becomes unstable and two new stable states emerge at $x = \pm\sqrt{\alpha}$, with these

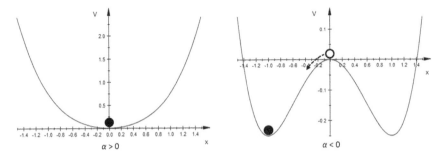

Figure 5.2 Eq. 1 plotted as a potential function for (left) α > 0 and (right) α < 0.

Notes: When α > 0 the system has one stable solution at x = 0, which is invariant with respect to the system's global (higher order) symmetry, namely inversion (i.e., x to –x). When α < 0 the system's local symmetry is broken with two stable solutions at x = ±√a that are not invariant with respect to the higher order symmetry of Eq. 1.

two minimal energy states having less symmetry than the system itself – the system is still invariant to inversion, even though at a local level the system has two differential solutions.

So how does a system that undergoes this kind of spontaneous symmetry break 'choose' which one of the many possible asymmetric states that could be actualized? That is, what determines whether the system defined by Eq. 1 moves from the previously stable state at $x = 0$ to one of the new stable solutions at $x = +\sqrt{\alpha}$ or $x = -\sqrt{\alpha}$? The answer is simple: some form of asymmetry is required in order for the system to 'choose'. This asymmetry could be extremely small and in most physical systems is determined stochastically via a small random fluctuation (i.e. noise). Of more potential interest here, however, is the possibility that the asymmetries that determine one behavioural state or pattern over another can also be non-random perturbations or pre-existing asymmetries or biases.

Explicit symmetry-breaking

Sometimes referred to as induced symmetry-breaking (Golubitsky & Stewart, 2002; Turvey, 2007), explicit symmetry-breaking occurs when the break in symmetry is not entailed by the higher order symmetry of the system (Castellani, 2003). Such symmetry-breaking can originate for different reasons and in some instances can break the symmetry of a system in such a way that it makes it impossible to know what underlying symmetry has been broken (the higher order symmetry becomes 'hidden'). The outcome of such symmetry-breaking, however, is essentially the same as spontaneous symmetry-breaking in that explicit symmetry-breaking can also create phenomena and bring about greater levels of behavioural order.

As a prototypical example of explicit symmetry-breaking, consider the situation in which some imperfection parameter is imposed on a system. For instance, take a pair of coupled oscillators whose collective behaviour is defined by the equation

$$\dot{\phi} = \Delta\omega + \beta \sin(\phi), \tag{Eq. 2}$$

where ϕ and $\dot{\phi}$ are the relative phase and change in relative phase of the two oscillators, $\beta\sin(\phi)$ defines the strength of the between-oscillator coupling, and $\Delta\omega$ is an imperfection parameter that equals the difference in the natural frequency of the two oscillators (Cohen, Holmes & Rand, 1982). As can be seen from an inspection of Figure 5.3, in which Eq. 2 is plotted as a potential function for $-4 < \phi < 4$, when $\Delta\omega = 0$ the system is symmetric with a stable solution at $\phi = 0$ – in degrees this corresponds to a stable 0° or inphase relative phase relationship. In other words, when the two oscillators have the same natural frequency (are symmetric) the solution to the system is symmetric and invariant with respect to inversion (i.e. $\phi \rightarrow -\phi$). However, when $\Delta\omega \neq 0$ the symmetry of the system is broken; the function tilts and the stable solution becomes weaker and moves away from $\phi = 0$ as $|\Delta\omega|$ increases, such that $+\phi \neq -\phi$. With regard to a physical system of coupled oscillators, the most significant effect of this explicit symmetry-break is a relative phase lead/lag, where the oscillator with the faster natural frequency leads the oscillator with the slower natural frequency. Accordingly, when $\Delta\omega \neq 0$ the deviation from perfect inphase coordination maps onto a differential relation that specifies an identity for each oscillator (i.e. oscillator 1 is different from oscillator 2). Of course, there is still a local symmetry operation that leaves the system invariant , which is essentially the identity transformation. Moreover, note that for this example the global symmetry that is broken when $\Delta\omega \neq 0$ can be identified by the fact that the (local) stable solutions for $\Delta\omega < 0$ and $\Delta\omega > 0$, when $|\Delta\omega| < 0 = \Delta\omega > 0$, are qualitatively equivalent.

Clearly, the increase in behavioural order that results from the imperfection parameter, $\Delta\omega$, is minimal. Later on this chapter, however, we will demonstrate how sufficiently large increases in $\Delta\omega$ can not only result in a differential leader-follower relationship between co-actors performing a rhythmic movement task (Schmidt & Turvey, 1994; Schmidt & Richardson, 2008), but can also result in the emergence of more complex patterns of behavioural coordination, including intermittent and polyrhythmic coordination. Before detailing this, there are three related points that need to be noted in regard to explicit symmetry-breaking.

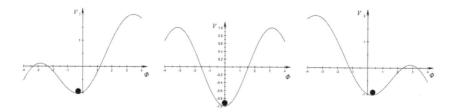

Figure 5.3 Eq. 2 plotted as a potential function when $\Delta\omega$=-.35, 0, and .35 from left to right, respectively (β=1).

Notes: When $\Delta\omega$=0 the system has one stable solution at ϕ=0, which is invariant with respect to the system's higher order symmetry, namely inversion (i.e. ϕ à -ϕ). When $|\Delta\omega|\neq$0, however, the system's local symmetry is broken with the function tilting and the stable solution moving away ϕ=0.

First, as the above example highlights, explicit symmetry-breaking can occur when an introduced or pre-existing asymmetry, imperfection or differential constraint operates to restructure the symmetry of a system. With this in mind, let us restate the point raised previously: *the (a)symmetry of an effect is defined by the (a)symmetries of the causes that bring about that effect.*

Second, nearly all natural or biological systems include asymmetries or imperfections, in that even those systems that one considers to be symmetrical are at best 'nearly symmetrical' or only symmetrical with regards to some idealized or abstract realization of the system. For instance, a human face or a butterfly's wings are never perfectly symmetrical, but are close enough that we can consider them as having bilateral or mirror symmetry. Why is this an important point? Well, not all of these imperfections or pre-existing asymmetries result in a functionally significant redistribution of symmetry and therefore they have little to no effect on the behavioural order expressed. The key is identifying what, when and how these induced or pre-existing asymmetries operate as explicitly symmetry-breaking factors to restructure or reorder the behaviour of a system in a behaviourally relevant manner.

Finally, the parameters that one considers to be explicit symmetry-breaking factors at one spatial or temporal scale of description may reflect the occurrence of a spontaneous symmetry-break at another scale of description. This implies that explicit symmetry-breaking can reflect a cascade of spontaneous symmetry-breaks. The highly differentiated and imperfect universe we exist in is a prime example of this. Such is the high diversity of human physical, cognitive and interpersonal capabilities and dispositions. Moreover, small asymmetries that result from spontaneous symmetry-breaking can significantly bias a system towards specific asymmetric states.

Symmetry-breaking and social coordination

So what do the principles of symmetry and symmetry-breaking have to do with understanding the dynamics of social coordination? The answer is twofold. First, the formal and conceptual language of symmetry provides a context-independent theory of phenomena creation and destruction, and as such provides a general framework for investigating the processes that operate to self-organize the dynamics of social coordination. Second, by identifying the symmetries and symmetry-breaking factors that define the behavioural order of a particular coordination task, one can better understand and identify the dynamical laws that underlie and shape the patterns of coordination that emerge.

Example 1. Detuning and multi-frequency rhythmic interpersonal coordination

As noted above, differences between the natural frequencies of coupled oscillators (i.e. $\Delta\omega \neq 0$), commonly referred to as "detuning", introduce an explicit break in the component symmetry of the system. With regard to the rhythmic coordination that naturally occurs between interacting individuals, say between the leg movements of two individuals walking and talking or the rocking movements of two individuals sitting side-by-side in rocking chairs, the general effects of this explicit symmetry-break can create a differential leader-follower relationship, with the individual with the slower movement frequency lagging behind the individual with the faster movement frequency. However, the effects of detuning on the potential emergence and stability of interpersonal rhythmic coordination are modulated by the strength of the coupling that links the movements of the co-actors. For example, decreases in coupling strength, brought about through a reduction in the visual or auditory information about a co-actor's movements, are known to significantly increase the effects of detuning (i.e. increase coordination variability and the lead/lag relationship between individuals), such that for sufficiently weak coupling the likelihood of even phase lagged 1:1, rhythmic coordination completely vanishes (Schmidt et al., 1998; Schmidt & Turvey, 1994; Richardson et al., 2007).

The 1:1 frequency locked coordination that is easily observed between the same or similar movements of interacting individuals is, however, just one of many possible patterns of rhythmic behavioural coordination: the one pattern defined by a specific combination of low component asymmetry and high coupling strength – high system symmetry. An intriguing question, therefore, is whether more complex patterns of rhythmic interpersonal coordination can be generated by the explicit and spontaneous symmetry breaks introduced through frequency detuning and coupling strength respectively. Previous research on intrapersonal rhythmic coordination supports this possibility, in that an explicit symmetry-break introduced by frequency detuning may actually facilitate the performance of coordination modes other than 1:1 frequency locked, inphase coordination (e.g. Sternad, Turvey & Saltzman, 1999; Treffner & Turvey, 1993). However, tasks explicitly requiring individuals to maintain complex oscillatory phase relationships are often very difficult to execute and require extensive practice (Fontaine, Lee & Swinnen, 1997; Zanone & Kelso, 1992). Here the necessity to produce a whole number frequency ratio between oscillatory movements which deviates from 1:1 (i.e. a multifrequency pattern of coordination, such as 1:2 or 2:3) introduces what has been referred to as a "task asymmetry" distinct from the inherent asymmetry of the component oscillatory subsystems that results from detuning (see Byblow & Goodman, 1994). The difficulty associated with successfully producing intrapersonal, multi-frequency coordination is due to high (biomechanical) coupling strength between limb movements. Indeed, at a very high coupling strength, 1:1 rhythmic coordination remains the most stable coordination mode even for a pair of oscillators that have very different natural frequencies (Peper et al., 1995). However, the introduction of detuning between oscillators with low or moderate coupling strength can facilitate the production of more complex patterns of coordination. Sternad et al. (1999) were able to demonstrate this effect by maintaining a consistent coupling strength between a participant's hands during a bimanual pendulum swinging task, and showed that participants exhibited significantly greater stability for 1:2

coordination when swinging two wrist-pendulums with distinctly different natural frequencies compared to when swinging two identical wrist-pendulums.

The possible *n:m* ratios of multifrequency coordination that can be achieved between two coupled oscillators is specified by the Farey Tree (see Figure 5.4), where movement down the tree is associated with an increase in the order of the performed frequency ratio between oscillators and a simultaneous decrease in the stability of coordination. Somewhat counterintuitively, then, when the coupling strength between a pair of oscillators with distinctively different natural frequencies is low, the possibility for multifrequency modes of coordination actually increases. In other words, lower coupling strength in the presence of frequency-based component asymmetries actually facilitates the successful production and emergence of multifrequency coordination. Extended to the circle map that is used to model rhythmic coordination as an emergent frequency relationship between two oscillators based on the ratio of their natural frequencies and coupling strength (Epstein, 1990; Peper et al., 1995; Pikovsky et al., 2003), the relative stabilities of possible multifrequency coordination modes correspond to an explicitly moderated, spontaneous symmetry-break. The stability predictions of the circle map illustrated as Arnold tongues (see Figure 5.4 right) allow one to easily identify which combinations of frequency detuning and coupling strength afford favourable conditions for the emergence of specific multifrequency patterns. Here, the black areas are the Arnold tongues and represent the specific intersection of component asymmetry and coupling strength between oscillators that supports the emergence of a number of multifrequency coordination patterns, as well as the differentially sized basins of attraction for each n:m mode.

So what does all this mean for the question we raised above – namely, can more complex patterns of rhythmic interpersonal coordination be generated by the explicit and spontaneous symmetry breaks introduced through frequency

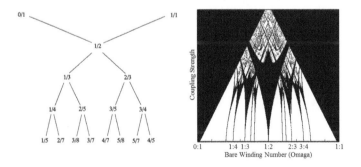

Figure 5.4 Left: Farey Tree (Peper et al., 1995). Higher order ratios are less stable than lower order ratios, and thus are less likely to occur or be sustained over time. Right: Arnold tongues for the stable frequency ratios (captured by the bare winding number, omega) predicted by the circle map as a function of coupling strength. Solid areas correspond to stable n:m behaviour.

detuning and coupling strength, respectively? In short, recent research indicates that the answer to this question is yes! Most notably, the research by Washburn, Coey, Romero, and Richardson (2014), in which participants comfortably swung a fixed-period wrist-pendulum while observing an oscillating, computer-generated stimulus that moved at a frequency ranging from well below to well above the resonant frequency of the pendulum. Results demonstrated that the participants became spontaneously and unintentionally entrained with the oscillatory movements of the stimuli at a variety of multifrequency patterns, namely 1:2, 2:3 and 3:4. Moreover, the occurrence of these $n{:}m$ patterns was completely consistent with the (a)symmetry predictions inherent to the Farey Tree, as well as the stability predictions of the circle map, in that they only occurred if (i) the ratio of inherent frequency difference was close to a rational number that characterizes a specific $n{:}m$ pattern and (ii) the between-movement coupling was weak (i.e. visual and not task defined). Thus, the behavioural order observed was a direct consequence of the symmetry-breaking factors that defined the system components and component interactions, in this case one explicit and one spontaneous.

One should appreciate that because the multifrequency coordination observed by Washburn et al. (2014) was exhibited in a task in which no coordination was explicitly required, the emergence of multifrequency coordination as a function of component asymmetries is not limited to contexts in which coordination is intentional. In other words, even for circumstances in which people's natural dynamics are largely dissimilar to the rhythms of those around them, these asymmetries create opportunities for different and more complex spontaneous entrainment and coordination to occur. One can imagine this might be the case when a young child walks alongside their parent (Kelso, DeGuzman & Holroyd, 1991), whereby the degree of pre-existing asymmetry between the physical characteristics of the parent and child can be seen as predictably prescribing their stable, coordinated behaviour to a small range of multifrequency modes. It is also worth noting that physical component asymmetries are not the only way in which the symmetry of rhythmic interpersonal coordination can be explicitly broken. Numerous social psychological asymmetries, including social competence and group membership or identity (i.e. in-group vs. out-group status) can also result in more complex patterns of rhythmic coordination, including lead/lag relationships and intermittent and potentially polyrhythmic modes of coordination (e.g. Miles, Lumsden, Richardson & Macrae, 2011; Schmidt, Christianson, Carello & Baron, 1994).

Example 2. Gait symmetry and the dynamics of "horsing around".

The potential power of symmetry arguments, when applied to the challenge of modeling the behavioural dynamics of coordinated activity, can also be exemplified by means of legged locomotion. To understand how, let us first briefly review the group symmetry approach developed by Golubitsky, Stewart and Collins (e.g. Collins & Stewart, 1993; Golubitsky & Stewart, 2003), in which modelling the gait patterns observed in nature proceeds by seeking the minimal network architecture of coupled oscillators that produce periodic solutions with matching symmetries. What makes this approach so noteworthy is that it is not concerned with the specifics of component oscillator dynamics; rather, at issue is the correspondence of the symmetry of the observed gait pattern (i.e. the symmetry of the effect) with respect to the symmetry of the inter-component dynamics and couplings hypothesized to underlie the gait system (i.e. the symmetry of the hypothesized cause; Golubitsky, Stewart, Buono & Collins, 1999).

Specifically, Golubitsky, Stewart and Collins (GSC) have demonstrated that the spontaneous spatiotemporal patterning of bipedal and quadrupedal locomotion can be encompassed by a network of coupled oscillators with oscillators numbering twice the number of legs (Golubitsky et al., 1999). This number is motivated by the assumption that each leg is associated with two oscillators, reflecting a distinction between functional extensor and flexor muscle groupings, and that each oscillator is associated with the output dynamics of only one leg. Thus, two symmetry constraints are imposed on GSC models of gait dynamics, namely leg number and the number of functional phases of limb action. The importance of these two symmetry constraints has been revealed in investigations that have shown that: (i) networks with less than twice as many oscillators as legs do not produce all of the periodic solutions observed in nature; and (ii) networks with more than twice as many oscillators as legs produce periodic solutions with unnatural phase shifts (Golubitsky & Stewart, 2003).

Figure 5.5 depicts the architecture of a network shown to have robust periodic solutions that correspond to the symmetries of bipedal gait patterns observed in nature (Pinto & Golubitsky, 2006). This network is a graph whose nodes are oscillators and whose edges indicate which oscillators are coupled to each other.

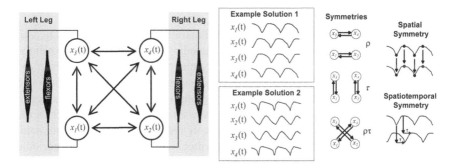

Figure 5.5 The left panel depicts a network of four mutually coupled oscillators used to model the symmetries of bipedal gait patterns. The middle panel shows examples of two stable periodic solutions produced by such a network. The two solutions differ with respect to both their spatial and spatiotemporal symmetries. The right panel shows the permutation symmetries of this network and the distinction between spatial and spatiotemporal symmetry.

The four identical second order oscillators and couplings are given by

$$\dot{x}_1 = F(x_1, x_2, x_3, x_4)$$
$$\dot{x}_2 = F(x_2, x_1, x_3, x_4)$$
$$\dot{x}_3 = F(x_3, x_1, x_2, x_4)$$
$$\dot{x}_4 = F(x_4, x_1, x_2, x_3) \qquad\qquad\qquad\qquad \text{(Eq. 3)}$$

The symmetry of the periodic solutions to Equation 3 is revealed through the transformations that map solutions to solutions – more specifically, where a gait pattern solution is invariant with respect is a specific (sub)group of transformations. Two symmetry subgroups can be considered: the subgroup of spatial symmetries, and the subgroup of spatiotemporal symmetries (Figure 5.5). Spatial symmetries map solutions to solutions point- or state-wise. Spatiotemporal symmetries map solutions to solutions via a phase shift. With reference to Figure 5.5, the spatiotemporal symmetry group of example solution 1 shown in the middle panel is D_2. It consists of the transpositions $\rho = (12)(34)$, $\tau = (13)(24)$ and $\rho\tau = (14)(23)$. The spatial symmetry group, however, is Z_2 as it only includes the transposition $\rho\tau$.

The example solution 1 in Figure 5.5 has symmetries that correspond to those of a bipedal walk, where during walking the gastrocnemius and tibialis anterior, both ankle flexors, are activated out-of-phase. Accordingly, a simple classification of gait patterns can be achieved through group symmetry analysis between primary gaits and secondary gaits. Classification is based upon the spatiotemporal symmetry group. Gaits with D_2 spatiotemporal symmetry are classified as primary gaits while gaits with less than D_2 spatiotemporal symmetry are classified as secondary gaits. As such, primary gaits include walking, running and two-legged hopping, and secondary gaits include skipping, gallop-walking, and hesitation-walking (i.e. a wedding march). Example solution 2 shown in Figure 5.5 is a secondary gait with Z_2 as both its spatial symmetry group and its spatiotemporal symmetry group.

The symmetry group of the network illustrated in Figure 5.5 is D_2. The symmetry of the network identifies the periodic solutions (and consequently, gaits) that are permitted given particular symmetry constraints (i.e. given the constraints of two legs and two functional phases of limb action). As mentioned above, the identification of the symmetry group is ambiguous with respect to which solution should be observed in a given context and with respect to the specific dynamics of component oscillators. If the symmetry constraints hold, however, what the symmetry group does predict is the symmetry of the periodic solutions that can be observed. A minimal application of this GSC symmetry group perspective in interpersonal coordination can be made with respect to the experimentally created interpersonal locomotion system investigated by Harrison and Richardson (2009) depicted in Figure 5.6. Here two participants were tasked with walking one behind the other while maintaining a fixed distance of separation. A priori, without knowledge of the specific form of interpersonal coupling in this system, it is not possible to predict the emergent form of

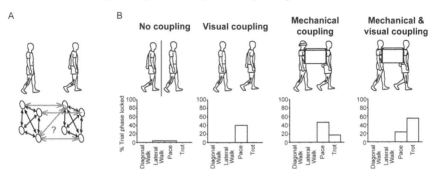

Figure 5.6 Coupling the bipedal locomotion dynamics of two walking participants.

Notes: A) The challenge of modelling the behavioural dynamics of unintentional interpersonal locomotion. B) Percentage of trials in which phase locked unintentional coordination was observed between participants for the full duration of the trial. Observed coordination patterns are those common to quadrupeds in nature, and vary as a function of interpersonal coupling.

coordination that should arise in this system. However, with knowledge of the relevant symmetry constraints in the system, such as the number of participating limbs (four), and the number of functional phases of each limb (two), it can be hypothesized that this system should have robust periodic solutions and, more importantly, that the solutions that arise from this system should have symmetries comparable to those of quadrupedal gaits. Consistent with this expectation, stable periodic solutions were found to emerge between the leg movements of the two participants that were consistent with quadrupedal gaits commonly observed in nature, and as predicted by the GSC group symmetry model.

Another important aspect of the study by Harrison and Richardson (2009) was the manipulation of the form and strength of inter-agent coupling. The different coupling conditions are illustrated in Figure 5.6b and resulted in different gait patterns. To understand why, one simply needs to examine the symmetry of visual informational couplings that these different conditions produced, which were all largely unidirectional (i.e. the back person was visually coupled to the front person, but not vice versa). For the visual coupling condition, the most salient visual information for the actor at the back was the leg movements of the person at the front. The symmetry of the flexion and extension phase of the front person's leg movement therefore defined a symmetry relation for the back person's leg movements, which resulted in a clear preference for in-phase coordination and the emergence of a quadrupedal pace gait. For the mechanical and the mechanical and visual coupling conditions, however, the actors tended to produce a trot pattern, rather than a pace pattern, with the movements of the legs of the two actors becoming coordinated in anti-phase. This was due to the fact that the most salient information in these conditions was the rhythmic oscillations of the front person's shoulders, which move in an anti-phase pattern with regard to the motion of the legs during walking. Again, the symmetry of the effect (i.e. the behavioural order of the gait pattern produced) was determined by the symmetry of the cause (i.e. the visually information observed).

Note that the symmetry arguments defined with respect to the study of Harrison and Richardson (2009) and GSC models of quadrupedal locomotion are not restricted to a back-front arrangement of co-actors. On the contrary, they can be applied to any situation in which the leg movements of co-actors become coordinated via visual information. For instance, the same symmetry argument also explains the patterning of the unintentional coordination that occurs between actors walking side-by-side (van Ulzen et al., 2008). In this situation, the between-agent visual coupling is side-on. From this point of observation, transformations that differentiate the left and right limbs are attenuated (i.e. information about right and left is less differentiated). The result is a coupling function that is constrained by the symmetry of limb action, but not by the symmetry of limb identity. Accordingly, both in-phase and anti-phase modes of coordination are equally likely; the symmetry of both effects is equivalent with respect to the symmetry of the information detected.

Conclusion

Our aim in this chapter was to demonstrate that the dynamic stabilities and behavioural patterning of coordinated social motor activity are defined by the symmetries of the physical and informational properties of the coordination context. Although we only focused on physical joint-action and movement-based forms of social coordination, it is important to appreciate that the symmetry arguments raised here are by no means restricted to these simple forms of behaviour and can be applied at any level of behavioural description, from the micro level organization of the human nervous and perceptual-motor systems to the macroscopic organization of societies and cultural systems (Richardson & Kallen, 2015). Indeed, the beauty of understanding behavioural order from a symmetry perspective is that one can develop a common language of explanation, one that highlights the casual similitude of behavioural organization across all scales of nature. Of course, determining whether the symmetry approach detailed here will ultimately lead to a unifying set of task independent laws – the hidden symmetries – common to all coordinated human behaviour is dependent on better understanding the mutual and reciprocal relationship between symmetry-breaking and the lawful dynamics that emerge during ongoing human and social behaviour.

Acknowledgements

We would like to thank Charles Coey, Anthony Chemero, Heidi Kloos, Dilip Kondepudi, Lynden Miles, Elliot Saltzman, Jeffery Kinsella-Shaw, Robert Shaw, Richard C. Schmidt, Michael Turvey and Ashley Walton. This work was supported by the National Institutes of Health (R01GM105045).

Note

1   The "whose glass is whose" dilemma was first described to the first and second authors of this chapter by Robert Shaw (see also Shaw, Mcintyre, & Mace, 1974; Warren & Shaw, 1985).

References

Brading, K. & Castellani, E. (2003) S*ymmetries in physics: Philosophical reflections.* Cambridge University Press: Cambridge.

Byblow, W.D. & Goodman, D. (1994) Performance asymmetries in multifrequency coordination. *Human Movement Science, 13*(2), 147–174.

Castellani, E. (2003) On the meaning of symmetry breaking. In K. Brading & E. Castellani (Eds.), *Symmetries in physics: Philosophical reflections* (pp. 321–334). Cambridge University Press: Cambridge.

Cohen, A.H., Holmes, P. & Rand, R.H. (1982) The nature of the coupling between segmental oscillators of the lamprey spinal generator for locomotion: A mathematical model. *Journal of Mathematical Biology, 13*, 345–369.

Collins, J.J. and Stewart, I.N. (1993) Coupled nonlinear oscillators and the symmetries of animal gaits. *Journal of Nonlinear Science, 3*, 349–392.

Epstein, I.R. (1990) Chemical oscillators and nonlinear chemical dynamics. In E. Jen (Ed.), *1989 lectures in complex systems, SFI studies in the sciences of complexity* (Vol. 2, pp. 213–269). Addison-Wesley: Reading, MA.

Fontaine, R.J., Lee, T.D. & Swinnen, S.P. (1997) Learning a new bimanual coordination pattern: reciprocal influences of intrinsic and to-be-learned patterns. *Can. J. Exp. Psych. 51*, 1–9.

Golubitsky, M. & Stewart, I. (2002) Patterns of oscillation in coupled cell systems. In P. Newton, P. Holmes & A. Weinstein (Eds.) *Geometry, mechanics, and dynamics* (pp. 243–286). Springer: New York.

Golubitsky, M. & Stewart, I. (2003) *The symmetry perspective*. Birkhauser Verlag: Basel, Switzerland.

Golubitsky, M., Stewart, I., Buono, P.-L. & Collins, J.J. (1999) Symmetry in locomotor central pattern generators and animal gaits. *Nature, 401*, 693–695.

Harrison, S.J. & Richardson, M.J. (2009) Horsing around: spontaneous four-legged coordination. *Journal of Motor Behavior, 41*(6), 519–524.

Hegstrom, R.A. & Kondepudi, D.K. (1990) The handedness of the universe. *Scientific American, 262*(1), 108–115.

Kelso, J.A.S. (1995) *Dynamic patterns*. MIT Press: Cambridge, MA.

Kelso, J.A.S., DeGuzman, G.C. & Holroyd, T. (1991) The self-organized phase attractive dynamics of coordination. In A. Babloyantz (Ed.), *Self-organization, emerging properties and learning* (pp.41–62). New York: Plenum Press.

Kugler, P.N. & Shaw. R.E. (1990) Symmetry and Symmetry-Breaking in Thermodynamic and Epistemic Engines: A Coupling ·of First and Second Laws. In H. Haken & M. Stadler (Eds.), *Synergetics of Cognition* (pp. 296–331). Springer-Verlag: Berlin.

Miles, L.K., Lumsden, J., Richardson, M.J. & Macrae, N.C. (2011) Do Birds of a Feather Move Together? Group Membership and Behavioral Synchrony. *Experimental Brain Research, 211*, 495–503.

Peper, C.E., Beek, P.J. & van Wieringen, P.W. (1995) Multifrequency coordination in bimanual tapping: Asymmetrical coupling and signs of supercritically. *Journal of Experimental Psychology: Human Perception and Performance, 21*, 1117–1138.

Pikovsky, A., Rosenblum, M. & Kurths, J. (2003) *Synchronization: a universal concept in nonlinear sciences.* Cambridge University Press: Cambridge.

Pinto, C.M. & Golubitsky, M. (2006) Central pattern generators for bipedal locomotion. *Journal of Mathematical Biology, 53(3)*, 474–489.

Richardson, M.J. & Kallen, R.W. (2015) Symmetry-Breaking and the Contextual Emergence of Human Multiagent Coordination and Social Activity. In E. Dzhafarov, S. Jordan, R. Zhang & V. Cervantes (Eds.), *Contextuality from Quantum Physics to Psychology* (pp. 229–286). World Scientific: Singapore.

Richardson, M.J., Marsh, K.L., Isenhower, R., Goodman, J. & Schmidt, R.C. (2007) Rocking Together: Dynamics of Intentional and Unintentional Interpersonal Coordination. *Human Movement Science, 26*, 867–891.

Rosen, J. (1995) *Symmetry in Science: An Introduction to the General Theory.* Springer-Verlag: New York.

Schmidt, R.C., Bienvenu, M., Fitzpatrick, P. & Amazeen, P. (1998) A comparison of intra- and interpersonal interlimb coordination: coordination breakdowns and coupling strength. *Journal of Experimental Psychology: Human Perception and Performance, 24*, 884–900.

Schmidt, R.C., Christianson, N., Carello, C. & Baron, R. (1994) Effects of social and physical variables on between-person visual coordination. *Ecological Psychology, 6*, 159–183.

Schmidt, R.C. & Richardson, M.J. (2008) Dynamics of Interpersonal Coordination. In A. Fuchs & V. Jirsa (Eds.), *Coordination: Neural, Behavioral and Social Dynamics* (pp. 281–308). Springer-Verlag: Heidelberg.

Schmidt, R.C. & Turvey, M.T. (1994) Phase-entrainment dynamics of visually coupled rhythmic movements. *Biological cybernetics, 70*(4), 369–376.

Shaw, R.E., Kugler, P. & Kinsella-Shaw, J. (1990) Reciprocities of Intentional Systems. In R. Warren & A.H. Wertheim, (Eds.), *Perception & control of self-motion* (pp. 579–619). Lawrence Erlbaum Associates, Inc.: Hillsdale, NJ.

Shaw, R.E., McIntyre, M. & Mace, W.M. (1974) The Role of Symmetry in Event Perception. In R. B. McLeod & H. Pick, Jr. (Eds.), *Perception: Essays in Honor of James J. Gibson*, pp. 276–310. Cornell University Press: Ithaca, NY.

Shaw, R.E. & Turvey, M.T. (1999) Ecological Foundations of Cognition: II. Degrees of Freedom and Conserved Quantities in Animal – Environment Systems. *Journal of Consciousness Studies, 6.*

Sternad, D., Turvey, M.T. & Saltzman, E.L. (1999) Dynamics of 1:2 coordination: Generalizing relative phase to n:m rhythms. *Journal of Motor Behavior, 31*(3), 207–223.

Stewart, I. & Golubitsky, M. (1992) *Fearful symmetry: Is God a geometer?* Penguin Books: London.

Treffner, P.J. & Turvey, M.T. (1993) Resonance constraints on rhythmic movement. *Journal of Experimental Psychology: Human Perception & Performance, 19*, 1221–1237.

Turvey, M.T. (2007) Action and perception at the level of synergies. *Human Movement Science, 26*(4), 657–697.

van Ulzen, N.R., Lamoth, C.J., Daffertshofer, A., Semin, G.R. & Beek, P.J. (2008). Characteristics of instructed and uninstructed interpersonal coordination while walking side-by-side. *Neuroscience Letters, 432*(2), 88–93.

Warren, W.H., Jr. & Shaw, R.E. (1985) Events and Encounters as Units of Analysis for Ecological Psychology. In W.H. Warren, Jr. & R E. Shaw (Eds.), *Persistence and Change: Proceedings of the First International Conference on Event Perception* (pp. 1–27). Lawrence Erlbaum Associates, Inc.: Hillsdale, NJ.

Washburn, A., Coey, C.A., Romero, V. & Richardson, M.J. (2014) Unintentional Polyrhythmic Entrainment: Can 1:2, 2:3 and 3:4 Patterns of Visual Coordination Occur Spontaneously? *Journal of Motor Control, 46*, 247–57.

Weyl, H. (1952) *Symmetry*. Princeton University Press: Princeton.

Zanone, P.G. & Kelso, J.A.S. (1992) Evolution of behavioral attractors with learning: nonequilibrium phase transitions. *Journal of Experimental Psychology: Human Perception and Performance, 18*, 403.

Part II

Interpersonal coordination in competitive and cooperative performance contexts

6 Interpersonal coordination in performing arts

Inspiration or constraint

Carlota Torrents, Natália Balagué and Robert Hristovski

Introduction

Performing arts are the forms of creative activity that are performed in front of an audience and where performers use their voice, body or objects to convey artistic expression. Theatre, music, puppetry, magic, musical theatre or activities that are primarily based on human movement, such as dance or circus, exemplify different forms of performing arts. Some sports also have an artistic component, especially those performed with music such as synchronized swimming, rhythmic gymnastics or figure skating. In all of these activities coordination of actions with the music and/or with partners is essential for success. Nevertheless, research related to this topic has been more focused on performing arts, and especially in music, more so than in artistic sports, although results and conclusions from research studies can be also applied to these sports.

Interpersonal coordination in performing arts is relevant when artists have to coordinate their performance with other artists at the same time. Solo dances, monologues or an actor using puppets do not require coordination of behaviours with other individuals on stage, but they have to control other elements related particularly to their use of space, time or effort (Laban & Ullman, 1960). Even in those cases, interpersonal coordination can appear in order to respond to the reaction of the audience and respect the timing that they are demanding with their applause, silences or attention. In Table 6.1 we show the most relevant elements that are used in some of the most common performing arts, based on human movement and artistic sports. Nevertheless, contemporary art approaches tend to mix these arts, and often it is difficult to define what is dance or what is theatre (see examples of physical theatre, such as pieces by the company Dv8), what is circus and what is theatre (e.g. the Cirque Éloize) or even what is dance and what is music (e.g. the shows of the companies STOMP or MAYUMANA).

Due to the specific rules that constrain the evolution of sport, artistic sports have not evolved in that direction, and they maintain constant and very well defined characteristics. In artistic sports and traditional art forms space coordination is needed, except in music, although puppetry or theatre is not so

Table 6.1 Categories that are combined during the performance of different performing arts and artistic sports. An X means that an element can be present, and XX means that it is very important. We understand the substrate as the component that is used to express (e.g. for a painter, the substrate will be the colour, and for the sculptor the material). (Adapted from Laban & Ullman, 1960 and Mateu, 2010.)

Art or sport		Substrate	Space	Time	Effort	Use of objects	Interaction with others	Text
Performing Movement arts	Dance	Body	XX	XX	XX	X	X	
	Mime	Body	XX	X	XX	X	X	
	Circus	Body	XX	X	XX	XX	X	
Performing Arts	Music	Sound		XX		XX	X	
	Puppetry	Objects	X	X		XX	X	XX
	Theatre	Body	X	X	X	X	X	XX
Artistic Sports	Rhythmic gymnastics	Body	XX	XX	XX	XX	XX	
	Synchronized swimming	Body	XX	XX	XX		XX	
	Artistic Skating	Body	XX	XX	XX		XX	

determinant as dance or artistic sports. On the contrary, time coordination is the essential characteristic of music, and it is also very important in other art or sport modalities that are performed with a music or external rhythmic background. The effort or quality of movement is important when human movement is the substrate of the expression. Objects are used in most of these arts and in rhythmic gymnastics, and text is particularly popular in traditional theatre and puppetry. When dialogues are included in the performance another type of interpersonal synergy emerges, as dialogues can also be considered as an emergent, self-organizing, interpersonal system capable of functional coordination (Fusaroli, Raczaszek-Leonardi & Tylén, 2014). Finally, interaction with others will be instrumental when more than one performer is acting at the same time. This interaction is evaluated in artistic sports where judges appreciate the synchronization between the gymnasts, swimmers or figure skaters.

Synchronization is a very well known phenomenon in music or dance, based on coordinated behaviour that emerges from ongoing interactions between the performer and the music or between different performers through a process of self-organization. When different rhythms interact with each other and co-adjust towards to a common phase and/or periodicity, and consequently synchronize, the concept of entrainment is used (Clayton, Sager & Will, 2005). Entrainment has been applied to music research, but dancers also need to coordinate different rhythms that interact with the music or with other dancers, or a trapeze artist needs to create a rhythm with their partners to be able to perform the most impressive skills in the aerial phase between different trapezes. The dynamic systems approach describes musical rhythmic entrainment as an active, self-sustained, periodic oscillation at multiple time scales, enabling the listener to use predictive timing to maintain a stable multi-periodicity pattern and synchronize movements or produce coordinated movements at metrical levels (Large, 2000; Loehr, Large & Palmer, 2011; Phillips-Silver, Aktipis & Bryant, 2010). Entrainment does not necessarily involve an external stimulus. Self-entrainment describes the coordination between the body's oscillatory systems, such as respiration and heart rhythm patterns. Social entrainment requires the response of the system to rhythmic information generated by other individuals. This can occur when an organism interacts with other organisms or with an external rhythm (auditory or not). It was assumed that the tendency to move in rhythmic synchrony with a musical beat was uniquely human, as no primate can dance or collaboratively clap to the beat of the music. However, there is evidence that some parrots have the ability to display genuine dance, including changing their movements to a change in tempo (Patel, Iversen, Bregman & Schulz, 2009).

The emergence of innovative behaviours is related to the interpersonal coordination that has been studied in sports (Hristovski, Davids, Passos, & Araújo 2012) and dance (Torrents, Castañer, Dinusôva, M. & Teresa Anguera, M., Torrents, Castañer, Dinusôva & Anguera, 2010), and it could also be applied to other performing arts. For example, in music, improvisation gives rise to the spontaneous production of novel musical material, including novel melodic, harmonic and rhythmic musical elements. It has been suggested that spontaneous improvisation is characterized by widespread deactivation of lateral portions of the prefrontal cortex together with focal activation of the medial prefrontal cortex. This would mean that the creative process can occur outside of conscious awareness and beyond volitional control (Limb & Braun, 2008). This is not trivial, since, whether in sport or art, the creative process is usually related to cognition and an off-line process that is driven by 'possibility thinking', a process that begins with imaginatively asking 'what if?' and which then moves on to performing an action. Although this approach, in which the act of asking perturbs, destabilizes and constrains the performer to explore other possible solutions, can be useful for many situations related to dance or music composition, it is insufficient to explain how dancers or musicians create online new configurations of movements during improvisation. Nevertheless, it has been shown that the spontaneous coordination with other musicians can also give rise to the production of novel musical material.

Interpersonal coordination in music

In recent years much research on synchronization has focused on music performance and entrainment. This phenomenon is especially interesting in music, as combinations and possibilities of synchronisms and asynchronisms give expressiveness and aesthetic richness to musical pieces. As Keller describes

(2014), timing deviations in ensemble performance affect horizontal (timing of successive sounds) and vertical (degree of synchronization of sounds) relations between sounds. He defended the idea that strict simultaneity is not only impossible, but also undesirable in human music making. In fact, some styles look for these vertical timing deviations to define the expressive character, such as jazz. Ensemble musicians have to achieve precision in interpersonal coordination without sacrificing the flexibility required. That is to say, entrainment needs unstable modes of coordination in order to be able to adapt the musicians' performance to the aesthetic and expressive needs of the music.

Synchronization deviations have been studied quantitatively, but also some researchers have addressed the reasons apart from perceptual and motor limitations. In his review Keller points out that synchronization can be constrained by the degree to which ensemble members can see and hear one another during the performance (Goebl & Palmer, 2009), mechanical aspects (Moore & Chen, 2010), differences between instruments (Rasch, 1988), or the complexity of the musical structure (Repp, 2006). He also points out that dynamical systems approaches can help to study unstable coordination modes, such as polyrhythms. Pedagogical proposals that have appeared in recent years due to coordination dynamics research can also help musicians to improve this unstable coordination mode that characterizes musical entrainment (Laroche & Kaddouch, 2015).

Synchronization in music has also been studied in brain activity. Lindenberger, Li, Gruber and Müller (2009) recorded EEGs from the brains of each of eight pairs of guitarists and found that phase synchronization both within and between brains increased significantly during the periods of preparatory metronome tempo setting and during coordinated play together.

Interpersonal coordination in dance

Synchronization of body movements with external rhythm has been studied from multiple perspectives (see Repp, 2005 for a review). Nevertheless, few researchers have studied the synchronization of whole-body movements with music (Burger et al., 2012; Eerola, Luck & Toiviainen, 2006; Toiviainen, Luck & Thompson, 2010; Zentner & Eerola, 2010). It is interesting to note that research could study the opposite phenomenon, how musicians can adapt their music to dancers' movements, or even how the improvisation of music and dance can give rise to variations and attunement of dancers' movements, but also to musical characteristics. This way of understanding the relation between music and dance was started with modern dance, and especially with the works of Merce Cunningham and his collaborator and partner John Cage. They made radical innovations, and in this field they concluded that music and dance may occur in the same time and space, but should be created independently of one another. In dance jams (dance meetings for improvising) it is common to join musicians and dancers and they improvise in the same timing, creating a bank of emergent interpersonal multi-dimensional coordination that would be really interesting to study.

As with the research on music-movement synchronization, a large body of research has been conducted on interpersonal movement synchronization. It is generally accepted that this follows principles of self-organization and can be mathematically modelled as a dynamical process (see Schmidt and Richardson, 2008 for a review). However, to our knowledge, synchronization in dance has not been studied from this perspective. Studying simple dance tasks, Wuyts and Buekers (1995) investigated the effect of visual and auditory concurrent models on the acquisition of a rhythmical synchronization task. Results revealed that auditory models, combined with auditory rhythms, produced significantly lower error scores during acquisition than learning without models, with visual models, or with visual models but without auditory rhythm. Himberg and Thompson (2009) compared the synchronisation and coordination of choreographed movements to music among expert and novice music choirs and found that the expert group had more group synchronization and cohesion, as expected.

There remains a need to study synchronization in dance companies or groups, since results for music could probably also be applied to dance. Dancers also need precision in interpersonal coordination, along with flexibility in order to introduce variations in the dance.

Another topic related to interpersonal coordination in dance is the emergence of movement behaviour. Dance interaction with partners gives rise to different behaviours depending on the partners involved, as well as task or environmental constraints will vary the behaviour emerging from the interaction. In the next chapter, some experiments will show how specific instructional constraints, as well as the level of the dancers, had a significant effect on the type of configurations performed by the dancers and also on the variability of their outcomes.

Applications to other disciplines and artistic sports

To our knowledge, little work has been done related to interpersonal coordination or synchronization in circus, mime or even in artistic sports. However, these disciplines require the entrainment of performers with the rhythms of their partners or with the music. In the case of artistic sports, rules force the performers to synchronize their movements as much as possible with the partners and with the music, allowing traditional variations, such as canons, or simple combinations of rhythms.

In rhythmic gymnastics, aerobic gymnastics, figure skating or synchronized swimming, the musicality or timing and synchronization with the partners is specifically evaluated. The relation of the performers with the music is strictly constrained by the rules. For example, in rhythmic gymnastics, synchronicity between movements or the handling of the apparatus and musical phrases or the relation of the musical intensity with the gymnast's expression have clearly changed over the last thirty years due to changes in the points code (Chiat & Ying, 2012).

Interpersonal synchronization in artistic sports is required with more precision than in dance (with the exception of some dance styles, such as some traditional dances). Training methods to improve this synchronization are mainly based on the repetition of the routines, using mirrors, video recordings and the feedback of the coach. As reported in music, these methods probably could be improved by new pedagogical proposals emerging from the coordination dynamics approach.

Emotional interpersonal coordination

Artistic activities are not only characterized by the efficacy of the technique but also by shared interpersonal emotions which underpin the aesthetics of body expression and music. The communication between the agents in performing arts and artistic sports, i.e. between performers and spectators or judges, is emotion laden. Thus, emotion entrainment is a key factor for the success of the performance, as in many other types of activities. Emotionally induced entrainment was detected in violin players (Varni, Camurri, Coletta & Volpe, 2010) as measured through phase synchronization of the head motions of the players. Under different emotional states and perceptual coupling strengths, non-verbal expressive gestural interactions played a significant role in entraining musicians' head movements. In similar vein, listening to music, watching movies and listening to narratives forms time-locked functionally selective activity dynamics in a vast number of brain areas, including limbic emotion systems (Chapin, Jantzen, Kelso, Steinberg & Large, 2010; Nummenmaaa, Glerean, Viinikainen, Jääskeläinen, Haria & Sams, 2012; Hatfield, Cacioppo & Rapson, 1994). Hence, shared emotions mould and amplify interpersonal synchronization of brain activity across individuals and consequently underpin similar perceptions, experiencing of and acting in the world. These time dependent interpersonal dynamic emotion systems consist of components of emotion which interact not only within but across the agents as well (Butler, 2011).

It has recently been hypothesized that a stable and continuous relation between the brain and heart dynamics exists (Shaffer, McCraty & Zerr, 2014), and that the level of their mutual coherence is a significant measure of an individual's personal emotional well being. In connection to this, some researchers have found finely-tuned synchronizations of cardiac rhythm during fire-walking rituals between fire-walkers and spectators who were family members or who had a close relationship, but not between fire-walkers and unrelated spectators (Konvalinka et al., 2011). Dynamic couplings between individuals have been shown to produce coherence not only in heart rhythms but also in neural rhythms, respiration, autonomic systems, posture and behaviour (see Konvalinka & Roepstorff, 2012 for a review). In this sense, emotions play the role of a crucial constraint and simultaneously of an instrumental dynamic variable in interpersonal coordination in performing arts and artistic sports. This is a young realm of research and many fascinating cases still wait to be researched on the path to a comprehensive understanding of the role of emotional coherence in these areas. In general, sociological coordination such as synchronization and entrainment may be traced down to cellular (Massie et al., 2010), organic and organizmic levels (Roenneberg, Serge & Merrow, 2003) and coordination in performing arts and artistic sports may be considered as a highly evolved human artistic refinement of these basic processes. Yet not all forms of performing arts and artistic sports form strongly entrained and long lasting modes of behaviour, as in some forms of music, folk and classical dances, where the stable rhythm synchronizes the performers. Performing arts and artistic sports may show a large spectrum of interpersonal coordination and levels of entrainment to external sources in which subtle and intermittent coherencies form and decay, as in contact improvisation or in theatre performance. However, what is common to the high levels of performance in all of these various domains of artistic expression is the emotional inter-personal coordination and coherence between all participants. Research in this rich area of phenomena is at its beginning, and it is our hunch that focusing on it will be more than rewarding.

References

Burger, B., Thompson, M.R., Luck, G., Saarikallio, S. & Toiviainen, P. (2012) Music moves us: beat-related musical features influence regularity of music-induced movement. In E. Cambouropoulos, C. Tsougras, P. Mavromatis & K. Pastiadis (Eds.), *Proceedings of the 12th International Conference on Music Perception and Cognition and the 8th Triennial Conference of the European Society for the Cognitive Sciences of Music*, Thessaloniki, Greece.

Butler, E. (2011) Temporal Interpersonal Emotion Systems: The "TIES" That Form Relationships. *Personality and Social Psychology Review*, *15*(4), 367–393.

Chapin, H., Jantzen, K., Kelso, J.A.S., Steinberg, F. & Large, E. (2010) Dynamic Emotional and Neural Responses to Music Depend on Performance Expression and Listener Experience. *PLoS ONE*, 5(12), e13812. doi:10.1371/journal.pone.0013812

Chiat, L.F. & Ying, L.F. (2012) Importance of music learning and musicality in rhythmic gymnastics. *Procedia – Social and Behavioral Sciences*, *46*, 3202–3208.

Clayton, M., Sager, R. & Will, U. (2005) In time with the music: the concept of entrainment and its significance for ethnomusicology. *European Meetings in Ethnomusicology*, *11*, 3–142.

Eerola, T., Luck, G. & Toiviainen, P. (2006) An investigation of preschoolers' corporeal synchronization with music. In M. Baroni, A.R. Addessi, R. Caterina & M. Costa (Eds.), *9th International Conference on Music Perception and Cognition* (pp. 472–476), Bologna, Italy.

Fusarolia, R., Rączaszek-Leonardi, J. & Tylén, K. (2014) Dialog as interpersonal synergy. *New Ideas in Psychology*, *32*, 147–157. doi:10.1016/j.newideapsych.2013.03.005

Goebl, W. & Palmer, C. (2009) Synchronization of timing and motion among performing musicians. *Music Perception*, *26*, 427–438.

Hatfield, E., Cacioppo, J. & Rapson R.L. (1994) *Emotional Contagion*. Cambridge University Press: New York.

Himberg, T. & Thompson, M. (2009) Group synchronization of coordinated movements in a cross-cultural choir workshop. In J. Louhivuori, T. Eerola, S. Saarikallio, T. Himberg & P.S. Eerola (Eds.), *Proceedings of the 7th Triennial Conference of European Society for the Cognitive Sciences of Music (ESCOM 2009)*. Jyväskylä, Finland.

Hristovski, R., Davids, K., Passos, P. & Araújo, D. (2012) Sport performance as a domain of creative problem solving for self-organizing performer-environment systems. *The Open Sports Science Journal*, 5, (Suppl 1-M4), 26–35.

Keller, P.E. (2014) Ensemble performance: Interpersonal alignment of musical expression. In D. Fabian & R. Timmers (Eds.), *Expressiveness in music performance: Empirical approaches across styles and cultures.* Oxford University Press: Oxford.

Konvalinka, I. & Roepstorff, A. (2012) The two-brain approach: how can mutually interacting brains teach us something about social interaction? *Frontiers in Human Neuroscience, 6*, 215.

Konvalinka, I., Xygalatas, D., Bulbulia, J., Schjodt, U., Jegindo, E.M., Wallot, S., Van Orden, G. & Roepstorff, A. (2011) Synchronized arousal between performers and related spectators in a fire-walking ritual. *Proceedings of the National Academy of Sciences, 108*, 8514–8519.

Laban, R. & Ullman, L. (1960) *The mastery of movement (revised). 3rd Ed.* Macdonald & Evans: London.

Large, E.W. (2000) On synchronizing movements to music. *Human Movement Science, 19*, 527–566.

Laroche, J. & Kaddouch, I. (2015) Spontaneous preferences and core tastes: embodied musical personality and dynamics of interaction in a pedagogical method of improvisation. *Frontiers in Psychology, 6*, 522. doi: 10.3389/fpsyg.2015.00522

Loehr, J.D., Large, E.W. & Palmer, C. (2011) Temporal Coordination and Adaptation to Rate Change in Music Performance. *Journal of Experimental Psychology: Human Perception and Performance, 37*, 1292–1309. doi: 10.1037/a0023102

Limb, C.J. & Braun, A.R. (2008) Neural substrates of spontaneous musical performance improvisation. *PLOSOne, 3*, e1679. doi:10.1371/journal.pone.0001679

Lindenberger, U., Li, S.C., Gruber, W. & Müller, V. (2009) Brains swinging in concert: cortical phase synchronization while playing guitar. *BMC Neuroscience, 10*, 22. doi:10.1186/1471-2202-10-22

Massie, T.M., Blasius, B., Weithoff, G., Gaedke, U. & Fussmann, G.F. (2010) Cycles, phase synchronization, and entrainment in single-species phytoplankton populations. *PNAS, 107*, 9, 4236–4241. doi: 10.1073/pnas.0908725107

Mateu, M. (2010) *Observación y análisis de la expresión motriz escénica. Lógica interna de espectáculos artísticos profesionales: Cirque du Soleil 1986–2005.* Doctoral thesis. Universitat de Barcelona: Barcelona.

Moore, G. & Chen, J. (2010) Timing and interactions of skilled musicians. *Biological Cybernetics, 103*, 401–414.

Nummenmaaa, L., Glerean, E., Viinikainen, M., Jääskeläinen, I.P., Haria, R. & Sams, M. (2012) Emotions promote social interaction by synchronizing brain activity across individuals. *PNAS, 109*(24), 9599–9604.

Patel, A.D., Iversen, J.R., Bregman, M.R. & Schulz, I. (2009). Experimental evidence for synchronization to a musical beat in a nonhuman animal. *Current Biology, 19*, 827–830. DOI 10.1016/j.cub.2009.03.038

Phillips-Silver, J., Aktipis, C.A. & Bryant, G.-A. (2010) The ecology of entrainment: Foundations of coordinated rhythmic movement. *Music Perception, 28*, 3–14. doi:10.1525/mp.2010.28.1.3

Rasch, R.A. (1988) Timing and synchronization in ensemble performance. In J.A. Sloboda (Ed.), *Generative processes in music: The psychology of performance, improvisation, and composition* (pp. 70–90). Clarendon Press: Oxford.

Repp, B.H. (2005) Sensorimotor synchronization: a review of the zapping literature. *Psychonomic Bulletin & Review, 12*, 969–992.

Repp, B.H. (2006) Rate limits of sensorimotor synchronization. *Advances in Cognitive Psychology, 2*, 163–181.

Roenneberg, T., Serge, D. & Merrow, M. (2003) The Art of Entrainment. *Journal of Biological Rhythms, 18*(3), 183–194 .doi:10.1177/0748730403253393

Schmidt, R.C. & Richardson, M.J. (2008) Dynamics of interpersonal coordination. In A. Fuchs & V.K. Jirsa (Eds.), *Coordination: Neural, Behavioral and Social Dynamics* (pp.281–308). Springer: Berlin Heidelberg. doi: 10.1007/978-3-540-74479-5_14

Shaffer, F., McCraty, R. & Zerr, C.L. (2014) A healthy heart is not a metronome: an integrative review of the heart's anatomy and heart rate variability. *Frontiers of Psychol*ogy, 5, 1040. doi: 10.3389/fpsyg.2014.01040

Toiviainen, P., Luck, G. & Thompson, M. (2010) Embodied meter: Hierarchical eigenmodes in music-induced movement. *Music Perception, 28*, 59–70.

Torrents, C., Castañer, M., Dinusôva, M. & Teresa Anguera, M. (2010) Discovering New Ways of Moving: Observational Analysis of Motor Creativity While Dancing Contact Improvisation and the Influence of the Partner. *The Journal of Creative Behavior, 44*(1), 45–61.

Varni, G., Camurri, A., Coletta, P. & Volpe G. (2008) Emotional Entrainment in Music Performance. *Proc. 8th IEEE Intl Conf on Automatic Face and Gesture Recognition*, Sept. 17–19, Amsterdam.

Wuyts, I.J. & Buekers, M.J. (1995) The effects of visual and auditory models on the learning of a rhythmical synchronization dance skill. *Research Quarterly for Exercise and Sport, 66*, 105–115. doi:10.1080/02701367.1995.10762218

Zentner, M. & Eerola, T. (2010) Rhythmic engagement with music in infancy. *Proceedings of the National Academy of Sciences of the United States of America, 107*, 5768–5773.

7 Interpersonal coordination in Contact Improvisation dance

Carlota Torrents, Robert Hristovski, Javier Coterón and Ángel Ric

Introduction

Dance can consist of following a musical rhythm with body movements, the movements of a partner, the movement of a group of dancers or the rhythm of the dancer's own feelings and emotions. In all cases, dance requires the coordination of the dancer with an internal or external rhythm, as well as with the space, which is transformed with each movement. It is possible to dance alone, in a *solo* on a stage, or as a recreational activity in a group or at a party, but many types of dance need partners to be performed. That is the case for many folk dances, ballroom dances, duets in classical ballet, or any style danced in a group, as a company of modern or contemporary dance. Dancers play with the coordination with their partners, time and space, changing from synchrony to asynchrony, using temporal variations as canons or counterpoints, varying the position of the participants in the scenic space, or answering the shapes or movements proposed by other partners. Therefore, interpersonal coordination can be considered as a skill that needs to be practiced in the learning process of many dance styles, but it is also something that emerges spontaneously, especially when dancers are allowed to improvise.

Dance improvisation plays with the spontaneity of the dancer or group of dancers but particularly with listening to what is happening in the moment of the dance, what is called the *flow*. The inspiration to move can come from the dancer's own body or from an external stimulus, and can be completely free (except for the physical forces that govern the movement) or follow some task constraints or rules. Improvisation is used as part of the creative process of making choreographies (Lavender & Preddok-Linnell, 2001), during dance education (Biasutti, 2013) or because some dance styles are based on improvisation. That is the case in social ballroom dances, some folk dances such as flamenco, or some variations of contemporary dance such as Contact Improvisation (CI). CI can be defined as a movement form that is improvisational in nature and which involves two bodies in contact (Sidall, 1997), or as a spontaneous mutual investigation of the energy and inertia paths created when two people engage actively and dance freely, using their sensitivity to guide and safeguard them (Paxton, 1997). The improvisational characteristics of CI are such that the generation of movements is not based on fixed and standardised movements or techniques, since this dance form requires a body that responds to the physical exchange of weight and contact (Albright, 2003; Novack, 1990). These situations can be thought as a bank of emergent human movement forms and an expression of the continuous exploration and discovery of idiosyncratic gestures, postures and actions supported by immediate affordances, i.e. opportunities for action or properties of the environment relative to a performer (Chemero, 2003; Gibson, 1979).

The unit of this dance is not the individual but the system formed by the two dancers. The dynamic and non-linear relationship between both dancers and the context gives rise to patterns or movements and transitions among them. As in all non-linear systems the nonlinearity arises as a consequence of the self-interaction of the perceptual-motor systems of performers. Self-interaction is manifested as a co-adaptive change between a system's components: the behaviour of a certain component changes itself through influencing another component, which in turn influences the first one. As a consequence of these non-linear interactions, movement configurations, i.e. patterns, arise following a process of self-organization, without the need to be consciously controlled by one of the dancers or imposed, prescribed, by an external agency. The feeling of flowing depends on the ability of the dancers to let the movements emerge, without forcing the system to perform a movement following aesthetic criteria, for instance. Some movements attract the system formed by the two dancers and their specific characteristics, while some other movements will be statistically rare and constitute fluctuations. Following a dynamical systems approach, exploration involves enhancing reconfigurations, reflected by larger variability in behaviour. These reconfigurations enable the dancers to discover new movement possibilities and facilitate the emergence of new skills.

The variability of situations occurs from the interaction between the constraints of the dancers (physical characteristics, skill level, emotional state), the environment (open space, stage, specific place for improvising, ground properties) and the task (instructions, dancing with material, presence of music) (Newell, 1986). In a previous study, Torrents, Castañer and Anguera (2011) showed how task constraints varied the behaviour of a group of seven dancers improvising during a series of four five-minute trials. The variations were: free movement; free movement, but previously they had spent one minute breathing at the same time; only walking and moving one's arms; free movement, but if one dancer stops, they must all stop at the same time. Systematic observational methodology revealed that very strong constraints limited the emergence of varied patterns, and that a previous synchronisation of breathing enhanced the group's synchronisation and the use of physical contact between members while dancing. In another study, also using a systematic observational methodology, Torrents et al. (2010) showed how the partner influenced and varied the movements of a duet CI.

Due to some very particular characteristics of CI, it is a perfect scenario to analyse spontaneous interpersonal coordination in dance. When two people dance CI they generate an evolving system of movement based on the communication between two moving bodies and their combined relationship with the physical laws that govern their motion, such as gravity, momentum or inertia (Warshaw, 1997). Movement is amplified by the contact, and very little goes unregistered, even unconsciously. Participants move as a result of their partner's movements, since partnering is not just an addition of one movement to another but rather the change of these movements in the midst of an improvisational duet (Albright, 2003, 2004; Stark Smith, 1997). These principles would explain the feeling of dancers during improvisation, when they are surprised by the new movement patterns that suddenly appear. Moreover, in CI the sense of touch is more important than the sense of sight, which is the most widely-used sense in our everyday life. Communication is predominantly based on this contact rather than on verbal and non-verbal communication signs, and social distance and organisational rules are re-structured, thus giving us an opportunity to analyse an activity that is not structured according to the most common cultural rules.

What is being explored in Contact Improvisation? The landscape of affordances

Kugler and Turvey (1987) defined the perceptual-motor workspace as a dynamical interface between the information flows coming from perception and kinetic flows occurring from action (see also Newell et al., 1991). The main properties of the perceptual-motor workspace are the affordances. Affordances have been defined as opportunities (Gibson, 1979), invitations (Withagen, de Poel, Araújo & Pepping, 2012) or solicitations (Bruineberg & Rietveld, 2014) for action, but always as a scaled relation between certain properties of the environment and a performer's effectivities, i.e. abilities (Turvey, 1992; Chemero, 2003). It is the affordances that channel the exploratory activity of the performer. They map the environmental information on the movement behaviour within the perceptual-motor workspace consistent with the task goal demands.

CI differs from a vast majority of motor learning tasks because of the absence of a particular and fixed goal movement coordination to be learned or prescribed in order for a performance goal or sub-goal to be achieved. Hence, one cannot expect a convergence of the exploratory behaviour towards a well defined configuration. The main task goal of CI is the continuity of whole body movements and postures, a flow of movement forms, satisfying the general task constraint of being in permanent physical contact with the partner. These properties of CI entail and enable the vast potential richness of the perceptual-motor workspace of performers. The perceptual (dominantly proprioceptive, tactile and visual) information, encompassing the partner's movements and postures, channels a performer's exploratory behaviour. However, this channelling function of the perceptual information does not lead to a sole predetermined gradient flow of movement forms. If that was the case, the same initial position of dancers would produce the same movement sequence. As has been shown during boxing improvisation (e.g. Hristovski, Davids & Araújo, 2006; Hristovski, Davids, Passos & Araújo, 2012) the same body-scaled information may invite or solicit more than one action. The probability of occurrence for the same constraint value depended on the perceived action effectiveness which suppressed the concurrent action probabilities. Also, it was shown that a minuscule change in body-scaled constraints, as well as sufficient variability in the environment, may create and/ or annihilate radically novel affordances. In other words, opportunities, solicitations or invitations for action, i.e. affordances, may be simultaneously concurrent to one another (see Czisek, 2007; Czisek & Pastor-Bernier, 2014 for neurophysiological evidence of this effect), but may also suffer qualitative changes as a result of minor environmental shifts. This common situation allows very subtle and consequently unpredictable (by the performer or the experimenter) interactions between the environmental information flow and the performer's organizmic constraints to decide which concrete affordance, from the field of affordances (Bruineberg & Rietveld, 2014), will be realized at each moment. The interaction of deterministic and random influences is the main characteristic of the ecological dynamics of decision making (Araújo, Davids & Hristovski, 2006). Hence, the randomness plays a substantial role in the exploratory behaviour of the perceptual-motor workspace of dancers. In other words, deterministic gradients and random fluctuations may cooperate or compete in the exploration. In this sense the exploration of perceptual-motor workspace may be defined as an exploration of the landscape of affordances (Bruineberg & Rietveld, 2014). The main challenge, then, is to make available tools that would afford the analysis of the dynamical, *nested* structure and the performer's/learner's exploration dynamics of the landscape of affordances.

Formally, the exploration of the landscape of affordances is defined as a hopping dynamics among many metastable minima of the nested basins of attraction which form the landscape, or equivalently as a random walk on a tree (see Hristovski et al., 2012; Hristovski et al., 2013 for definitions in the area of performer-environment systems, or Sibani & Jensen, 2013, for more up-to-date general information). Since in CI there is no pre-determined configuration or fixed goal to be achieved or to which the exploration should converge, the exploration proceeds by hopping between local minima (locally stable patterns) without trying to find any globally minimized (optimal) state as a final task solution. In this sense, it may be defined as an exploration by winner-less competition among affordances where the only goal is the exploration of movement possibilities – the landscape of affordances itself. During this process each metastable minimum (temporally stable configuration) represents the discovery of a temporary task solution. The strength of attraction of local minima may be assessed by the associated dwell times and the barriers between them through Hamming distances (see Hristovski et al., 2013). For illustration, in Figure 7.1 we present a histogram of Hamming distances between consecutive configurations generated during a Contact Improvisation. One can see that the most frequent (most probable) barriers separating local minima are of small size, and there are only a few of large size (an order of magnitude larger). This gives a picture of a large number of shallow local basins of attraction, nested within larger basins of attraction separated by a small number of large barriers. Hence,

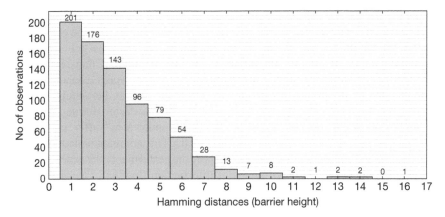

Figure 7.1 The distribution of Hamming distances, i.e. barrier heights, separating whole
body configurations during a contact improvisation.

the exploratory behaviour consists mostly of small reconfigurations (barrier crossings) and rare large reconfigurations (barrier crossings), after which, because they are more probable, further small reconfigurations take place.

The dynamics of exploration within the nested landscape of affordances is measured with the overlap order parameter (see Figure 7.2). This measure captures the average similarity of *whole-body coordination patterns* at ever increasing time distances (i.e. time lags) from each other. The stationary part, i.e. the static overlap, measures the width of the movement configuration space, i.e. the width of the general basin of attraction, that has been explored by the performer during a certain observation time (Figure 7.2). The smaller its value,

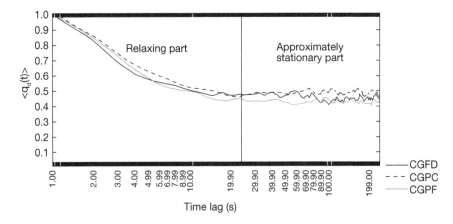

Figure 7.2 The dynamic overlap measure (vertical axis) as a function of time lag (horizontal axis) for three different conditions given in the legend. This may be divided into a relaxing part and an approximately stationary part. The stationary part measures the long-term exploratory breadth of movement forms.

the larger the general basin of attraction that has been explored (the larger the exploration that has been achieved) and vice versa.

How constraints influence exploratory behaviour

Constraint manipulation can help to determine how dynamic whole-body coordinative structures emerge from the more elementary classes of actions and postures given in Table 7.1. The main point here is to realize that there are classes of task and personal constraints that the experimenter can manipulate, and task constraints that spontaneously emerge and dissolve as a form of affordance on a shorter time scale, as a consequence of co-adaptive movements of partners. For example, the weight, height, flexibility, relative strength, skill level and mood of the performers may be seen as constant personal constraints on the observational time-scale of seven minutes; that is, they are quasi-stationary variables. Instructional task constraints that are not changed on the observational time scale also belong to this type of constraint; they are quasi-stationary variables too. These variables, which are constant over the whole observation time, act as non-specific control parameters. On the other hand, the quickly changing, short-time-scale task constraints are those which produce quick reconfigurations of movement patterns (so that constraints and movement forms emerge according to the same time-scale). For example, a momentary position of the partner affords certain actions and affords-not others. As a certain action is selected the movement configuration changes, and shortly afterwards another set or a sole affordance emerges. This co-adaptive, perception-action cycle occurs over much shorter time-scales (usually seconds) than the observation time-scale of seven minutes, and even on a shorter time scale than the control parameters that were mentioned above.

Our main aim was to investigate how different quasi-stationary parameters, which on the observational time scale of seven minutes may to a large extent be considered as constant – such as the instructional constraints, the partner and the skill level of performers – moulded the dynamic structure and the exploratory breadth of their whole-body movements. We expected that different sets of control parameters would have an effect on the properties of the general basin of attraction, such as its hierarchical dynamic structure of movement forms and its width. The dynamic structure was analysed using a hierarchical principal component analysis, and the width of the general basin of attraction by the stationary overlap order parameter.

We present two related studies in this section. In the first study, the aim was to analyse the hierarchical soft-assembled dynamic structure of three improvisers dancing CI under specific task constraint manipulations, examining the emergence of different action patterns (configurations) depending on these constraints (part of this study was published in Torrents, Ric and Hristovski, 2015). In view of the results obtained by this first study, the aim of the second study was to analyse differences in these parameters in different couples of improvisers with different backgrounds (expertise levels) dancing CI under

Table 7.1 Observational instrument defined on a coarse-grained scale based on elementary action or posture skills and movements (Torrents et al., 2015).

1-6: Support stability skills involving the floor: Actions of stability or balance using the floor that can be supported by…	Upper limbs (one or two) Lower limbs (one or two) Head Pelvis Torso Back
7-12: Support stability over a partner: Actions of stability or balance supported over the … of the partner	Upper limbs (one or two) Lower limbs (one or two) Head Pelvis Torso Back
13-16: Axial stability skills: Turns around the…	Longitudinal axis of the body Horizontal transverse axis of the body Horizontal anteroposterior axis of the body A combination of axes
17: Jumping 18: Being lifted or sustained by the partner 19-20: Level changing from…	Middle to down or falling Down to middle or rising from the floor
21-24: Locomotor skills involving…	Walking Movement on all fours Rolling Sliding
25: Receiving the partner 26: Colliding with the partner 27: Leading the partner 28: Lifting or sustaining the partner 29: Avoiding or escaping from the partner 30-47: Positions or actions of the different parts of the body:	One leg is bent Both legs are bent One leg is moving Both legs are moving One arm is bent Both arms are bent One arm is moving Both arms are moving One arm is relaxed, following gravity line Both arms are relaxed Body is almost aligned Body is bent forwards Body is bent backwards Body is bent to the right Body is bent to the left Body is moving Body is inverted (more than 45°) Head is moving
48: There is a change in the direction of the movement 49: There is a change in the global position of the body	

specific task constraint manipulations. In Table 7.1 an observational instrument with high inter-observer agreement metric characteristics (Cohen's kappa = 0.93) is presented, which was used in both investigations. The data obtained were subject to hierarchical principal component analysis (hPCA) and dynamic overlap analysis, as explained in more detail in Hristovski et al. (2013).

The role of task constraints

Three dancers with experience in CI participated voluntarily in the study. Participants warmed up individually. They danced together in duets lasting 480 seconds (the observation time), and in a limited square space measuring twelve metres along each side. The data acquisition, preparation and analysis were conducted as explained in Hristovski et al. (2013).

In the first duet dance there were no special instructional constraints (without instructional constraints – WIC) except those produced by contact with the partner (i.e. visual and haptic information, and gravity). Afterwards, the dancers danced twice more under two different instructional constraints:

- Instructional constraint 1: pelvis close – IC1: When dancing, try to keep your pelvis as close as possible to your partner
- Instructional constraint 2: pelvis far – IC2: When dancing, try to keep your pelvis as far away as possible from your partner

In general, the results from the component scores analysis showed that certain skills were persistent over time, while others were statistically rare and short-lived patterns that can be considered as fluctuations. More persistent elementary skills loaded predominantly on the first PC, capturing most of the variance, and they also had high component scores on several other PCs, playing the role of a 'bridge'; theses are mostly responsible for the correlations between first-level PCs (for more information about the dance skills performed, see Torrents, Ric & Hristovski, 2015). These long-term persistent elementary skills played the role of *seeds* from which more specific and shorter-lived movement forms emerged and dissolved due to more subtle and shorter-lived constraints (i.e. the state of the partner, inertia etc.).

The second-level PC structure was formed by salient correlated clusters of first-order PCs. A significant dimensional reduction was obtained, resulting in two (11%), three (78%) and four (11%) secondary PCs. This level's component scores revealed the more persistent skills by eliminating the short-term elementary skills. The extraction of two secondary PCs was only present under IC1, while the extraction of four secondary PCs was only achieved under IC2. These results indicated a stronger structural association of coordination patterns under IC1 and weaker structural association under IC2 constraints.

The secondary-level correlated structures systematically produced a sole third-level PC under all conditions. This level consisted of component scores of the most persistent elementary skills which, as expected, on the lower level PCs were projected on the first PC and in parallel on more than the first one. Hence, the HPCA enabled the analysis to view the dynamic structure of the CI dance at a different resolution. The lowest resolution related to the third-level PC reveals the elementary skills that form the skeleton of the dynamic structure with the lower order PCs adding more "meat" to it, revealing the elementary skills that were emerging and dissolving under subtle and quickly changing constraints. The co-relatedness of the first and second level PCs was a consequence of some properties of CI, and mostly due to the following:

1   The few, but most slowly changing (i.e. most frequently occurring), persistent elementary action classes, which played the role of a bridge between lower-level PCs.
2   The recurrence of small 'patches' of movement configurations of size L < L_{max} (where L_{max} = 49), which is the full size of the binary vectors.
3   The recurrence of whole configurations with L_{max}. This is not typical and occurred only a few times in all eighteen samples.
4   Rare but relatively long-lived (with a long dwell time) patterns with size L_{max}.

The results of the exploratory dynamics analysis are summed up in Figure 7.3. We present the averages of the stationary overlap order parameter $<q_{stat}>$ for each trial, comparing the values of each dancer in all conditions. In two of the three possible combinations of couples there was a significant effect of instructional constraints, being higher when dancing with IC2 and lower with IC1. Only one couple (S2 and S3) resulted in non-significant differences for IC1 and WIC. The instructional constraints WIC, IC1 and IC2 had no significant effects on the short-term exploratory breadth.

These results suggested that creative behaviour can be encouraged by manipulating instructional task constraints. Instructional constraints that produce a very different behaviour to that most commonly used by performers can lead

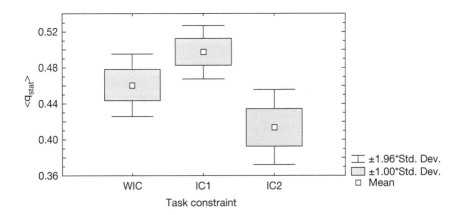

Figure 7.3 A typical relation of the average stationary dynamic overlap $<q_{stat}>$ values for different task constraints.

to greater exploration of movement possibilities. In contrast, constraints that reinforce the existing tendency of performers may limit the amount of exploratory behaviour.

The results of this study also show that it is possible to stimulate the practice of specific technical skills simply by modifying instructional constraints, without those skills having to be isolated from the normal practice of the activity.

The role of the partner

In order to examine the coupling dynamics of the duets we conducted a cross-correlation analysis of the Hamming distance (HD) time series. HD time series provide information about the size of individual reconfigurations at each time point, and hence, cross-correlating the time series told us about the time profile and the strength of duet couplings.

The cross-correlation of reconfiguration time series showed consistently weak but significant lag 0 (synchronous reconfigurations) and lag 1 (leading-following reconfigurations) correlations. Average lag 0 correlations had a value of $r = 0.35\pm$ 0.08, while lag 1 cross-correlations gave $r = 0.30 \pm 0.08$. Hamming distances revealed crossings over higher potential barriers between the different configurations when dancing with IC2. It is interesting to note that lag 1 cross-correlations were exclusively the property of one duet under different instructional constraints.

These results show that two duets preferentially adhered to synchronous reconfigurations, while one duet preferentially adhered to short-term delayed reconfigurations.

The role of skill level constraints (novice vs. expert)

This analysis was performed on one couple of contemporary dancers and one couple of Physical Activities and Sports Science students without a dance background. They danced under two different instructional constraints: pelvis close (PC) and pelvis far (PF).

The results again showed that certain classes of action had a persistent dominant role. These skills varied depending on the constraint and more strongly on the background of the dancers.

The highest level PC again contained the dominant long-term persistent actions, such as leading the partner using the foot-floor surface as support with bipedestrian locomotion, and support of the partner with the upper limbs. Nevertheless, when analysing the rest of the first-level PCs we were able to see that contemporary dancers were more influenced by the constraints imposed in the execution of specific skills. For instance, under the pelvis-close constraint lifts were more time-persistent, although novices explored fewer coordination patterns. The PCs of the novices under the pelvis-far constraint revealed that they dominantly used support with the partner using their upper limbs, while the experts performed lots of other combinations in all cases.

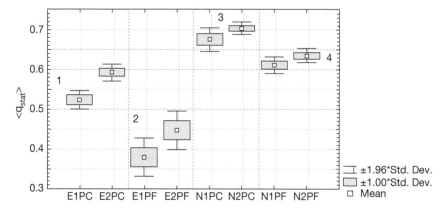

Figure 7.4 The profiles of the stationary values of the dynamic overlap $<q_{stat}>$ for expert (E1 and E2 dancers and novice (N1 and N2) dancers, under two instructional constraints – pelvis-close (PC) and pelvis-far (PF).

Figure 7.4 displays the average values obtained for all dancers and constraints for the stationary part of the dynamic overlap. The dynamic overlaps of pelvis-close constraints were always significantly higher than pelvis-far constraints (in both couples of different skill levels), while the exploratory breadths obtained by expert dancers were always significantly higher than for amateurs (in both constraints). Hence, a combination of skill level and instructional constraints revealed differences in the exploratory breadth of performers. It is interesting to note that the observational instrument coupled to the measure of dynamic overlap is able to detect the differences in the exploratory breadth within expert and novice couples as well as between them. For example, N2 showed lower exploratory breadth than N1 in both the pelvis-close and pelvis-far conditions. Similarly, E2 showed lower exploratory behaviour than E1 in both conditions.

What have we learned so far?

To summarize, the results corroborate the qualitative predictions of a nested dynamic structure of the landscape of affordances emerging during CI in dancers with different backgrounds (Hristovski et al., 2013). Instructional constraints, as well as the skill level of the dancers, had a significant effect on the dynamic structure of the landscape of affordances of dancers, and on their exploratory breadth. Dancing under IC1 constrained the emergence of some specific CI affordances in all the dancers with experience, and dancing under IC2 disabled the emergence of other affordances, but these instructional constraints were not enough to produce the emergence of those opportunities for action, and consequently movement forms, in novice dancers. Because affordances are always scaled relations between the immediate physical environment and a performer's abilities (effectivities), it is tempting to suppose that expert-novice differences in the exploration of the landscape of affordances are at least partly due to differences in their effectivities. Abilities as well as socio-cultural constraints may significantly limit or enhance the perceived opportunities and the associated readiness for certain action configurations.

In terms of their exploratory behaviour, all dancers produced a greater long-term exploratory breadth when dancing under pelvis-far constraints. For expert dancers, this constraint forced a different dance than they were used to, as CI is usually danced with a lot of contact with the partner. For novices, dancing under the pelvis-close constraint was really difficult. It created very repetitive behaviour based on the stereotyped movement classes of social dances. These results point to a significant reduction of the immediate field of perceived affordances and consequently the explored landscape of affordances as a result of constraining the dancers to dance so closely together. As Torrents et al. (2011) showed, very stringent constraints can limit the emergence of varied patterns. The average values of long-term exploratory breadth show that certain sub-configurations of movement classes are repeated during the dance, even in expert dancers. During free improvisation all configurations should be available to perform because there are no specific rules, but there are some specific dance constraints such as permanent physical contact with the partner, movements produced by the action of gravity when a dancer is lifted, turns produced by inertia, support with the floor, or the configurations that dancers repeat because of socio-cultural constraints (Bril et al. 2005).

The production of a sole third-level PC under all conditions in the first study means that the task constraints of CI generate a significant dimensional reduction in the configuration space of movement forms. In principle, under instructional and non-specific constraints the performers self-assembled movement and posture patterns in a dynamic hierarchical structure. This nested patterns-within-patterns dynamic is characterized by stronger relatedness at lower (elementary configuration) levels and less relatedness as one moves up the hierarchy.

The weak relatedness at higher patterning levels is due to the small number of slowly-varying, persistent elementary movement and posture classes which form the *skeleton* of the particular CI. They are outnumbered by more quickly changing elementary action classes which emerge to satisfy more immediate and specific task constraints. The long-term skeleton causally constrains the number and type of elementary action classes that can emerge and dissolve at shorter time-scales. For example, the long-term foot-floor support and upright stance action skeleton affords short-term actions like lifting the partner, falling and walking, and affords-not rolling or moving on all fours. This gives a picture of a nested landscape of affordances where the persistent, long-term afforded elementary movement and posture classes enable the perception of more immediate short-term emerging affordances that arise in the context of co-adaptive movements by the partner.

Furthermore, the fact that instructional constraints had an effect on the long-term exploratory breadth but not on short-term exploration (measured in seconds) suggests that in CI, instructional constraints may only have a significant influence on longer-term exploratory behaviour (measured in tens of seconds or minutes).

In order to significantly increase the exploratory breadth of the landscape of affordances in relation to short-term behaviour, a combination with other constraints, such as those coming from the partner, may prove to be essential.

The significant differences between the average values of the dynamic overlaps between experts and novices show that expert dancers use a more varied repertoire of patterns, as could be expected in an improvised dance. Dance teachers are usually worried about the creativity of their dancers (Chappell, 2007a, 2007b; Chen & Cone, 2003; Torrents et al., 2013), and the variety of the movements, i.e. the perceived field of affordances, is one of the criteria to take into account (Guilford, 1950). PF constraint seems to help novices to increase their rate of exploration, but they probably need to practice basic skills and develop abilities with other easier tasks in order to improve the richness of their dance.

From these results some applications are readily attainable. The first is connected to monitoring and possibly quantifying the evolution of the skill level of CI learners. By profiling the learner's nested dynamic structure and the exploratory breadth of her/his landscape of affordances, one can obtain information on the longer time-scales (months and years) of her/his learning dynamics. Scientifically guided modification of the constraints depending on the skill level of performers may be an obtainable challenge. A second application may be the analysis and quantifying of the effects of different constraints on the structure and dynamics of CI performance. In this sense the enhancement of exploratory and creative behaviour may be fostered.

References

Albright, A.C. (2003) Contact improvisation at twenty-five. In A.C. Albright & D. Gere (Eds.), *Taken by Surprise. A Dance Improvisation Reader.* Wesleyan University Press: Middletown.

Albright, A.C. (2003, 2004). Matters of tact: writing history from the inside out. *Dance Research Journal*, *35*(2) and *36*(1), 11–26.

Araújo, D., Davids, K. & Hristovski, R. (2006) The Ecological Dynamics of Decision Making in Sport. *Psychology of Sport and Exercise*, *7*, 653–676.

Biasutti, M. (2013) Improvisation in dance education: Teacher views. *Research in Dance Education*, *14*, 120–140.

Bril, B., Roux, V. & Dietrich, G. (2005) Stone knapping: khambhat (India), a unique opportunity? In V. Roux and B. Bril (Eds.), *Stone knapping, the necessary conditions for a uniquely hominid behaviour* (pp. 53-72). McDonald Institute for archaeological research: Cambridge.

Bruineberg, J. & Rietveld, E. (2014) Self-organization, free energy minimization, and optimal grip on a field of affordances. *Frontiers in Human Neuroscience*, 8, 599. doi: 10.3389/fnhum.2014.00599

Castañer, M., Torrents, C., Anguera, M.T., Dinušová, M. & Johnson, G.K. (2009) Identifying and analysing motor skill responses in body movement and dance. *Behavior Research Methods*, *41*, 857–867. doi: 10.3758/BRM.41.3.857

Chappell, K. (2007a) Creativity in primary level dance education: Moving beyond assumption. *Research in Dance Education*, *8*(1), 27–52.

Chappell, K. (2007b) The dilemmas of teaching for creativity: Insights from expert specialist dance teachers. *Thinking skills and creativity*, 2, 39–56.

Chemero, A. (2003) An outline of a theory of affordances. *Ecological Psychology*, 15, 181–195.

Chen, W. & Cone, T. (2003) Links between children's use of critical thinking and an expert teacher's teaching in creative dance. *Journal of Teaching in Physical Education*, 22, 169–185.

Czisek, P. (2007) Cortical mechanisms of action selection: the affordance competition hypothesis. *Philosophical Transactions of Royal Society B*, 362(1485), 1585–1599.

Czisek, P. & Pastor-Bernier, A. (2014) On the challenges and mechanisms of embodied decisions. *Philosophical Transactions of Royal Society B*, 369(1655), 20130479.

Gibson, J.J. (1979). *The ecological approach to visual perception*. Houghton Mifflin: Boston.

Guilford, J.P. (1950) Creativity. *American Psychologist*, 5, 444–454.

Hristovski, R., Davids, K. & Araújo, D. (2006) Affordance-controlled Bifurcations of Action Patterns in Martial Arts. *Nonlinear Dynamics, Psychology & Life Sciences*, 10(4), 409–444.

Hristovski, R., Davids, K., Passos, P. & Araújo, D. (2012) Sport Performance as a Domain of Creative Problem Solving for Self-Organizing Performer-Environment Systems. *Open Sports Science Journal*, 5(Suppl 1-M4), 26–35.

Hristovski, R., Davids, K., Araújo, D., Passos, P., Torrents, C., Aceski, A. & Tufekcievski, A. (2013) Creativity in sport and dance: Ecological dynamics on a hierarchically soft-assembled perception-action landscape. In K. Davids, R. Hristovski, D. Araújo, N. Balagué, C. Button & P. Passos (Eds.), *Complex systems in sport* (pp. 261–274). Routledge: London.

Joliffe, I.T. (2002). *Principal Component Analysis*. Second Edition. Springer-Verlag, New York Inc.

Kugler, P.N. & Turvey, M.T. (1987) *Information, Natural Law, and the Self-Assembly of Rhythmic Movement*. Erlbaum: Hillsdale.

Lavender, L. & Predock-Linell, J. (2011) From improvisation to choreography: the critical bridge. *Research in Dance Education*, 2(2), 195–209.

Newell, K.M. (1986) Constraints on the Development of Coordination. In M. Wade & H.T.A. Whiting (Eds.), *Motor Development in Children: Aspects of Coordination and Control* (pp. 341–360). Martinus Nijhoff: Dordrecht.

Newell, K.M., McDonald, P.V. & Kugler, P.N. (1991) The perceptual-motor workspace and the acquisition of skill. In J. Requin & G.E. Stelmach (Eds.), *Tutorials in motor neuroscience* (pp. 95–108). Kluwer Academic Publishers: Dordrecht.

Novack, C. (1990) *Sharing the Dance: Contact Improvisation and American Culture*. University of Wisconsin Press: Madison.

Paxton, S. (1997) Contact improvisation views. Round up. In L. Nelson & N. Stark Smith (Eds.), *Contact Quarterly's Contact Improvisation Sourcebook* (p. 79). Contact Editions: Massachusetts.

Sibani, P. & Jensen, H.J. (2013) *Stochastic Dynamics of Complex Systems: From Glasses to Evolution*. Imperial College Press: London.

Sidall, L.C. (1997) Round up. To definition. Volume 5, 1979–80. In L. Nelson & N. Stark Smith (Eds.), *Contact Quarterly's Contact Improvisation Sourcebook* (p. 54). Contact Editions: Massachusetts.

Stark Smith, N. (1997) Back in time. Editor note Volume 11, 1986. In L. Nelson & N. Stark Smith (Eds.), *Contact Quarterly's Contact Improvisation Sourcebook.* (p. 105). Contact Editions: Massachusetts.

Torrents, C., Castañer, M. & Anguera, M.T. (2011) Dancing with complexity: observation of emergent patterns in dance improvisation. *Education, Physical Training, Sport*, 80, 76–81.

Torrents, C., Castañer, M., Dinušová, M. & Anguera, M.T. (2010) Discovering new ways of moving: Observational analysis of motor creativity while dancing contact improvisation and the influence of the partner. *Journal of Creative Behavior*, *44*(1), 45–61. doi: j.2162-6057.2010.tb01325.x

Torrents, C., Ric, A. & Hristovski, R. (2015) Creativity and emergence of specific dance movements using instructional constraints. *Psychology of Aesthetics, Creativity and the Arts*, 9, 65–74.

Turvey, M. (1992) Affordances and prospective control: An outline of the ontology. *Ecological Psychology*, *4*, 173–187.

Warshaw, R. (1997) Round up. A definition. Volume 5, 1979–80. In L. Nelson & N. Stark Smith (Eds.), *Contact Quarterly's Contact Improvisation Sourcebook.* (p. 52). Contact Editions: Massachusetts.

Westerhuis, J.A., Kourti, T. & MacGregor, J.F. (1998) Analysis of Multiblock and Hierarchical PCA and PLS Models. *Journal of Chemometrics*, *12*, 301–321.

Withagen, R., de Poel, H., Araújo, D. & Pepping, G.J. (2012) Affordances can invite behavior: Reconsidering the relationship between affordances and agency. *New Ideas in Psychology*, *30*(2), 250–258. doi:10.1016/j.newideapsych.2011.12.003

8 Mix of phenomenological and behavioural data to explore interpersonal coordination in outdoor activities

Examples in rowing and orienteering

Ludovic Seifert, David Adé, Jacques Saury,
Jérôme Bourbousson and Régis Thouvarecq

Introduction

Consider the following problem: similar interpersonal coordination patterns can be based on factors that differ in meaning according to the situation and perspective of the individual (e.g. in term of perceptions, intentions or any other information that an individual considers meaningful). For instance, two rowers in a coxless pair crew can exhibit a small gap of synchronisation concerning their propulsion phase, which could be perceived differently by the rowers: one rower may relate the lack of propulsion synchronisation to the wind against the boat direction, while the other rower may perceive a lack of fluency in arm action. This example highlights that the rowers might be not attuned to the same information when interacting with a set of constraints. Conversely, several individuals might be similarly attuned to a set of constraints but they may exhibit different adaptations, from which different interpersonal coordination patterns can emerge. For instance, waves could be perceived as disturbing by two coxless pair crews, but their perceptions may lead them to different behavioural adaptations: in one crew the two rowers continue rowing simultaneously against the additional drag, while in a second crew one of the two rowers might overcome the wave to avoid wasting energy, leading to a period of temporary unsynchronised propulsion between the two individual rowers.

Therefore, although constraint manipulation can be well controlled in lab contexts, it can be more challenging to determine how a *set of constraints* (e.g. individual, task and environment; Newell, 1986) *interact in the ecological context of performance* (Araújo, Davids & Passos, 2007), especially in outdoor activities or competition where it is particularly salient that a small change in the initial conditions (i.e. in one of the constraints) can lead to a global destabilisation of the interpersonal coordination (i.e. a "butterfly" effect). In fact, it could be argued that when a phenomenon is fundamentally embedded in a complex interacting set of multiple constraints the phenomenon cannot be captured by a single experimental simulation of this situation, and must observed through a range of constraints. Therefore, for race performance analysis where many constraints interact without being manipulated by the experimenter, phenomenological data can provide fruitful information about how participants experienced these constraints as a whole, which is meaningful to their own point of view (Poizat, Sève & Saury, 2013; Sève, Nordez, Poizat & Saury, 2013). Therefore, the use of an ecological dynamics framework could be enriched by the analysis of performers' experience because it gives "experiential meaning" to the detection of periods of instability and transition between behavioural patterns (Poizat et al., 2013).

Thus, the main goal of this chapter is to present a heuristic approach concerning human activity analysis using: (i) mixed data that are the articulation of phenomenological data (i.e. how participants experience their involvement in the world) with behavioural data (i.e. how they behave in this world); and (ii) mixed methods that are the articulation of the theoretical and methodological framework used in the course of action research programme (e.g. reconstruction of the course-of-experience from semiotic categories; for more details, see Theureau, 2003, 2006) with traditional tools used in ecological dynamics framework to study the behaviour of neurobiological complex dynamical systems to explore their dynamics (Davids, Bennett, & Newell, 2006; Seifert, Button, & Davids, 2013; Seifert & Davids, 2012). This mix can show convergence, divergence and complementarity in the interpretation of the data and methods, in order to get new interpretations of the data and/or new approaches to a problem. As suggested by Varela and Shear (1999), the central point is "*whether the phenomenal description can help us both validate and constrain the empirical correlates that are available with the modern tools of cognitive neuroscience and dynamical system theory*" (p. 6). Hence we consider that mixing phenomenological and behavioural data and mixing methods can help us to understand: (i) how changes in the constraints encountered by the individuals may destabilise the performance and/or disturb the dynamics of interpersonal coordination in the ecological context of performance; and (ii) how a destabilisation of interpersonal coordination is overcome to maintain stability in coordination and/or the level of performance over time, despite the uncertain and constantly changing environments that surround outdoor activities. One key point that we would like to address in this chapter through the mix of data, tools and methods from the ecological dynamics and course of action frameworks is the *non-linear dynamical coupling* between changes in the ecological context of performance, the interpersonal coordination and the performance outcome. The ecological dynamics framework attempts to understand the laws, principles and mechanisms that govern how patterns of coordination emerge, adapt, persist and change in time and space according to a set of constraints present in ecological context of performance (Davids et al., 2014, 2012, 2015). How these patterns are maintained or reorganized according to stable or changing constraints is a crucial question to understand the adaptive and functional role played by movement *variability* (Davids, Bennett, & Newell, 2006; Davids, Button, & Bennett, 2008; Seifert et al., 2013), where it has been postulated that a state of permanent behavioural *adaptability* results from a *circular causality* between the components of the system (e.g. between an individual and its environment). Moreover, this behavioural adaptability to a set of constraints occurs *non-linearly* (e.g. there is non-proportionality between changes in the constraints and changes in behaviour), so a microscopic change in the constraints can lead to a superficial adjustment of the behaviour as well as to a macroscopic reorganization of the behaviour (Chow, Davids, Hristovski, Araújo, & Passos, 2011). One goal of the ecological dynamics framework is to determine the critical threshold over which the change in the constraints corresponds to a perturbation, in order to detect the *transition* between stable states and more broadly to analyse the system *stability* (i.e. its robustness). Interestingly, these dynamic relations between actors and their environments are also analysed in the course of action framework because any perturbations perceived in the environment by an individual structurally change these actor-environment coupling (Theureau, 2003) – i.e. they construct and modify the organization of individual activity at every instant (Varela, 1989). Thus, for these two frameworks, it may be suggested that behavioural variability can support functional adaptive behaviours. Adaptability relates to an appropriate relationship between *stability* (i.e. persistent behaviours) and *flexibility* (i.e. variable behaviours) during performance (Davids, Glazier, Araújo, & Bartlett, 2003; van Emmerik & van Wegen, 2000; Warren, 2006). Therefore, skilled performers are able to exhibit stable patterns of interpersonal coordination when needed, but can vary their relationships depending on dynamic performance conditions (Davids et al., 2012). Although human movement systems have a tendency to become stable and more economical with experience and practice (Sparrow & Newell, 1998), the nature of the interactions between individuals are not fixed into performing a rigidly stable solution, but can adapt into an emergent pattern in order to maintain behavioural functionality.

In that context, the two next sections provide research results about the variability of interpersonal coordination in rowing and orienteering, for which phenomenological and behavioural data were mixed and in each case with a different starting point. In rowing, the starting point was to capture the interpersonal coordination with behavioural data, i.e. the computation of the continuous relative phase based on kinematic and kinetic features. The phenomenological data were then used to better understand the interpersonal coordination variability. In particular, they allow an analysis of the inter-cycle variability of the interpersonal coordination of an international coxless pair crew during a 3000m race against the clock, during which several environmental (e.g. wind, wave, a corner along the route), organismic (e.g. fatigue, race strategy) and task (e.g. bow *vs.* stroke rower) constraints can influence the stability of interactions between rowers. The rowers had extensive experience of rowing together, having performed together for many years at the top level of their sport, and therefore a high degree of interpersonal coordination adaptability could be expected cycle by cycle.

Conversely, in orienteering the starting point was to identify the interpersonal coordination with phenomenological data, i.e. the analysis of perceptions, actions and intentions from self-confrontation interviews for eight dyads of two twelve-year-old children during physical education lessons. The students were novices in orienteering and were placed together for the first time in order to search for tags. Tools used in an ecological dynamics framework, such as a transition matrix, switching ratios (i.e. rates of change between patterns) and recurrence analysis (i.e. distribution of the patterns) (Chow, Davids, Button, & Rein, 2008; Wimmers, Savelsbergh, Beek, & Hopkins, 1998) were good candidates to study the interpersonal coordination dynamics under different task constraints. Specifically we evaluated the links between the constraints of different tasks (e.g. one or two maps shared between two students) and the dynamics of the interactions between two students. It was hypothesized that these various task constraints would lead students to share searching strategies or undergo conflicting decision making, which in both cases would impact on the nature and stability of the interactions between the students. Therefore, in this second case we did not mix data, but we used tools and methods popular in ecological dynamics frameworks to explain and understand the semiological dynamics of the interpersonal coordination in orienteering.

Interpersonal coordination in an elite coxless pair crew

As well described in the literature, the kinematics and kinetics data of the stroke and bow rowers could be collected using the *Powerline* system (Peach Innovations, Cambridge, UK) (e.g. Saury, Nordez, & Sève, 2010; Sève et al., 2013). This system has a data acquisition and storage centre connected to several sensors: two sensors measure the forces applied at the pin of the oarlocks (in the direction of the longitudinal axis of the boat) and two sensors measure the changes in oar angles in the horizontal plane (the angle formed by the oar with the axis perpendicular to the longitudinal axis of the boat) with a sample frequency of 50 Hz. As commonly reported in the literature (Hill & Fahrig, 2009; Hill, 2002; Sève et al., 2013), elite rowers demonstrate a high degree of synchronisation between each other both in terms of kinematics (oar angles) and kinetics (forces at oarlocks). In this study, the calculation of continuous relative phase (ϕ_{rel}, in degrees) between oar of the stroke rower and of the bow rower (as previously done for interpersonal coordination by de Brouwer, de Poel, & Hofmijster, 2013; McGarry et al., 1999) confirmed an in-phase pattern of coordination with few intra-cycle variations. Indeed, Figure 8.1 shows a mean ϕ_{rel} = -2.1 ± 4.2° and mean difference of forces = 3.6 ± 2.2 N, suggesting that on average these two rowers are well synchronised.

The inter-cycle variability of the interpersonal coordination could be assessed by the Cauchy index and root mean square (RMS) (Chen, Liu, Mayer-Kress, & Newell, 2005; Rein, 2012). The Cauchy index is based on the Euclidian distance that separates two successive cycles during a trial, and therefore it gives an indication of variability in comparison to the previous cycle, but does not reveal the nature of the variability. Comparatively, the RMS informs on the deviations between two curves at successive time points of the curves. In rowing, the deviations between each cycle and the average cycle were calculated, squared, summed, and the square root was taken from the resulting deviation score. Both

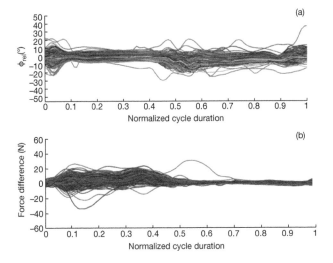

Figure 8.1 Intra-cycle variability of the interpersonal coordination based on the 350 cycles composing the 3000m race against the clock.

Notes: (A) continuous relative phase (φrel) for kinematics, (B) force difference.

for kinematics (Figure 8.2) and kinetics (Figure 8.3), the calculation of the Cauchy index and RMS for the ϕ_{rel} showed low values, with less than fifteen cycles out of a confidence interval (mean ± 2 standard deviations) for the 350 cycles performed during the race, indicating that these international-level rowers were able to maintain a low inter-cycle variability in their interpersonal coordination for 95% of the race.

Although the calculation of the continuous relative phase between rowers and the Cauchy index provided a good overview of the interpersonal coordination dynamics, they did not allow us to shed light on the instances of local destabilisation that were shown to appear regularly. Thus, it was hypothesized that phenomenological data reflecting the meaningful experience of the rowers can help to interpret interpersonal coordination variability. Indeed, the main point of collecting phenomenological data was to know whether rowers experienced (or not) these cycles as perturbed (or not). To do so, the rowers' behaviours and verbal communications were recorded during the entire race using two video cameras. Self-confrontation interviews were held immediately after the race. The self-confrontation interview has points in common with the stimulated recall interview technique, which was developed and tested by Cranach & Harré (1982). The interview techniques are based on video recordings of the participants during competition; as the participants then view these videotapes, they are invited to comment on their own lived experience step by step, as they may be able to report relevant perceptions (e.g. informational variables such as visual, kinaesthetic, haptic or acoustic variables), intentions and actions. Before each interview the researcher reminded the rower of the nature of the interview and the expectation that the participant needed to "re-live" and

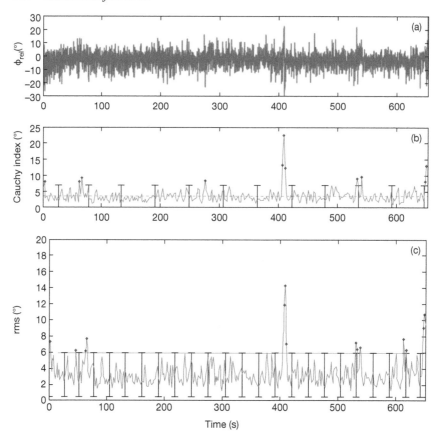

Figure 8.2 Continuous relative phase time series for kinematics and its related Cauchy
index and RMS time series for the 350 cycles.

Notes: (A) ϕrel between oar angles; (B) Cauchy index calculated on ϕrel from cycle i to i+1 as its mean
value and interval of confidence; (C) RMS calculated on ϕrel as its mean value and interval of
confidence. Stars indicate moments when the current cycle value is out of its interval of confidence.

describe his own experience during his race, without an *a posteriori* analysis,
rationalization or justification. This method is designed to reach the level of
activity that is meaningful for the rower: the goal is to encourage the rowers to
verbally report what they did, felt, thought and perceived during the race, as
naturally as possible. The self-confrontation interviews aimed to explore the
coupling between the perceptions, actions and intentions of each participant
based on a retrospective viewing of their performance. In the course of action
framework, activity gives rise to *pre-reflective consciousness*, which expresses
the actor's capacity to describe his/her activity by accounting for their experience
(according to the paradigm of enaction and the phenomenological approach;
Laroche et al., 2014; Legrand, 2007; Merleau-Ponty, 1945; Varela, Thompson &
Rosch, 1991). In this way, the researcher can further focus on three components

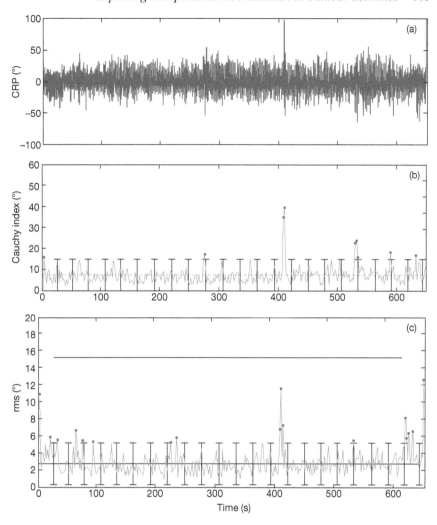

Figure 8.3 Continuous relative phase (CRP) time series for kinetics and its related RMS
and Cauchy index time series for the 350 cycles.

Notes: (A) CRP between oar angles; (B) Cauchy index calculated on CRP from cycle i to i+1 as its
mean value and interval of confidence; (C) RMS calculated on CRP as its mean value and interval of
confidence. Stars indicate moments when the current cycle value is out of its interval of confidence.

of the activity: the actor's dynamics of their concerns in the situation, the
dynamics of what is meaningful for the actor at each instant, and how elements
of knowledge are mobilized and broken with regards to past and current
experiences. This pre-reflective consciousness characterizes the immediate
experience for the actor – that is, the meaning that constitutes/emerges from their
action at each instant "*t*", for a given period, and in which the following action is
anchored. Interestingly, this awareness is the part of the consciousness that can

be described by actors; that is, they can *show it, tell it and comment on it* at each instant (Theureau, 2003, 2006), under certain methodological conditions of confrontation with the behavioural traces of their activity. Then, following the methodological framework of the course of action, the "course-of-experience" was reconstructed (Theureau, 2003, 2006). The course-of-experience consists in a semiological and dynamical description: it can be described as a chain of successive signs representing meaningful units of his/her ongoing activity from the actor's point of view, and which constitute actions, emotions, communications and interpretations (Theureau, 2003, 2006).

The results indicated that the cycles identified as perturbed with the Cauchy index were also reported as perturbing by the rowers. However, according to Millar, Oldham and Renshaw (2013), our results enabled us to identify two patterns of coordination between rowers' activities which were not detected from kinematics and kinetics analysis (Saury et al., 2014). The first pattern, with respect to the main perturbation occurring at around 400s in Figures 8.2 and 8.3, was identified in reference to coordination between the rowers' activities where one rower attempted to induce an adjustment by his partner; conversely, the other rower adjusted his behaviour to his partner's behaviour or communication. This pattern of coordination could be named *interpersonal* because each rower was focussed on his partner (Millar et al., 2013; Saury et al., 2014). In particular, an unexpected wave was responsible for the destabilisation of the rowers' coupling at around 400s in Figures 8.2 and 8.3, which led them to actively reorganize their interpersonal coordination at the kinematic and kinetic levels. The other perturbations occurring in the last part of the race have been identified as indirect interference between rowers' activities because both rowers' activities are oriented towards the boat, which induced the collective actions that Millar et al. (2013) called "rowing with the boat". This second pattern of coordination could be named *extrapersonal* because the rowers' perceptions of the boat mediated the coordination between each other (Saury et al., 2014). Typically, when a turn in the river path occurs, rowers focus on the crew (e.g. the speed and direction of the boat) and not on their partners. These suggestions are in accordance with those of Millar et al., (2013), who suggested that it is not the perceived actions of the respective rowers that determines their coupling but rather the perceived action of the boat. In other words, rowers attune to extrapersonal invariants as a primary source of information in order to achieve skilled interpersonal coordination.

Our results are in accordance with previous studies about sonification in rowing (i.e. acoustic feedback), which highlighted that rowers mainly focus on the boat speed rather than on the other rowers to organise their limb movements (Schaffert, Mattes & Effenberg, 2010, 2011). Finally, although behavioural data showed in-phase coupling between rowers, phenomenological data enabled the refinement of this first level of analysis by distinguishing direct and indirect interpersonal coordination to understand how the relationships between rowers could be temporarily destabilised without compromising the performance of the task-goal. This kind of mixed approach appeared particularly meaningful to understand interpersonal coordination dynamics in outdoor activities like rowing, and especially in the ecological context of a race. Indeed, as elite rowers only showed a few cycles perturbed (<5%), it seemed difficult to understand local perturbations exclusively from kinematic and kinetic data.

Interpersonal coordination in novice orienteering

Orienteering is usually performed individually; however, in the context of physical education lessons, teachers often pair students to increase the safety of the situation, to promote interpersonal coordination and to experience sharing in terms of social learning. In this context, the main research question is to identify the various patterns of interpersonal coordination and their dynamics according to the success or failure in finding tags. Although students can be equipped by a miniature HD video camera and microphone mounted on a pair of glasses and/or a GPS, using a metric (such as absolute distance between participants) does not necessarily enable the assessment of interpersonal coordination. In particular, metrics may be limited to identifying the various patterns of interpersonal coordination and their variability, and any shared experiences that might reflect cooperative learning may be missed. Indeed, as previously highlighted, Lafont (2012) outlined six patterns of cooperative learning in peer interaction (parallel behaviours, acquiescent mode co-elaboration, co-construction, confrontations without disagreement, contradictory confrontations and tutoring behaviours) that can be challenging to quantify from kinematical data collection. Therefore, using phenomenological data that express the course-of-experience of student dyads allows the identification of various interpersonal coordination patterns in orienteering and their dynamics during the task. Then, the tools and methods used in an ecological dynamics framework may help to locate and quantify temporary destabilisation of the interpersonal coordination in the semiological dynamics.

In the experiment described below, teenagers were equipped with an embedded microphone and camera on their head, and then immediately after their races self-confrontation interviews were organized on the basis of the video footage and the students' verbalizations in order to identify their actions, focus and concerns. The course-of-experience of each student was synchronized with their partner's in order to analyse the convergences/divergences of their actions, perceptions and intentions (Veyrunes & Yvon, 2013). This allowed us to distinguish three patterns of interpersonal coordination: *co-construction, confrontation* and *delegation* (Jourand, Adé, Sève, & Thouvarecq, 2014). *Co-construction* corresponded to convergence between two students' joint activities (e.g. often observed in the call room to prepare a searching strategy for tags), *confrontation* corresponded to divergence between two students' activities (e.g. often observed when each student proposed a different path for tag searching), and *delegation* corresponded to convergent actions and perceptions between the two teenagers but divergent intentions (e.g. often observed when one student took an initiative to search a certain path and the second one followed him with approval) (Jourand et al., 2014). This analysis has confirmed some patterns previously established by Lafont (2012), but also detected a new pattern called *delegation*.

From here, key questions emerged about the interpersonal coordination dynamics in orienteering: (i) is any one of the three patterns more prevalent? (ii) is there one pattern of interpersonal coordination that is more stable and always present, or could it be expected that one pattern leads to another one (e.g. can we expect that confrontation is not viable, and therefore confrontation may lead to co-construction or delegation)? (iii) finally, could it be expected that the beginning of the race in orienteering is dominated by a unique pattern (e.g. co-construction to define a searching strategy), or conversely, could we imagine that interpersonal coordination dynamics is very sensitive to task and environmental constraints? Following this questioning, the recurrence and variability of these three patterns of interpersonal coordination were evaluated under three manipulated teaching conditions: with or without time in the call room; with one or two maps; with or without "lure" tags. The first teaching situation consisted of offering (or not) to the two students one minute in a call room in order to prepare a search strategy for three tags in a forest during a running session lasting forty minutes. The second teaching situation consisted in providing either a map for each student or a single map for two students. When one map was provided, four tags were indicated and must be found in during the forty minute session. When two maps were provided, two tags were shown on the first map and the other two tags were shown on the second map; again, all four tags must be found within forty minutes. The third teaching situation consisted of providing one map for two students and placing some "lure" tags, which obliged the students to search carefully for the correct ones.

Similar to the previous example, the ecological dynamics framework can provide tools and methods to assess the semiological dynamics of the interpersonal coordination under the different task constraints; in this case this included the transition matrix, the switching ratio and the distribution of the patterns (Chow et al., 2008; Wimmers et al., 1998). The switching ratio corresponds to the number of observed changes divided by the theoretical maximal number of possible changes, knowing that a change of pattern can theoretically occur between two successive units of time – for instance, in a trial containing ten units of time, the total number of possible changes is nine. The distribution of the patterns corresponds to the percentage of occurrence of each pattern. In the different orienteering task constraints, the results showed that the confrontation pattern was present only 5% of the time, mainly during the second part or the end of the race (for an example, see Figure 8.4), whereas the other two patterns of interpersonal coordination (i.e. co-construction and delegation) appeared more often, and notably at the beginning of the race. In particular, when students had to share a map without any possibility of forming an anticipated searching strategy in the call room, the occurrence of co-construction and delegation patterns varied between 20% and 90% during the first half of the race, while the confrontation pattern occurred during the second part of the race (Figure 8.4).

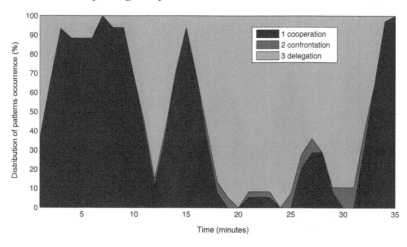

Figure 8.4 Distribution of pattern occurrence for dyads with one map and without time in a call room.

The confrontation pattern was more present at the beginning of the race when "lure" tags were introduced. According to the differential learning (Schöllhorn, Mayer-Kress, Newell & Michelbrink, 2009) and variable practice frameworks (Ranganathan & Newell, 2013), using "lure" tags with only one map per dyad and without time in a call room to prepare a searching strategy can create "noise" during practice that causes the confrontation pattern to emerge (Figure 8.5). Interestingly, Figure 8.5 also shows that "lure" tags induced high switching ratio values (>2.5) in the interpersonal coordination patterns: the delegation pattern was present for 51% of the time and alternated with co-construction, which was present for 44% of the time. In particular, the emergence of peak co-construction (65% of the time) occurred at the moment of finding a tag. In summary, "lure" tags induced a confrontation pattern of interpersonal coordination, which finally turned towards a co-construction pattern when the students approached the tag, and alternated with the delegation pattern when the students were running between two tags (Figure 8.5).

Finally, our interest in computing the switching ratio was to highlight changes in the patterns through time, i.e. to enable the analysis of interpersonal coordination dynamics. For instance, Figure 8.6 shows the same occurrence of the co-construction (46.5% of time) and delegation (49.5%) patterns, but with a different distribution through time: while delegation alternates with co-construction at the beginning of the race, the decrease of the switching ratio value indicated that the co-construction pattern became more present at the end of the race. This interpersonal coordination dynamics could be explained by the fact that the students had to share one map and could discuss their searching strategy in the call room.

This second example of mixed approaches, starting from phenomenological data and using the tools and methods of the ecological dynamics framework,

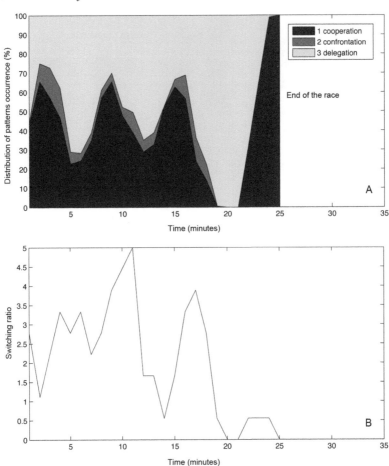

Figure 8.5 Distribution of pattern occurrence (A) and switching ratio (B) for dyad with "lure" tags, with one map and without time in a call room.

helped to analyse the semiological dynamics of the students' interpersonal coordination. The phenomenological data enabled the identification of interpersonal coordination patterns in a cooperative learning context, such as the co-construction and confrontation patterns (already emphasized by Gilly, Fraisse and Roux, 1988). However, Jourand et al., (2014) have also identified a third pattern named "delegation" that resembles a situation where one student accepted being "driven" by the other one (i.e. the "driver"). This driver/driven or actor/ reactor pattern of coordination has been already observed in competitive situations during tennis and squash games (Lames, 2006; McGarry et al., 1999; Palut & Zanone, 2005) but not in cooperative learning. Therefore, the use of phenomenological data appeared very fruitful in identifying interpersonal patterns. Indeed, although phenomenological data enabled the observation of transitions between patterns, the understanding of the interpersonal coordination dynamics

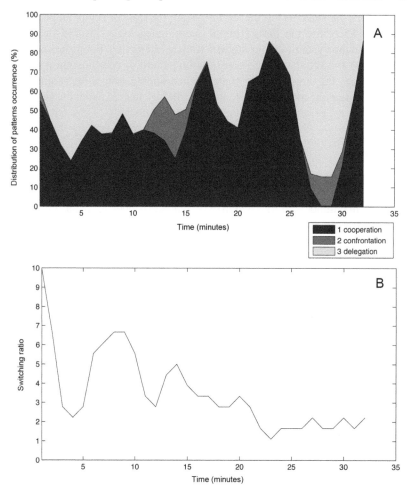

Figure 8.6 Distribution of pattern occurrence (A) and switching ratio (B) for dyad with one map and with time in a call room.

was enriched using the tools and methods employed in the Ecological dynamics framework by providing a quantification of variability, switching between patterns, and the identification of local destabilization. Finally, it was also shown that it was meaningful to come back to the phenomenological data in order to interpret this switching and local destabilization within the learning context. In other words, since it is sometimes difficult to precisely identify the emergence of a new pattern and/or switching among patterns from phenomenological data, the tools and methods used in the ecological dynamics framework enable an overview of the semiological dynamics. From a practical point of view, this mixed approach to the study of orienteering provided useful results about the interpersonal coordination dynamics that teachers cannot access due to the nature of this sport;

indeed, once the students left the starting point, the teacher could not provide any guidance. Therefore the design of the learning/teaching situation has a great influence on the dynamics of the student-student interaction. For instance, Jourand et al. (2014) showed that the co-construction pattern of coordination is favoured when a "lure" tag situation is designed: when approaching a tag, both students focus on reading the map in order to identify the key feature about the tag location and agree on the priority element that must be searched.

Conclusion

This chapter attempted to demonstrate how phenomenological and behavioural data can be mixed to explain and understand interpersonal coordination, especially in outdoor activities where the ecological properties of the environment cannot easily be isolated and controlled without decontextualizing the task. Starting the analysis from phenomenological data or from behavioural data and then mixing them, using the tools and methods of the ecological dynamics framework to understand the semiological dynamics, both approaches have demonstrated their usefulness in identifying the different patterns of interpersonal coordination and in understanding their dynamics (i.e. transition, switching patterns, local destabilization etc.). Therefore, this mixed approach provides guidance on which type of data should be collected to produce a meaningful description and a subsequent understanding of interpersonal coordination.

Acknowledgments

The authors would like to thank Julien Lardy, Antoine Nordez, John Komar, Clément Jourand and Dominic Orth for their advice and comments in the preparation of the manuscript. This project received support from the CPER/ GRR 1880 "Logistic, Mobility and Numeric" from the Normandy region, and the support of the ANOPACy convention from the Pays de Loire region.

References

Araújo, D., Davids, K. & Passos, P. (2007) Ecological validity, representative design, and correspondence between experimental task constraints and behavioral setting: Comment on Rogers, Kadar, and Costall (2005). *Ecological Psychology*, *19*(1), 69–78.

Chen, H.-H., Liu, Y.T., Mayer-Kress, G. & Newell, K.M. (2005) Learning the pedalo locomotion task. *Journal of Motor Behavior*, *37*(3), 247–56. doi:10.3200/ JMBR.37.3.247-256

Chow, J.Y., Davids, K., Button, C. & Rein, R. (2008) Dynamics of movement patterning in learning a discrete multiarticular action. *Motor Control*, *12*(3), 219–40.

Chow, J.Y., Davids, K., Hristovski, R., Araújo, D. & Passos, P. (2011) Nonlinear pedagogy: Learning design for self-organizing neurobiological systems. *New Ideas in Psychology*, *29*(2), 189–200. doi:10.1016/j.newideapsych.2010.10.001

Cranach, M. von & Harré, R. (1982) *The Analysis of Action: Recent Theoretical and Empirical Advances*. Cambridge University Press: London.

Davids, K., Araújo, D., Hristovski, R., Passos, P. & Chow, J.Y. (2012) Ecological dynamics and motor learning design in sport. In N.J. Hodges & A.M. Williams (Eds.), *Skill acquisition in Sport: research, theory and practice* (2nd Edition, pp. 112–130). Routledge: New York.

Davids, K., Araújo, D., Seifert, L. & Orth, D. (2015) Expert performance in sport: An ecological dynamics perspective. In J. Baker & D. Farrow (Eds.), *Handbook of Sport Expertise* (pp. 273–303). Taylor & Francis: London.

Davids, K., Bennett, S.J. & Newell, K.M. (Eds.) (2006) *Movement System Variability*. Human Kinetics: Champaign, IL.

Davids, K., Button, C. & Bennett, S.J. (Eds.) (2008) *Dynamics of skill acquisition: A Constraints-led approach*. Human Kinetics: Champaign, IL.

Davids, K., Glazier, P.S., Araújo, D. & Bartlett, R.M. (2003) Movement systems as dynamical systems: the functional role of variability and its implications for sports medicine. *Sports Medicine (Auckland, N.Z.)*, *33*(4), 245–60.

Davids, K., Hristovski, R., Araújo, D., Balague Serre, N., Button, C. & Passos, P. (2014) *Complex systems in sport*. Routledge: London.

de Brouwer, A.J., de Poel, H.J. & Hofmijster, M.J. (2013). Don't rock the boat: how antiphase crew coordination affects rowing. *PloS One*, *8*(1), e54996.

Gilly, M., Fraisse, J. & Roux, J. (1988) Résolution de problèmes en dyades et progrès cognitifs chez des enfants de 11 à 13 ans: dynamiques interactives et socio-cognitives. In A. Perret-Clermont & M. Nicolet (Eds.), *Interagir et connaître: Enjeux et régulations sociales dans le développement cognitifs* (pp. 73–92). DelVal: Fribourg.

Hill, H. (2002). Dynamics of coordination within elite rowing crews: evidence from force pattern analysis. *Journal of Sports Sciences*, *20*(2), 101–17.

Hill, H. & Fahrig, S. (2009). The impact of fluctuations in boat velocity during the rowing cycle on race time. *Scandinavian Journal of Medicine & Science in Sports*, *19*(4), 585–94.

Jourand, C., Adé, D., Sève, C. & Thouvarecq, R. (2014) Dynamique des modalités d'interaction entre élèves en course d'orientation: une étude comparative entre deux formats de travail lors de leçons d'EPS [Modalities of interaction between students dynamics in orienteering: a comparative study between two formats for PE lessons]. In *8ème biennale de l'ARIS* (p. 113). Geneva, Swizterland: Univeristy of Geneva (UNIGE).

Lafont, L. (2012) Cooperative learning and tutoring in sports and physicals activities. In B. Dyson & A. Casey (Eds.), *Cooperative learning in physical education* (pp. 136–149). Routledge: Oxon, Canada.

Lames, M. (2006) Modelling the interaction in game sports – relative phase and moving correlations. *Journal of Sports Science & Medicine*, *5*(4), 556–60.

Laroche, J., Berardi, A. & Brangier, E. (2014) Embodiment of intersubjective time: relational dynamics as attractors in the temporal coordination of interpersonal behaviors and experiences. *Frontiers in Psychology*, *5*(1180). doi:10.3389/fpsyg.2014.01180

Legrand, D. (2007) Pre-reflective self-as-subject from experiential and empirical perspectives. *Consciousness and Cognition*, *16*(3), 583–99.

McGarry, T., Khan, M.A. & Franks, I.M. (1999) On the presence and absence of behavioural traits in sport: an example from championship squash match-play. *Journal of Sports Sciences*, *17*(4), 297–311.

Merleau-Ponty, M. (1945) *Phénoménologie de la perception [Phenomenology of perception]*. Gallimard: Paris.

Millar, S.-K., Oldham, A.R.H. & Renshaw, I. (2013) Interpersonal, intrapersonal, extrapersonal? Qualitatively investigating coordinative couplings between rowers in

Olympic sculling. *Nonlinear Dynamics, Psychology, and Life Sciences, 17*(3), 425–43.

Newell, K.M. (1986) Constraints on the development of coordination. In M.G. Wade & H.T.A. Whiting (Eds.), *Motor development in children. Aspects of coordination and control* (pp. 341–360). Martinus Nijhoff: Dordrecht, Netherlands.

Palut, Y. & Zanone, P.G. (2005) A dynamical analysis of tennis: concepts and data. *Journal of Sports Sciences, 23*(10), 1021–32. doi:10.1080/02640410400021682

Poizat, G., Sève, C. & Saury, J. (2013) Qualitative aspects in performance analysis. In T. McGarry, P. O'Donoghue & J. Sampaio (Eds.), *Routledge Handbook of Sports Performance Analysis* (pp. 309–320). Routledge: London, UK.

Ranganathan, R. & Newell, K.M. (2013) Changing up the routine: intervention-induced variability in motor learning. *Exercise and Sport Sciences Reviews, 41*(1), 64–70.

Rein, R. (2012) Measurement methods to analyze changes in coordination during motor learning from a non-linear perspective. *The Open Sports Sciences Journal, 5*(1), 36–48.

Saury, J., Adé, D., Lardy, J., Nordez, A., Seifert, L., Thouvarecq, R. & Bourbousson, J. (2014) Experiential and mechanical data crossed analysis for a better understanding of sports performance. Case study on expert interpersonal coordination in rowing. In F. D'Arripe-Longueville & K. Corrion (Eds.), *The 5th International Congress of the French Society of Sport Psychology: Sport Psychology for Performance and Health across the lifespan* (p. 134). Nice, France.

Saury, J., Nordez, A. & Sève, C. (2010) Coordination interindividuelle et performance en aviron: apports d'une analyse conjointe du cours d'expérience des rameurs et de paramètres mécaniques. *Activités, 7,* 2–27.

Schaffert, N., Mattes, K. & Effenberg, A. (2010) Listen to the boat motion: acoustic information for elite rowers. In *3rd Interactive Sonification Workshop* (pp. 31–37). KTH: Stockholm.

Schaffert, N., Mattes, K. & Effenberg, A. (2011) An investigation of online acoustic information for elite rowers in on-water training conditions. *Journal of Human Sport and Exercise, 6*(2), 392–405.

Schöllhorn, W.I., Mayer-Kress, G., Newell, K.M. & Michelbrink, M. (2009) Time scales of adaptive behavior and motor learning in the presence of stochastic perturbations. *Human Movement Science, 28*(3), 319–33. doi:10.1016/j.humov.2008.10.005

Seifert, L., Button, C. & Davids, K. (2013) Key properties of expert movement systems in sport: an ecological dynamics perspective. *Sports Medicine, 43*(3), 167–78. doi:10.1007/s40279-012-0011-z

Seifert, L. & Davids, K. (2012). Intentions , Perceptions and Actions Constrain Functional Intra- and Inter- Individual Variability in the Acquisition of Expertise in Individual Sports. *The Open Sports Sciences Journal, 5*(8), 68–75.

Sève, C., Nordez, A., Poizat, G. & Saury, J. (2013) Performance analysis in sport: Contributions from a joint analysis of athletes' experience and biomechanical indicators. *Scandinavian Journal of Medicine & Science in Sports, 23*(5), 576–584.

Sparrow, W.A. & Newell, K.M. (1998) Metabolic energy expenditure and the regulation of movement economy. *Psychonomic Bulletin & Review, 5*(2), 173–196. doi:10.3758/BF03212943

Theureau, J. (2003) Course of action analysis and course of action centred design. In E. Hollnagel (Ed.), *Handbook of cognitive task design* (pp. 55–81). Lawrence Erlbaum Associates :Mahwah, NJ.

Theureau, J. (2006). *Le cours d'action: Méthode développée [The course of action: developments in methods]*. Octarès: Toulouse, FR.

Van Emmerik, R.E.A. & van Wegen, E.E.H. (2000) On Variability and Stability in Human Movement. *Journal of Applied Biomechanics, 16,* 394–406.

Varela, F. (1989) *Autonomie et connaissance. Essai sur le vivant [Autonomy and knowledge: Essay about living system].* Seuil: Paris.

Varela, F. & Shear, J. (1999) First-person methodologies: what, why, how? *Journal of Consciousness Studies, 6*(2–3), 1–14.

Varela, F., Thompson, E. & Rosch, E. (1991) *The embodied mind: Cognitive science and human experience.* MIT Press: Cambridge, MA.

Veyrunes, P. & Yvon, F. (2013) Stability and transformation in configurations of activity: the case of school teaching in France and Mexico. *International Journal of Lifelong Education, 32*(1), 80-92.

Warren, W.H. (2006) The dynamics of perception and action. *Psychological Review, 113*(2), 358–89. doi:10.1037/0033-295X.113.2.358

Wimmers, R.H., Savelsbergh, G.J.P., Beek, P.J. & Hopkins, B. (1998) Evidence for a phase transition in the early development of prehension. *Develomental Psychology, 32,* 235–248.

9 Theoretical perspectives on interpersonal coordination for team behaviour

Duarte Araújo and Jérôme Bourbousson

Introduction

How is one to understand the structure of interpersonal coordination, particularly in the case of team coordination? This chapter describes how there can be three possible theoretical perspectives to answer this question: the social-cognitive, the enactive and the ecological dynamics approaches. Team sports, as representative social contexts, are used as a task vehicle for presenting these theoretical frameworks. We also discuss their strengths and weaknesses, and further implications.

The social-cognitive approach

The ability to coordinate actions with those of others is a key factor for succeeding in collective performance contexts, such as team sports (Eccles & Tenenbaum, 2004). A traditional approach to understanding the mastery of group processes is predicated on the notion of team cognition. The concept of team cognition has been addressed in cognitive and social psychology, and a key aim has been to understand how shared knowledge can be represented in groups of coordinating individuals (Cannon-Bowers, Salas & Converse, 1993). These ideas are rooted in a key principle of cognitive science that performance (whether individual or collective) is underpinned by the existence of a representation or schema, responsible for the organization and regulation of behaviours (e.g. Rentch & Davenport, 2006).

The assumption of shared knowledge results from the possession by team members of complementary goals, strategies and relevant tactics that provide a basic shared understanding of desired performance outcomes, which underpins how each team member individually, and the team globally, aims to achieve performance goals (Ward & Eccles, 2006). Team members form clear expectations about each other's actions, allowing them to coordinate quickly and efficiently in adapting to the dynamic changes and demands of competitive performance environments, like sport, by selecting appropriate goal-directed actions to execute at appropriate times (Eccles, 2010; Salas et al., 1997). In this context, the processing of information is considered to play a crucial role in understanding how shared cognitive entities provide the basis of players' decision making in team sports (Reimer et al., 2006).

Previous reviews addressing social cognition models have emphasized shared knowledge believed to be associated with team effectiveness (Reimer et al., 2006). In this case, team efficacy may increase when a sophisticated, global and comprehensive representation of a collective action, linked to a mental representation of a performance context, is somehow shared by all players and put into practice. An asynchrony between the goals of individual performers and those of the team imply that a shared state has yet to be achieved, with resulting difficulties in coordination between performers (Eccles, 2010).

The role of memory is emphasized in each individual player for successful team functioning. Practice and experience are deemed important for enhancing the encoding of domain-specific information in, and retrieval from, long-term memory structures (Eccles & Tenenbaum, 2004). They are also relevant for the formation of new and more elaborated mental representations, models or schemas, developed by performers for regulating behaviours in task-specific situations (Eccles, 2010; Eccles & Tenenbaum, 2004).

Consequently, this framework provides a description of the knowledge possessed by team members, how this knowledge is shared/distributed, and how this knowledge is updated during a given game (Eccles & Tran-Turner, 2014).

In this view, it is the construction and updating of the individual's situation model that explains how multiple actors may simultaneously act together. The building of a mental representation of the current state of the setting implies that a given athlete is able to select, prepare and execute responses that will particularly allow for coordination with team members. As many team members are simultaneously coordinating, the amount of similarity within individuals' representations as they act together becomes a key feature, indicating a state of shared understanding within the team.

Shared knowledge thus constitutes an emerging state that teams can reach, and the amount of such shared knowledge is assumed to discriminate between novice and expert teams. A key benefit of shared knowledge is that each team member can similarly figure out what other team members are doing and generate expectations about the behaviours that will occur within the team, such that coordination can be achieved. These expectations help each team member to anticipate the operations of their team members. Each team member thus selects and undertakes appropriate operations at appropriate times in response to a given task. The concept of shared knowledge can be associated with two main types of content (Cannon-Bowers et al., 1993): i) taskwork shared knowledge includes content thats help to predict environmental events; while ii) teamwork shared knowledge supports expectations related to teammate's activities. Moreover, two main ways of establishing shared knowledge have been distinguished: i) time-consuming procedures can be involved in building shared knowledge prior to a given game, and ii) shared knowledge-building should be routinized enough to succeed during a given game. Accordingly, it has been suggested that team members can acquire shared knowledge by explicit planning or by actual play

(Eccles & Tran-Turner, 2014). Explicit planning means that coaches (or team leaders) communicate plans by introducing them verbally or by presenting them graphically. This process often occurs in an off-the-field location, and aims at planning various levels of action (i.e. various levels of abstraction, such as planning outcomes, designs, procedures, or operations). Actual play leads to shared knowledge by playing the sport generally (i.e. practicing the sport regardless of the team in which they perform), and by playing within a particular team (i.e. allowing for acquiring knowledge specific to the team under consideration) (e.g. Blickensderfer et al., 2010). This last process accounts for a dynamic and situated form of establishing shared knowledge.

Regardless of how shared knowledge is established, it can be updated over time. Because new and unexpected events are expected to occur in dynamic and uncertain complex tasks, new states of shared knowledge may happen, and thus converge and/or compete with the state of the knowledge shared prior to the game. Thus, an updating of the individuals' representations is needed to maintain the accuracy of the knowledge that was shared prior to the performance (Uitdewilligen, Waller & Pitariu, 2013). This updating can occur by two processes: incidentally, and/or by deliberate means (Eccles, 2010; Eccles & Tran-Turner, 2014). Incidental means include how team members adapt to the unfolding game by changing some part of what was planned prior to the game. These changes occur without explicitly intended communication within the team, and are thus facilitated when shared knowledge existed prior to the game that set the range of strategies available to the team. When such an updating occur via deliberate means, it refers to a process by which team members deliberately communicate (i.e. verbally or not) with their teammates about how the plan has to be updated to better face upcoming events.

Being grounded in the mainstream human information processing approach of team cognition, the social-cognitive framework can make use of methodologies elaborated by this paradigm. Questionnaires, verbal protocols and various elicitation methods are the common methods (e.g. Blickensderfer et al., 2010).

Although shared knowledge has tended to dominate research on mental models in team behaviour, and is still accepted as a necessary pre-condition for the existence of team coordination, some investigators claim that it needs to be conceptually reformulated and much more carefully defined (Ward & Eccles, 2006; Mohammed & Dumville, 2001). Further, informational cues are likely to be used differently by each individual, according to their characteristics and each phase of play (Ward & Eccles, 2006). Thus, knowing *"who knows what"* at each moment of a match would involve an immense mental load.

Particularly, the mechanism to explain re-formulations of a team member's schema, when changes occur in the content of another member's schema, has proved difficult to verify (Mohammed, Klimoski & Rentsch, 2000). Moreover, several studies have failed to find significant relationships between measures of convergence of mental models and various dimensions of team performance (Mohammed & Dumville, 2001). Finally, it is hard to consider that representations exist beyond the boundaries of an individual organism and can be somehow shared (Riley, Richardson, Shockley & Ramenzoni, 2011; Silva et al., 2013).

The enactive approach

The theoretical and methodological framework known as course-of-action (Theureau, 2003), grounded in the enactive approach (Varela, Thompson & Rosch, 1991), has been recently used to study interpersonal coordination in teams. The enactive approach suggests that team coordination processes should be investigated by reconstructing how individual 'cognitions' articulate during performance environments and by determining how these articulations are step-by-step adjusted over time. Cognitions are described with respect to a phenomenological approach to humans, which assumes the sense-making process to highly contribute to the human/environment coupling. Thus, the method used in this approach tries to capture through verbal description the lived experience of team members involved in a dynamic cooperative effort.

Importantly, this framework aims to contribute to a paradigm shift in cognitive science: course-of-action researchers joined the enactive approach in calling for a rejection of the dominant positivist computationalist (i.e. cognitivist) paradigm (e.g. Varela, 1979, Varela et al., 1991). Lived experience is not conceived as an epiphenomenal issue, but it is central to the situated achievement of human activity. Therefore, research follows a careful phenomenological paradigm. To go further, the "own-world" of the living being is primarily made by (a) the nature of its sensory apparatus, that is genetically inherited, (b) the history of the actor/environment coupling (e.g. recurrent patterns of perception and action building during individual development), and (c) the way the individual experiences his or her coupling with the environment in the moment (Thompson, 2011). This last assumption tends to make the situated experience as lived by the performers the *sine qua non* condition for describing how individuals operate in their real-time activity.

Focused in the claim that a living being possesses an interiority that escapes any objectivist approach, this interiority has been the starting point of the descriptions of the actor/environment coupling. This framework has thus given a subsequent special attention to pre-reflective consciousness – that is, the implicit ways in which a given actor experiences his/her ongoing activity. To capture this interiority, the course-of-action framework includes a methodology that makes use of phenomenological forms of *a posteriori* interviews (e.g. self-confrontation interview techniques) to elicit subjective data. This primary source of data can be combined with secondary behavioural data (e.g. ethnomethodological data: Poizat, Bourbousson, Saury & Sève, 2012; biomechanical data: Sève, Nordez, Poizat & Saury, 2013) to deepen the understanding of the activity under study (see also Seifert et al.'s chapter in this volume). Together, the subjective and the observational data are combined to build units of activity that are, *a posteriori*, chained together sequentially, to describe any given individual's activity (see Theureau, 2003 for details). After detailing the sense-making activity embedded in its context, as it was reported by the participants, researchers make use of additional processing regarding the particular object of their research (e.g. Braun & Clarke, 2006).

In line with the enactive approach, when investigating team coordination phenomena the team cognition concept refers to the extent to which individual actors are mutually coupled – that is, the extent to which individual activities contribute to or perturb the activity of the others (De Jaegher & Di Paolo, 2007; Laroche, Berardi & Brangier, 2014). From this perspective, it is assumed that a novel domain of phenomena emerges from the cooperative effort, the so-called 'participatory sense-making' (De Jaegher & Di Paolo, 2007). Participatory sense-making processes refer to how social agents coordinate their sense-making during interaction (i.e. how the meanings that each individual builds in his/her activity corroborate with the meanings simultaneously built by co-performers) and how this participation in sense-making is experienced (De Jaegher & Di Paolo, 2007; Froese & Di Paolo, 2011). The experience of the mutual engagement, in phenomenological terms, is thus assumed to be not prescribed by elements of knowledge or shared habits observable in the members' culture prior to the activity, but dynamically achieved in real-time by actors working together. This achievement can be facilitated by various elements that support it, such as previous shared experiences of performers (see Froese & Di Paolo, 2011 for details about the role of cultural factors in general). However, the particular perspective that each participant builds from the common setting does not ensure that the sense-making will be participatory. Any divergence in how each member experiences the situation leads to various degrees of participation in sense-making, various feelings of connectedness with the other, and various expectancies for actions from others. In line with this, the extent to which team members experience the same context has been a key issue of empirical studies, assuming that interpersonal coordination depends primarily on participatory sense-making. For example, the dyadic coordination in tennis doubles generally occurs in a context where the teammates do not necessarily build a similar perspective on the world (Poizat, Bourbousson, Saury & Sève, 2009). When conducted in a basketball team, such an analysis contributes to observing how several singular perspectives on the world (i.e. many team members developing their activities individually) can harmoniously function together, despite the difficulty of drawing a stable common picture (e.g. Bourbousson, Poizat, Saury & Sève, 2012).

An enactive phenomenological analysis starts from a temporal account of how each individual activity unfolds in the ongoing team coordination; and after this, it describes how individually perceived situations are arranged together by means of the adaptive activities of participants and the mediation of their physical environment. For this second step, the analysis points out what is shared and what it is not depending on individuals' personal perspectives, rather than presupposing sharedness. In short, an enactive approach, and consequently the course-of-action framework that operationalizes it, always gives primacy to the individual subjective analysis, and only then, as a second step, describes the team coordination. For analyzing team coordination there is a procedure called the 'synchronization of the participants' courses of action'. In this analysis, for a given team the participants' individual courses of action are "connected" by presenting them side-by-side in chronological order. The connection is achieved by scrutinizing correspondences between individual activities thanks to an objective description of the unfolding game, obtained, for example, by video recording. Once the individual participants' courses of action are step-by-step connected, each connection is considered as a collective unit – that is, a reference unit to which the individual meaningful units at a given moment of the game are related. Then, it is possible to analyze the temporal chain of collective units. Once identified, collective units facilitate the characterization of the relationships between individual courses of action according to the specific research question. For instance, the researcher may ask how the individual activities interplay according to the sharedness of their situated concerns (Bourbousson, Poizat, Saury & Sève, 2012) or the degree of sharedness of the contextual information taken into account by each teammate (Poizat et al., 2009). Subsequently, the convergence/divergence of individual perspectives across time can be described (Sève, Nordez, Poizat & Saury, 2013). This step is mainly conducted with respect to a qualitative methodology.

Moreover, a quantitative behavioural analysis can be used to complement the description of how the lived situation is experienced. Such a complementary combination of first- (the starting point of the analysis) and third-person approaches is needed to describe the actor/world coupling by considering the complex nature of the constraints that give rise to a perceived setting (i.e. physiological, cultural), and to consider the nature of the effects that a given experience of an action produces in the dynamics of the collective task. For example, Bourbousson and colleagues discussed the way in which most points of subjective interplays between players occurred at the local level (between two or three players). These local interplays may alternatively be seen in terms of the players' concerns (Bourbousson et al., 2012), the players' mutual awareness (Bourbousson et al., 2010, 2015), the players' elements of knowledge (i.e. expectations for action for himself and for others, and recurrent expected patterns of perception and action) (Bourbousson et al., 2011), expectations and judgements about the situation. They found that the situated interplays between team members were constantly being modified; that the local participatory sense-making was dynamically created; and that the temporal interaction of the local instances of sharedness permitted the players to regularly revise their own perspective on the situation, update the sharing process, and even adopt a perspective similar to others.

Conceived as such, the enactive approach goes beyond the positivist approach to cognitive science that tries to capture the objective world (independent of the perceiver, and by means that are neutral to the individual), to include analyses of the subjective experiences of the individual as the starting point. The enactive approach tries to avoid representationalism, but by being grounded in the "interiority" of individual it needs to operationally define what is the internal sense-making process and contrast it with representations of the lived situations stored in memory. The risk of falling into a representational realm has to be thoroughly considered and managed by researchers in their procedures. To develop studies that contribute to a genuine enactive approach of human activity, the concept of (individual/environment) coupling also needs further elaboration and empirical description: it is inadequate to understand the coupling by analysing only the individual side of it. Research scientists often describe the asymmetry of the coupling as a consequence of the methods and not as a principle that has to be empirically described. For instance, it is presupposed that the subjective perspective of the actor defines the perturbations of the environment that contribute to the coupling, but it is less well described how this subjective experience is finely nested in the behavioural side of the coupling, or is produced by this behavioural dynamics. It is true that there is a recent effort to combine first- with third-person approaches (e.g. Seifert et al. in this volume), but this combination is not yet an embedded form of capturing and expressing the phenomena under analysis. How (individual or participatory) sense-making could be the starting point of how interpersonal interactions are shaped needs to be contrasted with studies that also consider the subjective facet as a consequence of such interaction. Such studies should thus be able to question the accuracy of the verbalization about the individual experiences that emerged from the dynamics of interpersonal coordination.

In general the enactive approach to team coordination relies on the interview techniques used to collect the empirical data. Importantly, the expertise of the researcher is needed because it is through his/her understanding of the theory and the method that the data are combined – the portions of verbalizations, and also the links with the "objective" data. Such criticisms are anchored in indirect access to the phenomenon under study, and further methodological research is needed to guarantee that participatory sense-making data-capture is less constrained by the *medium* than by the *thing* being researched.

The ecological dynamics approach

Understanding interpersonal coordination from the perspective of ecological dynamics means investigating the dynamical principles that form the patterns of interpersonal coordination in daily contexts (Araújo et al., 2006, after Kelso, 1995; Warren, 2006; Schmidt et al., 1999). The interpersonal patterns at the behavioural scale can be understood and modelled in terms of their own dynamics, without investigating the micro level of the individual's interior (Schmidt et al., 2011). This dynamics comprehends the interaction of system components to form stable behavioural patterns which can be characterized as attractor states of the system's dynamics (Schmidt et al., 1990). Important for the study of the dynamics of interpersonal systems is the fact that synchronization process (i.e. the temporal coordination of unfolding events in a system) has been found to occur between biological rhythmic units that are connected not only mechanically but also informationally. Much research has demonstrated how this dynamical synchronization process can operate across perceptual media to produce a coordinated timing of interpersonal coordination (Schmidt et al., 1990). In these studies, the only possible way that the rhythmic units could be interacting is via the information available to the visual systems of the human participants.

Systematically exploring the informational basis for the emergence of interpersonal coordination and the dynamics that can allow the self-organization of coordinated action to occur is important to study, not because it implies humans do not have intentions, expectations and mental states, but rather because research demonstrates how integrally intertwined are body (e.g. nervous, physiological, psychological, movement) and contextual (e.g. social, cultural, climate, altitude) sub-systems when engaging in interpersonal coordination (Marsh et al., 2009), and that a theory of the dynamics of self-organization is needed to understand the complexity of such interacting components (e.g. Kelso, 1995). Ecological dynamics takes seriously the embodiment of cognition and the embeddedness in the environment (Richardson et al., 2008). With motion sensors, it is possible to examine the near-sinusoidal interpersonal rhythmic coordination of movement that is described by a motion equation known as the HKB model (Haken, Kelso & Bunz, 1985; also see Kelso, 1995). Related differential equations modelled dyadic interactions in team sports (Araújo et al., 2014), making predictions about the kinds of coordination that could and could not be sustained. Following an environment-individual system perspective (Järvilehto, 2009; Richardson et al., 2008), an ecological approach proposes that knowledge of the world is based upon recurrent processes of perception and action through which humans perceive affordances (i.e. opportunities for action; see Gibson, 1979). The concept of affordances presupposes that the environment is directly perceived in terms of what actions a performer can achieve within a performance environment (i.e. it is not dependent on a perceiver's expectations, nor mental representations, nor interpretations linked to specific performance solutions) (Richardson, Shockley, Fajen, Riley & Turvey, 2008). Importantly, the ecological perspective posits that meanings, the value possibilities for action in the environment, are directly perceived by individuals in that environment. Extending this idea to an interpersonal level, ecological dynamics predicts that the presence of others extends the action possibilities ("affordances") that are realizable by individuals to action possibilities realizable by groups. Therefore an interpersonal system can be seen as a new entity with new abilities, and that the emergence of such a social synergy should be predictable using affordance theory (Gibson, 1979; Marsh et al., 2009). Indeed, affordances can be perceived by a group of individuals trained to become perceptually attuned to them (Silva et al., 2013). In collective sports, both teams have the same objective (i.e. to overcome the opposition and win). Hence, the perception of collective affordances acts as a selection pressure for overcoming opponents and achieving successful performances (Reed, 1996). In this sense, collective affordances are sustained by the common goals of team members who cooperate to achieve group success. From this perspective, team coordination depends on the team's collective attunement to shared affordances founded on a prior platform of (mainly non-verbal) communication or information exchange (Silva et al., 2013). Through practice, players become perceptually attuned to affordances *of* others and affordances *for* others during competitive performance, and refine their actions (Fajen, Riley & Turvey, 2009) by adjusting behaviours to functionally adapt to those of other teammates and opponents (see also Chapter 12 in this volume). This process enables them to act synergistically with respect to specific team goals (Araújo et al., 2015).

An important feature of a synergy is the ability of one of its components (e.g. a player) to lead changes in others (Riley et al., 2011). The decisions and actions of players forming a synergy should not be viewed as independent, and can explain how multiple players act in accordance with dynamic performance environments in fractions of a second (Silva et al., 2015/in press). Therefore, the coupling of players as independent degrees of freedom into integrated synergies is based upon a social perception-action system supported by perception of shared affordances (Silva et al., 2013). In line with this, an ecological dynamics perspective provides predictions about the conditions under which individuals will be better able to coordinate their movement with others, and what features of a situation facilitate/perturb interpersonal coordination in completing some task.

This view has major implications for designing experimental research for studying team performance behaviours, as well as for practice and training design. Brunswik (1956) was probably the first psychologist advocating theoretical principles for sampling the features of a task, using the concept of 'representative experimental design'. Perceptual variables incorporated into experimental designs should be sampled from the performer's typical performance environment, to which behaviour is intended to be generalized. Experimental designs, following an ecological dynamics perspective, need to focus on *player-player-environmental* interactions that can be elucidated in compound variables specifying functional collective behaviours of teams (e.g. geometrical centres), underpinned by interpersonal synergies created between performers (see Araújo et al., 2015 for a review).

In sport training and practice design, 'representative learning design' implies the generalization of learning task constraints to the task constraints encountered during performance (Pinder et al., 2011), for example when perceiving the actions of a 'live' opponent in a study of anticipation (Araújo et al., 2006; Travassos et al., 2012). An important pedagogical principle in sport is the need to ensure that there is adequate 'sampling' of informational variables from the performance environment in a practice task, ensuring that modified tasks will correspond to an actual competition context so that important sources of information are present (Araújo & Davids, 2015/in press). Therefore, the tendency to design simplistic and controlled tasks, both in experiments and in practice, may not accomplish a key principle in ecological dynamics (Davids et al., 2012).

Although ecological dynamics is a useful way to conceptualize and investigate interpersonal dynamics, it can also be seen more fundamentally as a manifestation of an integrative conceptual framework and research paradigm for psychological processes at all levels of human functioning, from brain dynamics to societal dynamics. However, different aspects of human experience have distinct constraints, information and processes. Therefore, it is important to avoid contamination of the specificities of one psychological phenomenon for other psychological phenomena.

For example, social interaction in one context (e.g. interaction among friends) may be expressed differently from social interaction in a different context (e.g. interaction with adversaries), but the foundational laws of dynamics and self-organization are presumably the same in both contexts. More importantly, emergence via self-organization is a general phenomenon that can be manifested in different ways. It is important to take ecologies seriously and try to understand the specific form that interpersonal coordination takes in various social contexts (Vallacher & Jackson, 2009).

Another point that needs further research is the study of how the intrinsic dynamics of each individual create favourable or unfavourable conditions for interpersonal entanglement (Phillips et al., 2010). Finally, an important issue is the explanation of how affordances for interpersonal interaction are selected from a world full of other possibilities (Beek, 2009; Withagen et al., 2012). Indeed, the fact that performers are coupled does not clarify what the perceived affordance was, or which source of information is constraining the link. These and other questions, if clarified, may open the avenue for extensive embodied-embedded research on human functioning.

Conclusion

There are many possibilities to study interpersonal coordination. In this chapter, we discussed the three approaches that have underpinned many studies of team coordination in sport. Briefly, the socio-cognitive approach to understanding team coordination in sports involves the idea of team cognition grounded on the premise of shared knowledge of the performance environment, internalised among all team members (Cannon-Bowers et al., 1993). It is derived from the information-processing rationale that argues that performance (whether individual or collective) is based on the existence of a representation, programme, script or schema, responsible for the organization and regulation of behaviours. The course-of-action framework is predicated on the enactive approach to activity (Varela et al., 1991). This perspective supports the use of phenomenological techniques to study interpersonal coordination, to address the extent to which the team members share their situated understanding of the unfolding activity. Recent efforts attempt to describe the actor/world and actor/actor coupling by combining first- and third-person approaches. Finally, an ecological dynamics perspective on team coordination focuses on the available information offering possibilities for regulating goal-directed activity in individuals, often with others (Gibson, 1979) in representative contexts (Araújo et al., 2006), and how these self-organized patterns of behaviour take place and evolve over time (Kelso, 1995).

Instead of a dominant paradigm that supplants the others, we argue in favour of explanatory pluralism in (social) psychology (Dale, Dietrich & Chemero, 2009). Human behaviour is extraordinarily complex. This phenomenon is sufficiently rich to admit several goals and levels of analysis that require pluralism to approach them all. This is not to say that there cannot be genuine competition between different theories about similar phenomena; we are not defending a forced complementarity between approaches. These three perspectives have very different philosophical roots, different traditions and different, even contradictory, assumptions. Instead, we are arguing that theoretical assumptions need not force scientists into any kind of fundamentalism, but rather serve to define explanatory boundary conditions. Possibly, correspondence to the innumerable phenomena (Hammond, 1996) would be more helpful than the search for a coherent monism to explain all the phenomena. Human beings, and the behaviour and structure of organisms in general, may not subscribe to one single theoretical paradigm for explanation. Therefore, we agree with Dale et al. (2009) that it is worth trying to figure out how and why certain frameworks work when and where.

References

Araújo, D. & Davids, K. (2015) Towards a theoretically-driven model of correspondence between behaviours in one context to another: Implications for studying sport performance. *International Journal of Sport Psychology, 46*, 268–280.

Araújo, D., Davids, K. & Hristovski, R. (2006) The ecological dynamics of decision making in sport. *Psychology of Sport and Exercise, 7*, 653–676.

Araújo, D., Diniz, A., Passos, P. & Davids, K. (2014) Decision making in social neurobiological systems modeled as transitions in dynamic pattern formation. *Adaptive Behaviour, 22*(1), 21–30.

Araújo, D., Silva, P. & Davids, K. (2015) Capturing group tactical behaviours in expert team players. In J. Baker & D. Farrow (Eds.), *Routledge Handbook of Sport Expertise* (pp. 209–220). Routledge: New York.

Beek, P.J. (2009) Ecological approaches to sport psychology: prospects and challenges. *International Journal of Sport Psychology, 40*(1), 144–151.

Blickensderfer, E., Reynolds, R., Salas, E. & Cannon-Bowers, J. (2010) Shared Expectations and Implicit Coordination in Tennis Doubles Teams. *Journal of Applied Sport Psychology, 22*(4), 486–99.

Bourbousson, J., Poizat, G., Saury, J. & Sève, C. (2010) Team coordination in basketball: Description of the cognitive connections between teammates. *Journal of Applied Sport Psychology, 22*, 150–166.

Bourbousson, J., Poizat, G., Saury, J. & Sève, C. (2011) Description of dynamic shared knowledge: An exploratory study during a competitive team sports interaction. *Ergonomics, 54*, 120–138.

Bourbousson, J., Poizat, G., Saury, J. & Sève, C. (2012) Temporal aspects of team cognition: A case study on concerns sharing within basketball. *Journal of Applied Sport Psychology, 24*, 224–241.

Bourbousson, J., R'Kiouak, M. & Eccles, D.W. (2015) The Dynamics of Team Coordination: A Social Network Analysis as a Window to Shared Awareness. *European Journal of Work and Organizational Psychology, 24*, 742–760.

Braun, V. & Clarke, V. (2006). Using thematic analysis in psychology. *Qualitative Research in Psychology*, *3*, 77–101

Brunswik, E. (1956) *Perception and the representative design of psychological experiments* (2nd ed.). University of California Press: Berkeley.

Cannon-Bowers, J.A., Salas, E. & Converse, S. (1993) Shared mental models in expert team decision making. In J.J. Castellan (Ed.), *Individual and group decision making*. Erlbaum: Hillsdale, NJ.

Dale, R., Dietrich, E. & Chemero, A. (2009) Explanatory Pluralism in Cognitive Science. *Cognitive Science*, *33*, 739–742.

Davids, K., Araújo, D., Hristovski, R., Passos, P. & Chow, J.Y. (2012) Ecological dynamics and motor learning design in sport. In N. Hodges & M. Williams (Eds.), *Skill acquisition in sport: Research, theory and practice* (2nd ed., pp. 112–130). Routledge: Abingdon.

De Jaegher, H. & Di Paolo, E. (2007) Participatory sense-making. *Phenomenology and the cognitive sciences*, *6*, 485–507.

Eccles. D.W. & Tenenbaum, G. (2004) Why an expert team is more than a team of experts: A social-cognitive conceptualization of team coordination and communication in sport. *Journal of Sport & Exercise Psychology*, *26*, 542–60.

Eccles, D.W. (2010) The coordination of labour in sports teams. *International Review of Sport and Exercise Psychology*, *3*, 154–170.

Eccles, D.W. & Tran-Turner, K. (2014) Coordination in sports teams. In R. Beauchamp & M.A. Eys (Eds.), *Group dynamics in exercise and sport psychology: Contemporary themes* (pp. 240–255). Routledge: Abingdon.

Fajen, B., Riley, M. & Turvey, M. (2009) Information, affordances, and the control of action in sport. *International Journal of Sport Psychology*, *40*, 79–107.

Froese, T. & Di Paolo, E. (2011) The enactive approach: Theoretical sketches from cell to society. *Pragmatics & Cognition*, *19*, 1–36.

Gibson, J. (1979) *The Ecological Approach to Visual Perception*. Lawrence Erlbaum Associates: Hillsdale, NJ.

Haken, H., Kelso, J.A.S. & Bunz, H. (1985) A theoretical model of phase transitions in human hand movements. *Biological Cybernetics*, *51*, 347–356.

Hammond, K.R. (1996) *Human judgment and social policy: Irreducible uncertainty, inevitable error, unavailable injustice*. Oxford University Press: New York.

Järvilehto, T. (2009) The theory of the organism-environment system as a basis of experimental work in psychology. *Ecological Psychology*, *21*(2), 112–120.

Kelso, J.A.S. (1995) *Dynamic Patterns*. MIT Press: New York.

Laroche, J., Berardi, A.M. & Brangier, E. (2014) Embodiment of intersubjective time: Relational dynamics as attractors in the temporal coordination of interpersonal behaviours and experiences. *Frontiers in Psychology*, *5*, 1180.

Marsh, K., Richardson, M. & Schmidt, R. (2009) Social Connection Through Joint Action and Interpersonal Coordination. *Topics in Cognitive Science*, *1*, 320–339.

Mohammed, S. & Dumville, B. (2001) Team mental models in a team knowledge framework: Expanding theory and measurement across disciplinary boundaries. *Journal of Organizational Behaviour*, *22*, 89–106.

Mohammed, S., Klimoski, R. & Rentsch, J. (2000) The measurement of team mental models: We have no shared schema. *Organizational Research Methods*, *3*(2), 123–165.

Phillips, E., Davids, K., Renshaw, I. & Portus, M. (2010) Expert performance in sport and the dynamics of talent development. *Sports Medicine*, *40*(4), 271–283.

Pinder, R.A., Davids, K., Renshaw, I. & Araújo, D. (2011) Representative Learning Design and Functionality of Research and Practice in Sport. *Journal of Sport & Exercise Psychology*, *33*, 146–155.

Poizat, G., Bourbousson, J., Saury, J. & Sève, C. (2009) Analysis of contextual information sharing during table tennis matches: An empirical study on coordination in sports. *International Journal of Sport and Exercise Psychology*, *7*, 465–487.

Poizat, G., Bourbousson, J., Saury, J. & Sève, C. (2012) Understanding team coordination in doubles table tennis: Joint analysis of first- and third-person data. *Psychology of Sport & Exercise*, *13*, 630–639.

Reed, E. (1996) *Encountering the world: Toward an ecological psychology*. Oxford University Press: Oxford.

Reimer, T., Park, E. & Hinsz, V. (2006) Shared and coordinated cognition in competitive and dynamic task environments: an information-processing perspective for team sports. *International Journal of Sport and Exercise Psychology*, *4*, 376–400.

Rentch, J,. & Davenport, S. (2006) Sporting a new view: team member schema similarity in sports. *International Journal of Sport and Exercise Psychology, 4,* 401–21.

Richardson, M.J., Shockley, K., Fajen, B.R., Riley, M.A. & Turvey, M.T. (2008) Ecological psychology: Six principles for an embodied-embedded approach to behaviour. In P. Calvo & T. Gomila (Eds.), *Elsevier Handbook of New Directions in Cognitive Science.* Elsevier: New York.

Riley, M., Richardson, M., Shockley, K. & Ramenzoni, V. (2011) Interpersonal synergies. *Frontiers in Psychology*, *2*(38), 1–7.

Salas, E., Cannon-Bowers, J.A. & Johnston, J.H. (1997) How can you turn a team of experts into an expert team? Emerging training strategies. In C. Zsambok & G. Klein (Eds.), *Naturalistic decision making: Where are we now?* (pp. 359–370). Erlbaum: Hillsdale, NJ.

Schmidt, R.C., Carello, C. & Turvey, M.T. (1990) Phase transitions and critical fluctuations in the visual coordination of rhythmic movements between people. *Journal of Experimental Psychology: Human Perception & Performance*, *16(*2), 227–247.

Schmidt, R.C., O'Brien, B. & Sysko, R. (1999) Self-organization of between-persons cooperative tasks and possible applications to sport. *International Journal of Sport Psychology*, *30*, 558–579.

Schmidt, R.C., Fitzpatrick, P., Caron, R. & Mergeche, J. (2011) Understanding social motor coordination. *Human Movement Science*, *30*, 834–845.

Sève, C., Nordez, A., Poizat, G. & Saury, S. (2013) Performance analysis in sport: Contributions from a joint analysis of athletes' experience and biomechanical indicators. *Scandinavian Journal of Medicine and Science in Sports*, *23*, 576–584.

Silva, P., Esteves, P., Correia, V., Davids, K., Araújo, D. & Garganta, J. (2015) Effects of manipulating player numbers vs. field dimensions on inter-individual coordination during youth football small-sided games. *International Journal of Performance Analysis of Sport*, *15*, 641–659.

Silva, P., Garganta, J., Araújo, D., Davids, K. & Aguiar, P. (2013) Shared knowledge or shared affordances? Insights from an ecological dynamics approach to team coordination in sports. *Sports Medicine*, *43*, 765–772.

Theureau, J. (2003). Course-of-action analysis and course-of-action centered design. In E. Hollnagel (Ed.), *Handbook of cognitive task design* (pp. 55–81). Lawrence Erlbaum: Mahwah, NJ.

Thompson, E. (2011) Précis of mind in life: Biology, phenomenology, and the sciences of mind. *Journal of Consciousness Studies*, *18*, 10–22.

Travassos, B., Duarte, R., Vilar, V., Davids, K. & Araújo, D. (2012) Practice task design in team sports: Representativeness enhanced by increasing opportunities for action, *Journal of Sports Sciences, 30*, 1447–1454.

Uitdewilligen, S., Waller, M.J. & Pitariu, A.R.H. (2013) Mental model updating and team adaptation. *Small Group Research, 44*, 127–158.

Vallacher, R.R. & Jackson, D. (2009) Thinking inside the box: Dynamical constraints on mind and action. *European Journal of Social Psychology, 39*, 1226–1229.

Varela, F.J. (1979) *Principles of Biological Autonomy*. Elsevier: New York.

Varela, F.J., Thompson, E. & Rosch, E. (1991) *The Embodied Mind: Cognitive Science and Human Experience*. MIT Press: Cambridge, MA.

Ward, P. & Eccles, D. (2006) A commentary on "team cognition and expert teams: emerging insights into performance for exceptional teams". *International Journal of Sport and Exercise Psychology, 4*, 463–83.

Warren, W.H. (2006) The dynamics of perception and action. *Psychological Review, 113*, 358–389.

Withagen, R., de Poel, H.J., Araújo, D. & Pepping, G.-J. (2012) Affordances can invite behaviour: Reconsidering the relationship between affordances and agency. *New Ideas in Psychology, 30*, 250–258.

10 Crew rowing

An archetype of interpersonal coordination dynamics

Harjo J. de Poel, Anouk J. De Brouwer and Laura S. Cuijpers

Introduction

Coordinating our actions with others is paramount in our daily lives. Between-person coordination is important for functional cooperative behaviour of a group of people, but also when the interaction between people is competitive. Regarding the group dynamics of a team of agents that has to cooperate (for instance, cooperation with colleagues at work or within a sports team), the behaviour of the multi-agent system as a whole emerges as a function of the cooperative interactions between the agents that constitute the system. For obvious reasons, team managers, coaches etc. seek to optimize team performance as well as the performance of each individual within the context of that multi-agent system. In that respect, an often-used model of perfect within-team tuning is the rowing crew; managers often proclaim to members of their team that in order to achieve optimal (productive) performance their work efforts should be ideally synchronized, just like in a rowing crew. Also in scientific literature, crew rowing is often adopted as an archetypical, natural example to illustrate, explain and examine joint action, interpersonal coordination dynamics and synchronization (e.g. Keller, 2008; Marsh, Richardson & Schmidt, 2009; Richardson, Marsh, Isenhower, Goodman & Schmidt, 2007) as well as group processes in general (e.g. Ingham, Levinger, Graves, & Peckham, 1974; King & De Rond, 2011). However, as direct scientific examination of crew coordination is limited, it appears that the example of crew rowing is often adopted merely as a metaphor.

To examine interpersonal interactions, the theoretical framework of coordination dynamics (for a general overview see e.g. Kelso, 1995) offers expedient analysis tools. Using models of coupled oscillators, this approach is particularly well-suited for investigating cyclical movement behaviour. The cyclical nature of the rowing act, and the fact that it has to be performed in unity with others in the same boat, suggest that crew rowing is arguably one of the most relevant and exemplary real-life tasks demonstrating interpersonal coordination dynamics. Hence, inspired by studies of inter-personal coordination dynamics (e.g. Richardson et al., 2007), the purpose of the current chapter is to outline crew rowing within the pertinent theoretical framework of coordination dynamics, alongside some recent empirical work in this context. In doing so, we aim to demonstrate the expediency of coordination dynamics for crew rowing and vice versa. First, we briefly introduce the essentials of the sport of crew rowing and address previous studies that investigated crew coordination in rowing. Second, relevant concepts and issues in coordination dynamics will be elaborated on, and then applied to crew rowing.

Crew rowing

Rowers perform a cyclic movement pattern in which the legs, trunk and arms take part in a synchronized fashion in the propulsion of the boat. Each rower faces backwards in the boat and uses a single oar (sweep rowing) or two oars (sculling) to propel the boat. It is important to recognize that each stroke cycle consists of a propulsion phase (or drive) and a recovery phase. In the drive phase, a rower (sitting on a sliding seat) pushes off against the footboard (attached to the boat) using primarily the strong leg extensor muscles, while pulling on the oar(s) with the oar blade(s) in the water. At the 'finish' of the drive, the arms release the blade(s) from the water. In the recovery phase the rower returns to the initial position by sliding forward on the seat with the blade(s) out of the water, to reposition the oar(s) for the next stroke. Then, at the so-called 'catch', the blade(s) enter(s) the water again so that the next stroke can be performed. As we will see later, the existence of a drive phase and a recovery phase, and, accordingly, the movement of the rowers' centre of mass relative to the boat due to the sliding seats, have a great influence on boat velocity.

In competitive rowing, the goal is to cover a course of 2000 metres as fast as possible, preferably faster than opponent crews. To achieve a high average boat velocity, maximizing power production is of course prerequisite, while the crew also needs to minimize the power *losses*. In other words, rowers produce power to propel the boat via the oars, but at same time they want to lose as little of the produced power as possible to, for instance, slippage of the blades in the water. As such, a proper and fluent technique is required for efficient power application.

In crew rowing, even a team of individually strong and technically skilled rowers will probably not win races if they do not properly coordinate their movements with each other (O'Brien, 2011). Hence, they also have to develop an efficient 'crew technique'. Both in rowing practice and rowing science, mutual synchronization is generally regarded as a main determinant for optimal performance among a given crew (e.g. Hill, 2002; Wing & Woodburn, 1995). The overall idea is that when the crew is moving perfectly in sync the power the rowers produce is optimally converted into forward velocity, mainly because it reduces unwanted boat movements (e.g. Hill & Fahrig, 2009). Indeed, if the overall forces at the blades are unbalanced and/or applied with imperfect mutual timing, this would cause net torques around the centre of the boat, which entails dispensable rolling, yawing and pitching of the boat (Baudouin & Hawkins, 2002). Poor synchrony would thus involve additional movements that disturb the balance of the boat, elongate the travelled distance of the boat, and create greater hydrodynamic drag.

Together, the general opinion of researchers and coaches is that to maximize average velocity, a crew must optimize their synchrony. The question arising then is how well do rowers in a boat synchronize, and how can we measure and analyse that? This knowledge could be highly relevant for optimizing performance. Obviously expert crews are expected to demonstrate better, more consistent synchronizing patterns than crews with less expertise. Besides, despite the well-known natural tendency for coupled systems towards synchronization (Kelso, 1995), it is also important to realize that rowing in sync does not simply come naturally. In fact, freshmen crews often need some months of training to achieve a common rhythm and require years of training to reach a desired expert level of synchronization. This calls for methods and concepts that allow for the examination of such crew synchronization processes, and how these change over time as a function of practice and other factors and interventions (e.g. crew member changes, boat velocity, stroke rate etc.).

One way is to analyse the crew's synchronization from force measurements, for instance the force applied to the oars or blades over time. Although studies showed that rowers display an individually characteristic force-time pattern, varying the combination of rowers within a crew led the force-time profiles to be correlated with those produced by the other crew members, clearly indicating that rowers adjusted their behaviour to the crew (Baudouin & Hawkins, 2004; Hill, 2002; Wing & Woodburn, 1995). Furthermore, from force data from 180 successive strokes (at a stroke rate of about 18 strokes/min) for four rowers in an eight, Wing and Woodburn (1995) determined stroke onset times (i.e. catch) and saw that the degree of correspondence in catch timing between all four rowers was remarkably accurate, within a range of 10 to 20 ms. Also, they analysed between-rower cross-correlations of peak forces, drive duration and recovery duration. Somewhat surprisingly, only the recovery durations positively correlated between rowers, while the drive durations did not, suggesting that crew synchronization varies within the stroke cycle. In line with this finding, based on force data, Hill (2002) found that timing differences between rowers in a boat (in this case coxless fours: crews of four sweep rowers without a coxswain) were generally smaller for the catch than for the finish for endurance intervals (at 23 to 25 strokes/min: 14.2 ms and 25.8 ms, respectively) and intensive intervals (at 31 to 41 strokes/min: 11.2 ms and 21.7 ms, respectively). In addition, this tentatively suggests that crew synchronization improved at higher stroke rates (see the section on *Movement frequency* for further discussion of the impact of stroke rate).

Another way is to analyse crew coordination based on movement data, such as trunk movement, oar angles, and/or displacement of the sliding seat. For instance, in a case study on a coxless pair with a self-indicated 'dysfunction of crew coordination', Sève, Nordez, Poizat and Saury (2011) used the turning points in the oar angles to determine the onset differences between the rowers for each stroke. With such analysis the authors were able to depict systematic differences in the interpersonal coordination of the catch times and entry angles of the oars, which allowed the pair's coaches to define new training objectives to remedy the imprecisions in the pair's coordination.

Apart from determining discrete features for each stroke, analysis of movement data implies that coordination can also be determined for the whole cycle and not solely for the drive phase, as is the case with force data. As such, an instantaneous and continuous measure of synchronization can be determined. Motivated from coordination dynamics, crew coordination can be displayed by the relative phase (ϕ) between rowers. In short, this measure depicts the difference between two rowers in terms of where they reside in their respective stroke cycle. If the relative phase equals zero, the rowers are exactly synchronous, whereas the amount of variation of the relative phase over time indicates the degree of consistency of crew coordination. Recently, this has been adopted in rowing experiments in the laboratory (De Brouwer, De Poel, & Hofmijster, 2013; Cuijpers, Zaal, & De Poel, 2015a; Varlet, Filippeschi, Ben-sadoun, Ratto, Marin, Ruffaldi & Bardy, 2013) and on water (Cuijpers, Passos, Hoogerheide, Lemmink, Murgia, & De Poel, 2016). We will involve these studies in the discussion of coordination dynamics related to crew rowing, which will be done in the subsequent paragraphs.

Coordination dynamics and crew rowing

In general, coordination dynamics encompasses the study of coordinative patterns – that is, how patterns of coordination form, adapt, persist and change over time. This approach (e.g. Haken, Kelso & Bunz, 1985; Kelso, 1995) offers an expedient framework for studying rhythmic coordination, in which the (in) stability of coordinative patterns is explained with reference to the coupling between the components that comprise the system. Most importantly for the purposes of the present chapter, it offers a well-established non-linear model of coupled oscillators, known as the Haken-Kelso-Bunz, or HKB model, that captures key properties and phenomena of isofrequency (i.e. identical movement rates) coordination (Haken et al., 1985). Although the HKB model was originally developed for rhythmic bimanual coordination (i.e. within-person coordination), to date many studies have underwritten that between-person coordination abides by similar coordinative phenomena and principles (for reviews, see Schmidt, Fitzpatrick, Caron & Mergeche, 2011; Schmidt & Richardson, 2008). Knowing this, crew rowing perfectly meets the conditions of this coupled oscillator model, in that it is a cyclical act of two (or more) coupled agents that are moving at equal movement rates. Moreover, the stroke cycles show behaviour that is near to harmonic (i.e. sinusoidal). In the following sections we will describe in a point by point fashion the main predictions derived from the model, how these predictions are supported by previous empirical findings (or not), and how they may apply to crew rowing.

Differential pattern stability

In a seminal paper, Kelso (1984) demonstrated that while tapping fingers in an antiphase pattern (i.e. perfectly alternating, $\phi = 180°$) an increase in movement frequency resulted in a spontaneous, involuntary transition towards in-phase coordination (i.e. perfectly coinciding, $\phi = 0°$). However, when starting in in-phase coordination, no shift towards antiphase occurred. To account for this phenomenon, Haken et al. (1985) formulated a model of two non-linearly coupled limit cycle oscillators that constituted an equation of motion that could describe the rate of change in relative phase angle between the two oscillating components, following

$$\phi = a \sin \phi - 2b \sin 2\phi \qquad \text{(Eq. 1)}$$

with a affecting the attractor strength of in-phase coordination and b affecting the attractor strength of both in- and antiphase coordination. The ratio b/a is directly related to the movement frequency. Given Eq. 1, at low frequencies (i.e. $b/a > .25$) the model has two stable attractors, namely in-phase and antiphase, while other patterns are intrinsically unstable. Also for between-person tasks, the difference in stability for in-phase and antiphase coordination has been consistently demonstrated (for an overview see Schmidt & Richardson, 2008).

This difference in the stability properties of coordinative patterns already poses the first challenge for crew rowing. At first glance this may seem a rather curious argument, since rowing crews only synchronize in an in-phase manner and other patterns are not and cannot be performed. However, perhaps somewhat surprisingly, the latter does not appear to be the case. That is, out-of-phase rowing *has* been considered in the past.

Rowing 90° out-of-phase

In fact, there is a long-standing 'myth' that out-of-phase crew rowing may be beneficial over the conventional in-phase crew coordination (see also *Steady-state antiphase crew rowing*). By 1930 there had already been several actual attempts to row out-of-phase on water, also termed 'syncopated rowing' or 'jazz rowing'. For instance, newspapers reported British crews rowing in a four-phase strategy ($\phi = 90°$; quarter-cycle-lag pattern) in an eight[1], and a three-phase strategy ($\phi = 120°$; third-cycle-lag pattern) with a crew of six rowers. Although some reports suggested that these attempts were successful, the syncopated boats received a lot of criticism. Stories mentioned that after some training the crews apparently managed to master the coordination, but this did not lead to sufficient gain in boat velocity and, more importantly, victories. In the end the critics won and these trials were aborted, probably due to the fact that a sufficient gain in boat velocity was not achieved.

With hindsight, we know from coordination dynamics that 90° and 120° coordination patterns are intrinsically unstable patterns, and even after considerable practice they are probably extremely difficult to maintain, especially at higher movement rates. For bimanual coordination, support for this prediction has already been provided, for instance in experiments in which subjects practiced 90° interlimb patterns (e.g. Zanone & Kelso, 1992), while for between-person coordination, to our knowledge no studies on learning new phase relations have been reported. Perhaps such studies might not exist because for two (or more) persons it is virtually impossible to (learn to) interact stably in a quarter cycle relation.

Therefore, we performed a single case experiment in which we had four experienced rowers row on ergometers (Concept2) that were positioned next to each other. The task was to row in a pace that was indicated by a sequence of beeps. The beeps had four different tone pitches that were presented in a 90° phase delay with respect to each other. Each rower was assigned to one of the pitches and instructed to align the catches with the incidence of the beeps. Starting at a tempo of 20 strokes/min, each minute the tempo of the beeps was increased in steps of 2 strokes/min to a maximum of 32 strokes/min. Initially the rowers were able to maintain the 90° interpersonal pattern, but a breakdown of the pattern had already occurred at 24 strokes/min. The breakdown involved a switch towards three rowers moving in in-phase relation, while the fourth rower was moving in antiphase relation to the other three. The subjects also performed a condition in which they were divided into two groups of two rowers. The two groups were instructed to row antiphase with respect to each other, while their pace was again indicated by beeps that increased in frequency. The rowers easily maintained the antiphase pattern until the end of the trial. This is not surprising, since the antiphase pattern is an intrinsically stable pattern. In subsequent studies we therefore considered the stable antiphase pattern. In sum, it is likely that at high stroke rates stable 90° crew coordination is extremely difficult to achieve. Although after practice 90° out-of-phase patterns can be mastered (Zanone & Kelso, 1992), the stroke rates at which this pattern can be performed are limited, and hence so is the velocity on water. As we will see, such problems exist to a much lesser extent for the antiphase pattern, because this is an intrinsically stable pattern.

Before we proceed, it is important to note that in ergometer rowing, rowers usually row on separate machines (as was also the case in the above experiment), whereas on water the rowers are linked in a mechanical way since they share the same boat. We therefore subsequently performed a series of studies in which we analysed dyads of rowers in the lab using a two-ergometer system. This involves two ergometers that are put on so-called 'slides' (Concept2), so that they can move freely with respect to the ground, and which also allows them to be physically linked so that they move as one 'boat' (see De Brouwer et al., 2013; Cuijpers et al., 2015).

Steady-state antiphase crew rowing

Although over the previous century some have suggested that the idea behind out-of-phase crew rowing was that it would be faster because of more continuous propulsion (much like a car engine, where the pistons do not ignite at the same time but with a mutual phase delay), others already recognized that there is something else that mediates the potential velocity-gaining mechanism behind out-of-phase rowing. In rowing, 5% to 6% of the total power produced by the rower(s) is lost to velocity fluctuations of the shell within each rowing cycle. Shell velocity fluctuates because propulsion is not continuous (viz. the drive and recovery phase) and the relatively heavy rower(s), seated on their sliding seat(s), push off with the feet against the relatively light boat, causing the shell to decelerate during the drive and accelerate during the recovery (Hill & Fahrig, 2009). As the power needed to overcome hydrodynamic drag is proportional to shell velocity cubed, minimizing velocity fluctuations of the boat while maintaining total power output will thus result in higher efficiency and hence, ceteris paribus, higher average boat velocity (see e.g. De Brouwer et al., 2013; Hill & Fahrig, 2009).

Theoretically, in the case of crew rowing this can be achieved by rowing in antiphase coordination, a strategy in which two (groups of) rowers within the boat perform their strokes in perfect alternation (Brearly, De Mestre & Watson, 1998; De Brouwer et al., 2013). In antiphase rowing, the movements of the rowers would almost perfectly counteract each other, resulting in a net centre of mass movement (CoM) of the crew that stays close to the movement of the boat. As such, boat velocity would remain close to constant over the whole rowing cycle. A recent study confirmed for dyads rowing at 36 strokes/min on coupled ergometers (see above) that antiphase crew coordination is indeed mechanically more efficient, in that the power loss was reduced by 5% compared to in-phase rowing (De Brouwer et al., 2013). Importantly, the crews produced similar amounts of total power during in-phase and antiphase rowing, resulting in a 5% greater amount of useful power for antiphase coordination. Furthermore, as expected, the coordination between the rowers, as measured by the relative phase between the rowers' CoM movements, was indeed less accurate and less consistent in antiphase as compared to in-phase rowing. Nevertheless, it was striking to see how little difficulty the rowers showed in rowing antiphase, although they were doing it for the first time ever. This supports the notion that for crew rowing, next to in-phase, antiphase is an intrinsically stable state. Still, one of the nine pairs in De Brouwer et al.'s study showed a breakdown of antiphase coordination towards in-phase rowing, which led to the following examination.

Involuntary pattern switches

It is essential that the stability of the crew coordination, whether in- or antiphase, remains sufficient to maintain at high stroke rates, because 2000-metre races are typically rowed at strokes rates above 30. Based on the predictions of the HKB model, one would expect the difference in stability between in- and antiphase crew coordination to increase with movement frequency. In fact, when rowers gradually increase their movement tempo, one might expect spontaneous switches from the less stable antiphase to the more stable in-phase pattern (Haken et al., 1985; Kelso, 1984) as was also shown in visually coupled humans (Schmidt, Carello & Turvey, 1990).

In a recent off-water experiment (Cuijpers et al., 2015), we tested whether rowing in antiphase coordination would show a tendency to break down into in-phase coordination when we increased the tempo. To this end, eleven experienced male rowing pairs rowed in-phase and antiphase on the two-ergometer system on slides in a steady state trial (2 min, 30 strokes/min) and a ramp trial during which the stroke rate was increased every 20 s from 30 strokes/min to as fast as possible in 2 strokes/min steps. There was sufficient recuperation time between the four trials carried out by each pair. Kinematics of rowers, handles and ergometers were captured (Vicon®, 200 Hz). Relative phase between rowers' trunk movements and between handles was determined. Continuous relative phase angle (CRP) was based on a procedure that took into account the fact that the recovery phases lasted longer than the propulsive phases (see also Varlet et al., 2013). Moreover, in a rowing stroke more time is spent around the finish than around the catch of the stroke. Therefore, we also determined a discrete measure of relative phase (DRP) that is not sensitive to such small though impactful deviations from perfect harmonicity (see also De Brouwer et al., 2013), based on the moments of the catch (Kelso, 1995).

First of all, for most pairs the highest achieved stroke rate was during in-phase, which was ascribed to the reduction of ergometer movement in antiphase, knowing that on dynamic ergometers higher stroke rates can be achieved than on static ergometers (Colloud, Bahuaud, Doriot, Champely, & Cheze, 2006). Furthermore, two of the eleven pairs showed a breakdown of antiphase into in-phase crew coordination. Notably, these breakdowns occurred at the very beginning of the ramp trial (around 32 strokes/min). After the transition occurred, one pair tried to restore the antiphase crew coordination but did not succeed (see Figure 10.1). The other pair had already showed difficulties maintaining antiphase coordination during the steady state trial (30 strokes/min); they indicated that they did not feel comfortable rowing in antiphase. Because transitions occurred at the initial tempo of the ramp trial and were also apparent in steady state trials (see also De Brouwer et al., 2013), we suspect that at these tempos the coordination of these two dyads might already have been too sensitive to perturbations (designating low stability), potentially related to a (temporary) loss of concentration or attention (e.g. Temprado & Laurent, 2004).

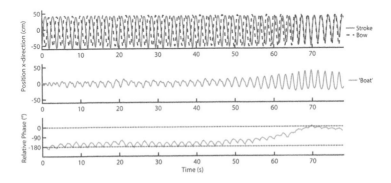

Figure 10.1 Transition from anti- to in-phase crew coordination at 32 strokes/min.

Notes: Movements of rowers (upper panel) and 'boat' (middle panel), and relative phase between the rowers (lower panel).

In any case, when antiphase rowing coordination is lost, it is difficult to regain. This is also due to the mechanical coupling, because once the boat starts oscillating (see Figure 10.1) it is difficult to counter it. Regarding mechanical coupling, Christiaan Huygens already observed in the seventeenth century that two pendulum clocks on a wall that were initially uncoordinated became coordinated over time in either an in-phase or antiphase pattern, because the clocks interact through the vibration in the wall. This was also demonstrated with mechanical metronomes that were jointly placed on a moving base (Bennet, Schatz, Rockwood & Wiesenfeld, 2002; Pantaleone, 2002). In this non-living system, due to mechanical coupling via the moving base the metronome pendulums are also attracted to in-phase synchronization when starting in antiphase.[2] Hence, in interpersonal coordination, a direct mechanical link forms a strong base for attraction to in-phase that is arguably more stringent than for perceptual coupling (Lagarde, 2013). In cases where two humans are mechanically coupled, it might require more 'mental effort' to stay coordinated in antiphase (see also previous paragraph), to prevent any attraction to in-phase from happening. For more discussion on this issue, see *Sources of coupling*.

Movement frequency

A second issue Cuijpers et al. (2015) could address was the predicted coordinative inconsistencies at increasing movement rate. Indeed, many interlimb coordination studies have confirmed that coordination deteriorates with movement tempo, and that for antiphase this effect is stronger than for in-phase (see Kelso, 1995). In rowing, stroke rates vary from 18 to 24 strokes/min during (endurance) training, up to 30 strokes/min for freshmen crews during racing, with Olympic crews often reaching up to 42 strokes/min (i.e. 0.7 Hz). It is therefore essential to know if the stability of the coordination pattern, whether in- or antiphase, remains sufficient at increasing stroke rates. The nine pairs that did not show a transition in the antiphase ramp trial were analysed in terms of the variability of DRP ('circular' standard deviation) and accuracy of CRP ('circular' absolute error) over steady state bins of each performed movement frequency.

Although the coordination was expected to deteriorate with increasing tempo, the results revealed no statistically significant effects of stroke rate on crew coordination (Cuijpers et al., 2015). On the other hand, as mentioned in the paragraph on *Crew rowing*, from on-water rowing studies there are some indications that crew synchronization might *improve* rather than deteriorate with increasing stroke rate (Hill, 2002). In fact, interpersonal pendulum swinging experiments demonstrated that at movement rates above 1.2 Hz (i.e. 72 cycles/min) coordinative variability increased with tempo, while for movement rates below 1 Hz (i.e. 60 cycles/min) coordinative variability *de*creased with tempo (Schmidt, Bienvenu, Fitzpatrick & Amazeen, 1998). The authors related the latter effect to the difficulty of moving at a rate lower than the preferred (or 'natural') movement rate (see also next paragraph). Importantly, as the ramp trials in Cuijpers et al.'s study were designed to invoke phase transitions, they already started at a reasonably high stroke rate, namely 30 strokes/min. It is conceivable that at rates lower than 30 stroke/min a movement frequency effect emerges and becomes more easily observable. In this respect, recent on-water measurements suggest that in-phase crew coordination indeed deteriorates for stroke rates below 26 strokes/min (Cuijpers, et al., 2016).

Detuning: within-crew individual differences

Differences between the oscillatory characteristics of the two components also affect coordination (e.g. De Poel, Peper, & Beek, 2009; Schmidt & Richardson, 2008), generally modelled as a detuning parameter (i.e. $\Delta\omega$; Kelso, DelColle, & Schöner, 1990). Implemented in the HKB model, the detuning parameter induces specific lead-lag relations and a decrease of coordinative stability (Kelso et al., 1990). It has commonly been inferred to reflect a difference between the eigenfrequencies (cf. 'intrinsically preferred movement rate') of the two oscillators (e.g. Schmidt & Richardson, 2008). As we consider rowers in terms of limit cycle oscillators, we may expect their movements to possess a characteristic amplitude (e.g. reflecting stroke length) and eigenfrequency (e.g. reflecting individually preferred stroke rate). For instance, in an ergometer rowing experiment, Sparrow, Hughes, Russell and Le Rossignol (1999) indicated that rowing at a preferred rate was metabolically more efficient. When increasing or decreasing stroke rate while maintaining the same power output, for instance, the rowers changed their stroke lengths (i.e. the excursion of the handle), which led to an increase in metabolic cost for both lower and higher rates. These results advocate that, for a given output level, each rower has his or her own individual 'optimal' stroke frequency, and hence the eigenfrequency varies over individuals. Therefore, if rowers with different eigenfrequencies and/or stroke amplitudes (cf. De Poel et al., 2009) are combined into a crew, not only the individual efficiency (see above) within the crew but, given the predictions from the HKB-model with detuning parameter, also the crew coordination may be compromised. As such, coupled oscillator dynamics provides an account for why it is beneficial to select rowers close to the same preferred movement frequency (and also in terms of other properties) into a crew.

Sources of coupling

In this paragraph, we briefly reflect on the merits that the task of crew coordination might have for examining (interpersonal) coordination dynamics in general, through delineating some issues regarding the sources that mediate interaction between rowers in a crew. The interaction, or coupling, may in general terms be considered as an information array between two rhythmically moving components. The majority of research on interpersonal coordination dynamics is done on movement synchronization mediated through visual coupling (e.g. Oullier, De Guzman, Jantzen, Lagarde & Kelso 2008; Peper, Stins & De Poel, 2013; Schmidt et al., 2011). Evident in rowing, though, is the direct mechanical coupling through the boat. This physical link also allows for the perception of haptic information about the others' movements through the movements of the boat that they share. Hitherto, mechanical and haptic coupling have received very limited attention in interpersonal coordination studies, but many relevant examples used in such studies are highly dependent on such coupling, like dancing the tango or, indeed, crew rowing. Laboratory experiments have shown that when dyads need to coordinate their actions on the basis of haptic information, they amplify their forces to generate a haptic information channel (Reed, Peshkin, Hartmann, Grabowecky, Patton & Vishton 2006; Van der Wel, Knoblich & Sebanz, 2011). This principle seems to hold for crew rowing as well, as Hill (2002) suggested that an increase in force output provides a better kinesthetic perception facilitating the adaption of force patterns.

In rowing the movements of each rower set the boat in motion, thereby moving the other crew members (similar to the coupled metronomes; Pantaleone, 2002). As such, mechanical interpersonal coupling may be considered as a source of perturbation requiring anticipatory movements (Bosga, Meulenbroek & Cuijpers, 2010) but can also be seen as a source of support that stabilizes coordination patterns by mutually constraining the movements of the mechanically coupled agents (Harrison & Richardson, 2010). Most importantly, mechanical coupling differs from perceptual coupling (visual, auditory and haptic/kineasthetic coupling) in that it is impossible to escape from: the body of each agent gets passively shaken by the movement of the other agent (Lagarde, 2013), whereas perceptual coupling is mediated by the degree to which an agent is sensitive to or able to detect the pertinent information, for instance by means of attention devoted to the information source (Meerhoff & De Poel, 2014; Richardson et al., 2007). This implies that the mechanical coupling is more stringent than perceptual coupling.

Application to on-water rowing

The theoretical analysis of crew rowing from a coordination dynamics perspective has already offered some new insights that can be directly applied to rowing practice. However, the lab is obviously not exactly the same as the real situation. For instance, in the lab studies of De Brouwer et al. (2013) and Cuijpers et al. (2015) there were no lateral and vertical (angular) movements of the 'boat', handles were used rather than oars (with a certain length and weight), oar handling technique and blade hydrodynamics were not present, and so on. Therefore, testing on water is a next important step. Commercial measurement systems are available for analyzing movements and forces in on-water crew rowing (e.g. Sève et al., 2013) and in recent on-water experiments with an Arduino-based measurement system (Cuijpers et al., 2016) we tested the hypothetical relation between the quality of crew coordination and unwanted, drag-increasing boat movements (as posed by Baudouin & Hawkins, 2002; Hill et al., 2009).

The case of antiphase rowing also offers quite a straightforward direct application for on-water rowing. Before it can be considered to implement in competitive practice, though, many research steps still have to be taken. This primarily involves biomechanical aspects that may cancel out the 5% velocity efficiency benefit, such as blade resistance and air friction, that remain to be further explored. Nevertheless, as delineated above, coordination dynamical examinations have already showed that performing the antiphase pattern per se is not a problem at all, not even at stroke rates as high as in a rowing race. This was also confirmed in recent exploratory on-water tryouts by ourselves, using two experienced rowers. With some extra space between the rowers (because otherwise the blades would clash and/or the bow rower would hit the stroke rower in the back), on-water antiphase rowing appears to be quite easy to perform, even without any practice. Whether it indeed leads to higher average velocity is not clear yet; as noted, this requires testing with measurement equipment and further biomechanical evaluation of the problem. However, that was not within the scope of this chapter.

Concluding remarks

In this chapter we illustrated the relevance of coordination dynamics for investigating crew rowing. Alongside issues such as the (differential) stability of coordinative patterns, (preferred) movement frequency and coupling in crew rowing, we showed how coordination dynamical research is particularly relevant in this context, how it may be applied to crew rowing, and also how the knowledge gained may be used for the benefit of improving performance. Together, this also further underscored crew rowing as an archetype of interpersonal coordination dynamics, and the research reviewed in this chapter provides means for the example of crew rowing to now surpass the stage of mere metaphor.

Notes

1 Footage of these attempts is available via British Pathe: http://www.britishpathe.com/video/syncopated-rowing
2 Many movies are available on-line in which synchronizing metronomes are demonstrated, of which arguably the most illustrative can be found here https://www.youtube.com/watch?v=yysnkY4WHyM and here https://www.youtube.com/watch?v=5v5eBf2KwF8.

References

Baudouin, A. & Hawkins, D. (2002) A biomechanical review of factors affecting rowing performance. *British Journal of Sports Medicine*, *36*, 396–402.
Baudouin, A. & Hawkins, D. (2004) Investigation of biomechanical factors affecting rowing performance. *Journal of Biomechanics*, 37, 969–976.

Bennet, M., Schatz, M.F., Rockwood, H. & Wiesenfeld, K. (2002) Huygens's clocks. *Proceedings: Mathematical, Physical and Engineering Sciences, 458*(2019), 563–579.

Bosga, J., Meulenbroek, R.G.J. & Cuijpers, R.H. (2010) Intra- and interpersonal movement coordination in jointly moving a rocking board. *Motor Control, 14,* 440–459.

Brearley, M.N., De Mestre, N.J. & Watson, D.R. (1998) Modelling the rowing stroke in racing shells. *Mathematical Gazette, 82,* 389–404.

Colloud, F., Bahuaud, P., Doriot, N., Champely, S. & Cheze, L. (2006) Fixed versus free-floating stretcher mechanism in rowing ergometers: mechanical aspects. *J Sports Sci, 24,* 479–493.

Cuijpers, L.S., Zaal, F.T.J.M. & De Poel, H.J. (2016) Rowing Crew Coordination Dynamics at Increasing Stroke Rates. *PloS ONE, 10*(7), e0133527.

Cuijpers, L.S., Passos, P., Hoogerheide, A., Lemmink, K.A.P.M., Murgia, A. & De Poel, H.J. (2016) Rocking the boat: rowing crew synchronisation and boat movements at different stroke rates. *Manuscript in preparation.*

De Brouwer, A.J., De Poel, H.J. & Hofmijster, M.J. (2013) Don't rock the boat: how antiphase crew coordination affects rowing. *PLoS ONE, 8*(1), e54996.

De Poel, H.J., Peper, C.E. & Beek, P.J. (2009) Disentangling the effects of attentional and amplitude asymmetries on relative phase dynamics. *Journal of Experimental Psychology: Human Perception and Performance, 35*(3), 762–777.

Haken, H., Kelso, J.A.S. & Bunz, H. (1985) A theoretical model of phase transitions in human hand movements. *Biological Cybernetics, 51,* 347–356.

Harrison, S.J. & Richardson, M.J. (2010) Horsing around: spontaneous four-legged coordination. *Journal of Motor Behavior, 41*(6), 519–524.

Hill, H. (2002) Dynamics of coordination within elite rowing crews: evidence from force pattern analysis. *Journal of Sport Sciences, 20*(2), 101–117.

Hill, H. & Fahrig, S. (2009) The impact of fluctuations in boat velocity during the rowing cycle on race time. *Scandinavian Journal of Medicine and Science in Sports, 19,* 585–594.

Ingham, A.G., Levinger, G., Graves, J. & Peckham, V. (1974) The Ringelmann effect: Studies of group size and group performance. *Journal of Experimental Social Psychology, 10*(4), 371–384.

Keller, P. (2008) Joint action in music performance. In F. Morganti, A. Carassa & G. Riva (Eds.), *Enacting intersubjectivity: A cognitive and social perspective on the study of interactions* (pp. 205–221). IOS Press: Amsterdam.

Kelso, J.A.S. (1984) Phase transitions and critical behavior in human bimanual coordination. *Am J Physiol, 246*(6 Pt 2), R1000–R1004.

Kelso, J.A.S. (1995) *Dynamic patterns. The self-organisation of brain and behaviour.* MIT Press: Champaign, IL.

Kelso, J.A.S., DelColle, J.D. & Schöner, G. (1990) Action-perception as a pattern formation process. In M. Jeannerod (Ed.), *Attention and performance 13: Motor representation and control* (pp. 139–169). Lawrence Erlbaum Associates: Hillsdale, NJ.

King, A. & de Rond, M. (2011) Boat race: rhythm and the possibility of collective performance. *The British Journal of Sociology, 62*(4), 565–585.

Lagarde, J. (2013) Challenges for the understanding of the dynamics of social coordination. *Frontiers in Neurorobotics, 7*(18), 1–9.

Marsh, K.L., Richardson, M.J. & Schmidt, R.C. (2009) Social connection through joint action and interpersonal coordination. *Topics in Cognitive Science, 1,* 320–339.

Meerhoff, L.A. & De Poel, H.J. (2014) Asymmetric interpersonal coupling in a cyclic sports-related movement task. *Human Movement Science, 35,* 66–79.

O'Brien, C. (2011) Effortless rowing. In V. Nolte (Ed.), *Rowing faster* (p. 173–182). Human Kinetics: Champaign, IL.

Oullier, O., De Guzman, G.C., Jantzen, K.J., Lagarde, J. & Kelso, J.A.S. (2008) Social coordination dynamics: Measuring human bonding. *Social Neuroscience, 3*(2), 178–192.

Pantaleone, J. (2002) Synchronization of metronomes. *American Journal of Physics, 70*(10), 992–1000.

Peper, C.E., Stins, J.F. & de Poel, H.J. (2013) Individual contributions to (re-) stabilizing interpersonal movement coordination. *Neuroscience Letters, 557*, 143–147.

Reed, K., Peshkin, M., Hartmann, M.J., Grabowecky, M., Patton, J. & Vishton, P.M. (2006) Haptically linked dyads. Are two motor-control systems better than one? *Psychological Science, 17*(5), 365–366.

Richardson, M.J., Marsh, K.L., Isenhower, R., Goodman, J. & Schmidt, R.C. (2007) Rocking together: dynamics of intentional and unintentional interpersonal coordination. *Human Movement Science, 26*, 867–891.

Schmidt, R.C., Carello, C. & Turvey, M.T. (1990) Phase transitions and critical fluctuations in the visual coordination of rhythmic movements between people. *Journal of Experimental Psychology: Human Perception and Performance, 16*(2), 227–247.

Schmidt, R.C., Bienvenu, M., Fitzpatrick, P.A. & Amazeen, P.G. (1998) A comparison of intra- and interpersonal interlimb coordination: coordination breakdowns and coupling strength. *Journal of Experimental Psychology, Human Perception and Performance, 24*, 884–900.

Schmidt, R.C., Fitzpatrick, P., Caron, R. & Mergeche, J. (2011) Understanding social motor coordination. *Human Movement Science, 30*, 834–845.

Schmidt, R.C. & Richardson, M.J. (2008) Dynamics of interpersonal coordination. In A. Fuchs & V.K. Jirsa (Eds.), *Coordination: neural, behavioural, and social dynamics* (pp. 281–308). Springer: Champaign, IL.

Sève, C., Nordez, A., Poizat, G. & Saury, J. (2013) Performance analysis in sport: Contributions from a joint analysis of athletes' experience and biomechanical indicators. *Scandinavian Journal of Medicine and Science in Sports, 23*(5), 576–584.

Sparrow, W.A., Hughes, K.M., Russell, A.P. & Le Rossignol, P.F. (1999) Effects of practice and preferred rate on perceived exertion, metabolic variables and movement control. *Human Movement Science, 18*, 137–153.

Temprado, J. & Laurent, M. (2004) Attentional load associated with performing and stabilizing a between-persons coordination of rhythmic limb movements. *Acta Psychologica, 115*(1), 1–16.

Van der Wel, R.P.R.D., Knoblich, G. & Sebanz, N. (2011) Let the force be with us: dyads exploit haptic coupling for coordination. *Journal of Experimental Psychology: Human Perception and Performance, 37*(5), 1420–1431.

Varlet, M., Filippeschi, A., Ben-sadoun, G., Ratto, M., Marin, L., Ruffaldi, E. & Bardy, B.G. (2013) Virtual Reality as a Tool to Learn Interpersonal Coordination: Example of Team Rowing. *PRESENCE: Teleoperators and Virtual Environments, 22*(3), 202–215.

Wing, A.M. & Woodburn, C. (1995) The coordination and consistency of rowers in a racing eight. *Journal of Sport Sciences, 13*, 187–197.

Zanone, P. G. & Kelso, J.A.S. (1992) Evolution of behavioral attractors with learning: nonequilibrium phase transitions. *Journal of Experimental Psychology: Human Perception and Performance, 18*(2), 403.

11 Interpersonal coordination in team sports

Pedro Passos and Jia Yi Chow

Introduction

In this chapter we discuss how team sports can be modelled as complex dynamical systems at an ecological scale of analysis. Interpersonal coordination between players in team sports (both within and between teams) can emerge through spontaneous self-organization processes, under the influence of specific task constraints such as field boundaries, location of the goal and players' relative position to each other. Self-organization is an intrinsic mechanism that exists in complex adaptive systems and explains how order emerges due to critical fluctuations in a system's intrinsic dynamics (Kelso, 1995). The presence of these critical fluctuations can be due to player adjustments to stabilize the performance of collective units (e.g. attacking subunits in team games). In other words, stable patterns of collective behaviours can emerge at a macro scale due to behavioural adjustments at a smaller, micro (e.g. dyadic) scale. These player adjustments are supported by relatively simple local interactional rules to create structures and patterns at a collective level that are more complex than the behaviour of each individual player considered separately (Couzin, Krause, Franks & Levin, 2005; Passos, Araújo & Davids, 2013). A feature of the local rules that specify the interactions among the players is focused on context dependency. This is important as it highlights the notion that interactions between players within a team, and between teams, are mainly sustained by information locally generated, without reference to a global pattern (Passos et al., 2013).

However, it should be emphasized that players' contextual dependency only emerges within *critical regions* of performance. Current evidence suggests that these regions are characterized by low values of interpersonal distances between attackers and defenders (Passos et al., 2008). Previous research suggests that within these critical regions, properties of complex dynamical systems can be observed in the interactions between attackers and defenders in team sports such as rugby union (Passos et al., 2008), basketball (Araújo, Davids & Hristovski, 2006) and football (Duarte et al., 2010). For example, findings from these studies have revealed that the coupled behaviours of a ball carrier and defender can be characterized by different coordination states. Transitions from one pattern of behaviour to another can emerge from inherent self-organisation tendencies and can depend on factors such as the position of a defender or proximity to a goal or target in relation to successful dribbling behaviours (Passos et al., 2013).

What is interpersonal coordination in team sports?

In this section we address two key questions with regards to players' interpersonal coordination within competitive environments. First, what is interpersonal coordination in team sports? Second, what sources of information are used by players to coordinate their actions with each other?

Coordination emerges when the components of a system (e.g. the players) are synchronized or adjusted to each other to perform as a coherent unit (Turvey, 1990). As stated by Richard C. Schmidt and Paula Fitzpatrick in Chapter 2 of this book, "Interpersonal coordination refers to how the behaviour of two or more individuals is brought into alignment". Thus, interpersonal coordination occurs when participant actions that are supported through informational fields (e.g. visual, acoustic) are coupled (Fajen, Riley & Turvey, 2009; Schmidt & O'Brien, 1997), which supports the achievement of levels of collective performance that are not afforded to individual parts of the system. Importantly, it is the players' movements that create the information fields which sustain the couplings among them. Information is seen as a cascade of energy flows across all spatial and temporal scales of an active biological system (Dixon, Holden, Mirman & Stephen, 2012; Vaz, 2015). The term 'active' deserves to be highlighted here because it is the players' actions that generate information which sustains subsequent actions, a cyclical process known as perception-action coupling (Gibson, 1979).

As players' couplings are sustained by information that is generated by their own movements, it is suggested that players' interactive behaviour should be analysed as a self-organized dynamical system, the behaviour of which is guided without an external controller (Davids, Handford & Williams, 1994; McGarry, Anderson, Wallace, Hughes & Franks, 2002; Passos et al., 2013).

With regards to the second question, the information that supports interpersonal coordination probably comes in the form of coaches' instructions, but also emerges from the dynamics of an ever-changing context (see Passos, Araújo & Davids, 2013). Previous research indicates that interpersonal coordination is usually achieved through the use of visual information that is locally generated due to the participants' movements (Schmidt, Bienvenu, Fitzpatrick & Amazeen, 1998). Performance contexts contain plenty of perceptual information (e.g. visual, acoustic) that players actively explore to use different perceptual-motor strategies to move towards a goal, receive a pass or tackle an opponent. The cyclical feature of perception and action is strengthened where actions play a critical role in perception and vice versa, perception is critical for players' assembly of decisions and actions (Fajen et al., 2009; Gibson, 1979). Data from previous research have revealed that players mainly use information that is locally created, such as from a ball's flight path or opposing players' movements, to regulate interceptive actions, which highlights the predominance of prospective information (Bastin, Craig & Montagne, 2006; Correia, Araújo, Craig & Passos, 2011). Thus for interpersonal coordination to be effective, the use of prospective control strategies, which entail anticipation of where a teammate or an opponent is likely to be (in time and space) at a future moment, will allow players to continuously co-adapt their behaviours to the behaviours of others in their surroundings (i.e. teammates or opponent players) (Fajen, 2005; Fajen & Devaney, 2006; Montagne, 2005; Passos & Davids, 2015).

To meet the time and space constraints required to succeed in interpersonal coordination tasks, some degree of perceptual flexibility (i.e. the use of different sources of information) is required. This is achieved with perceptual *attunement*, a mechanism which allows each individual player to use the most relevant information to perform a task (Fajen et al., 2009). Perceptual information is tightly coupled with *affordances*, which are opportunities for action provided by the environment to each individual (Gibson, 1979). This perceptual attunement influences the affordances that are available to each individual player (Turvey, 1992), but each player also has the ability to perceive affordances of other players.

Additionally, due to players' movements, information is dynamically generated for different players (e.g. attackers or defenders), or even for the same player at different moments in time (e.g. in the first half or second half of a match). This perspective suggests that affordances are dynamical functional relations embedded in ongoing player-environment interactions (Vaz, 2015).

Therefore, interactive behaviour is a crucial feature of success in team sports, which requires each player to continuously co-adapt to teammates and opposing players' actions (Passos et al., 2013). Players' co-adaptive behaviours are constrained by information emerging from task constraints, such as pitch locations, boundaries and rules etc. Importantly, the complementary association that exists during cooperative and competitive contexts in team sports requires that interpersonal coordination patterns should be analysed first from two different (or even complementary) perspectives. Specifically, a perspective on the complementary nature of team sports performance can be adopted at the intrateam and the interteam level of analysis.

Two levels of analysis of interpersonal coordination in team sports

Players' opportunities for action that may emerge in team sports are bounded by time and space constraints. The mutually exclusive objectives between opposing players are a key task constraint which will account for how a player may interact differently with teammates and opponents. Thus, interpersonal coordination should be analysed from two different perspectives: from an *intrateam* coordination perspective (i.e. between players of the same team) and from an *interteam* coordination perspective (i.e. between opposition players). The common ground for both perspectives is that players coordinate (intentionally or unintentionally) to achieve a dyadic or group level performance outcome.

Intrateam coordination

Intrateam coordination demands that players intentionally coordinate with each other to perform as a single entity to achieve performance outcomes; this single entity is usually called a synergy.[1] However, players are independent entities, which mean that there is no causal relation between them in the sense that the behaviour of one inevitably influences the behaviour of the other. Notwithstanding action under the influence of task constraints, the two 'independent' players may couple to behave as a single entity (Kugler & Turvey, 1987). These soft assembly functional synergies are emergent entities grounded on the reciprocal adjustments between system components, the players (at the lowest (micro) level), and bounded by previously defined task constraints (at the highest (macro) level) (e.g. pitch dimensions, players' relative positions). By soft assembly, we mean that players are temporarily coupled to fit a specific task goal such as the need to keep possession in soccer or the challenge to move towards the scoring area in rugby union (Eiler, Kallen, Harrison & Richardson, 2013). Changes in the dynamics of competitive performance force task goals to disappear that may be replaced by other more relevant task goals. When these task goals decay or emerge over time, the coupling between players can also disappear and the players who form it are free to form other couplings (with other players) that are more functional for other task goals.

The accomplishment of some collective behaviour requires intrateam coordination between players of the same team that must be sustained by local interaction rules – for example, keeping a 'functional' interpersonal distance from teammates (e.g. close enough to support the ball carrier, but not so close as to result in collisions). To satisfy this local rule other rules are required, like running at the same pace and in the same direction as the ball carrier (which in turn demands that support players adjust their running lines, speed and direction). It is this adjustable (co-adapted) behaviour between players on the same team, supported by local interaction rules, that allows the formation of an attacking subunit with a specific collective structure (e.g. a diamond-shaped structure). However, due to the proximity of the opponents (e.g. defenders), the stability of this intrateam coordinative structure is perturbed and players must adjust their relative positions to maintain the functional intrateam collective behaviour. There are two studies in rugby union that capture this interactive behaviour between attacking players in a competitive performance context (Passos et al., 2011; Rodrigues & Passos, 2013). The first study aims to describe how a set of players within an attacking subunit (forming a diamond-shaped structure) adjust their relative positions due to a decrease in interpersonal distances to defenders. Results reveal that on average, when the attacker's subunit faces the first defensive line, the attacking players are closer to each other than when they reorganize (spread out) to face the second defensive line (Passos et al., 2011). In a second study, the aim was to analyse how an intrateam coordination pattern influences successful performance in rugby union. Results revealed that intrateam coordination patterns may be crucial for successful performance when the opposition do not achieve the same levels of intrateam coordination. However, when players on opposing teams display the same level of intrateam coordination, there are other factors that impact on successful performance (Rodrigues & Passos, 2013).

Due to the complex nature of players' interactive behaviour within competitive matches, some analysis requires the use of variables which provide a focus for the most relevant information with regard to the behaviour of a set of players (i.e. a single coordinative/collective variable). For example, the centroids are most commonly used to describe intrateam collective behaviour. By definition, "a team's centroid is a geometrical configuration that represents the mean position of a group of points" (Frencken, De Poel, Visscher & Lemmink, 2012, pp. 1209). Therefore, centroids can be used as a collective variable which describes the interactive behaviour of a set of players (from the same team or from opposing teams) during a competitive performance (Folgado, Lemmink, Frencken & Sampaio, 2014; Frencken, et al., 2012; Frencken, Lemmink, Delleman & Visscher, 2011; Sampaio & Macas, 2012).

Interteam coordination

To analyse interteam coordination, the first challenge was to identify coordinative (i.e. collective) variables that could accurately describe the interactive behaviours between attackers and defenders. By definition, a coordinative variable is the "relational quantities that are created by the cooperation among the individual parts of a system. Yet they, in turn, govern the behaviour of the individual parts" (Kelso, 2009). One example of such a coordinative variable is the interpersonal angle which is calculated using a vector from the defender to the ball carrier with an imaginary horizontal line parallel to the score line or pitch end line (Figure 11.1) (Passos et al., 2009).

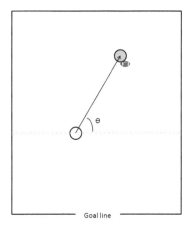

Figure 11.1 The interpersonal angle. The white circle represents the defender; the grey circle represents the ball carrier; the black arrow represents the vector from the defender to the ball carrier; the dashed grey line represents an imaginary line parallel to the goal line; Θ represents the interpersonal angle.

The angle values close to 90° signify that the ball carrier did not pass the defender, which indicates that the ball carrier's attempts to dribble past the defender were successfully counterbalanced (Passos et al., 2009; Passos et al., 2013). The zero crossing point is identified as the moment when the ball carrier dribbled past the defender, and negative angle values would signify that the ball carrier had become the player closest to the goal (e.g. basket, try line) (Figure 11.2).

Importantly, an increase in the fluctuations scale of the interpersonal angle signifies that the dyadic system is poised for a transition, i.e. a change in the system's structural organization. This in turn suggests that the attacker-defender dyadic system is poised to evolve towards regions of self-organized criticality (Jensen, 1998). This term is built upon two concepts: i) self-organization, which describes the ability of certain dynamical systems to create structures and patterns in the absence of control by an external agent; and ii) criticality, a concept usually associated with transitions in systems structural organization (Bak, 1996; Jensen, 1998). Criticality occurs when a certain parameter achieves a specific value – for instance when the ball carrier-defender relative velocity achieves a positive value. As a consequence, an abrupt transition in the structural organization of the dyadic system will occur, with the ball carrier transiting closer to the try line. After a transition period where one player usually gains an advantage over the opponent, the other system components (e.g. players) will likely reorganize their relative positions to settle into a balanced state between opposing teams. For all other values of ball carrier-defender relative velocity, the dyadic system will only be locally disturbed, usually seen as small fluctuations in the values of the ball carrier-defender interpersonal angle; these periods are usually characterized by system relative stability. The attacker-defender systems as team sports can alternate between periods of stability and variability, and this is why it was suggested that attacker-defender systems can move towards regions of self-organized criticality (Passos et al., 2009; Passos et al., 2013).

Interpersonal coordination under different task constraints

In this section, we highlight how interteam coordination can be influenced by different task constraints such as the values of player distances to the goal, the

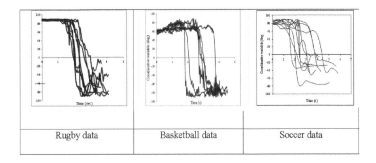

| Rugby data | Basketball data | Soccer data |

Figure 11.2 Interpersonal angle values.

defender's distance to the nearest sideline, pitch dimensions, or the numerical differences between competing players in a sub-phase of play.

Ball carrier-defender interactions are sensitive to task constraints, such as a dyad's distance to the defender's goal. For example, in a study that examined 1v1 situations in football, the defender's distance to the ball was used as a variable that describes the ball carrier-defender interactions in relation to pitch locations (Headrick et al., 2012). Results showed that when the ball is near to the defender's goal, the defender will be further away from the ball so as to reduce the attacker's affordances to shoot at goal. When the ball is further away from the defender's goal, the defender attempts to close the distance to the ball to regain possession of the ball (Headrick et al., 2012). To analyse the ball carrier's decisions and actions in a study with a 2v1 situation in rugby union, both ball carrier distance and tackler distance to the closest sideline were used to capture both players' relative positions (Passos et al., 2012). Results revealed that within a critical region of 2m bandwidth of interpersonal distance, the dynamics of the tackler's (defender's) relative position to the sideline was crucial for the emergence of the ball carrier's decisions and actions. On one hand, the ball carrier tended to perform a pass when the tackler was furthest from the sideline. On the other hand, the ball carrier's decision not to pass the ball but go forward towards the score line was influenced by the variability in the tackler's relative position (Passos et al., 2012). It is important to note here that it was not only the tackler's relative position to the nearest sideline that constrained ball carrier decisions and actions, but also the changes in the tackler's relative position.

A study on small-sided games of football which examined the role of different pitch dimensions in constraining player behaviour is a relevant example of the influence of task constraints (Vilar, Duarte, Silva, Chow & Davids, 2014). The variables created to analyse the ball carrier's decisions and actions were the *shooting interception point* and the *passing interception point*. Results revealed that ball carriers have fewer opportunities to retain ball possession on smaller pitches. However, different pitch dimensions did not influence affordances for the ball carrier to shoot at goal, or to perform passes to the support players. Importantly, the interpersonal coordination between players is dependent on various factors (e.g. distance between players, density of players within a pitch, player's distance to goal). In addition, the different phases of the match may also influence intrateam and interteam interpersonal coordination patterns (e.g. in futsal, from a study by Travassos, Araújo, Vilar, & McGarry, 2011; in basketball, from a study by Bourbousson, Sève & McGarry, 2010a). These studies highlight the need to be cautious in over-generalization of results since different task constraints present in various team sports afford the emergence of different behaviours.

Discussion and further research

The key message for further research is to move beyond a descriptive level of analysis. We suggest three different approaches to achieve this: i) identify possible control parameters[2] that can be progressively scaled to effect a change in overall system organization (see Kelso, 1995); ii) relate players' interactive behaviour with the concept of affordances; and iii) identify synergetic relationships which can explain players' behavioural adjustments with performance stabilization.

In relation to the first point, there is a lot that can be gained if we can successfully examine the role of control parameters. Practitioners can attempt to change practice conditions and accurately measure how these changes can eventually influence a player's interactive behaviour. For example, the rules of the game or equipment can be scaled to progressively push a player or the team to move towards a critical region of play where they can transit to a new pattern of behaviour. Such manipulation of control parameters (like rules, equipment or even task goals) that cause the system's behaviour to change qualitatively will be meaningful for practitioners to understand the functional strategies that can be adopted when designing practice conditions. Thus, further research should be focused on relating descriptive variables, such as interpersonal distances and angles, with parameters which add a temporal dimension to the player's interactive behaviour. Some formal models which express this relation between coordinative variables and control parameters have been successfully presented in the movement science literature (Araújo, Diniz, Passos & Davids, 2014; Diniz, Barreiros & Passos, 2014).

Interactive behaviour creates affordances that players can use to gain an advantage over their opponents. A suitable example of how players' interactive behaviour creates affordances is the work by Luís Vilar and colleagues with small-sided games in futsal. Their work provided valuable insights into the role of the ball carrier's and defenders' relative positions that quantify the *shooting interception point* (Sip) and the *passing interception point* (Pip). These two indicators can characterize the role of the ball carrier and the defender's affordances at any moment of a match (Vilar et al., 2014). In addition, the inclusion of a temporal component in the analysis of team play behaviours provided insights into the role of affordances. Undoubtedly, interactive behaviours change in space and time, and thus a spatial-temporal analysis may provide relevant information which can help us to better understand players' interactive behaviours. A study by Travassos and colleagues examined ball carriers' shooting and passing opportunities by calculating the velocity of each defender at each frame by the rate of change of the distance between his current position and the nearest point of the ball's projection trajectory. Results from the study revealed that defenders' ability to intercept the ball was influenced by continuous adaptations of the defenders' velocities relative to the ball trajectory (Travassos et al., 2012). Thus, the temporal component can provide additional information regarding how long an affordance is 'available'.

Last but not least, the synergies that are formed during a competitive match are also an area of work that requires further examination. A synergy occurs when a set of players adjust their behaviours to form a 'single entity' which stabilizes performance. It is this relation between players' behavioural adjustments and the performance components that need to be stabilized that requires a more formal (i.e. quantifiable, measurable) examination. The use of the Uncontrolled Manifold method (Scholz & Schoner, 1999) was suggested as a possible tool to examine interpersonal coordination in social systems (Riley, Richardson, Shockley, & Ramenzoni, 2011).

Conclusions

A dynamical systems approach to the study of interpersonal coordination in team sports provides insights into how the interactive nature of players and their behaviours can alter as a consequence of manipulating relevant constraints during practice. Clearly, there are still major challenges in this area of work that need to be addressed. For example, how to study forming synergies in team sports? How will such synergies influence team effectiveness? How do changes at dyadic level emerge and influence the entire team's play behaviour? These are non-trivial questions, and this chapter has provided an attempt to enhance our understanding of interpersonal coordination in team sports.

Notes

1   In the literature synergies may be also called coordinative structures.
2   "Control parameters refer to naturally occurring environmental conditions or intrinsic, endogenous factors that move the system through its repertoire of patterns and cause them to change" (Kelso, 2009, p. 3).

References

Araújo, D., Davids, K. & Hristovski, R. (2006) The ecological dynamics of decision making in sport. *Psychology of Sport and Exercise, 7*(6), 653–676.

Araújo, D., Diniz, A., Passos, P. & Davids, K. (2014) Decision making in social neurobiological systems modeled as transitions in dynamic pattern formation. *Adaptive Behavior, 22*(1), 21–30.

Bak, P. (1996) *How nature works. The science of self-organizing criticality.* Copernicus. Springer-Verlag NY. ISBN: 038798738

Bastin, J., Craig, C. & Montagne, G. (2006) Prospective strategies underlie the control of interceptive actions. *Hum Mov Sci, 25*(6), 718–732.

Bourbousson, J., Sève, C. & McGarry, T. (2010a) Space-time coordination dynamics in basketball: Part 1. Intra- and inter-couplings among player dyads. *J Sports Sci, 28*(3), 339–347.

Bourbousson, J., Sève, C. & McGarry, T. (2010b) Space-time coordination dynamics in basketball: Part 2. The interaction between the two teams. *J Sports Sci, 28*(3), 349–358.

Correia, V., Araújo, D., Craig, C. & Passos, P. (2011) Prospective information for pass decisional behavior in rugby union. *Hum Mov Sci, 5*, 984–997.

Couzin, I.D., Krause, J., Franks, N.R. & Levin, S.A. (2005) Effective leadership and decision-making in animal groups on the move. *Nature, 433*(7025), 513–516.

Davids, K., Handford, C. & Williams, A.M. (1994) The natural physical alternative to cognitive theories of motor behaviour: An invitation for interdisciplinary research in sports science? *Journal of Sport Sciences, 12*(6), 495–528.

Diniz, A., Barreiros, J. & Passos, P. (2014) To Pass or Not to Pass: A Mathematical Model for Competitive Interactions in Rugby Union. *Journal of Motor Behavior*, *46*(5), 293–302.

Dixon, J.A., Holden, J.G., Mirman, D. & Stephen, D.G. (2012) Multifractal Dynamics in the Emergence of Cognitive Structure. *Topics in Cognitive Science*, *4*(1), 51–62.

Duarte, R., Araújo, D., Fernandes, O., Fonseca, C., Correia, V., Gazimba, V. et al. (2010) Capturing complex human behaviors in representative sports contexts with a single camera. *Medicina (Kaunas)*, *46*(6), 408–414.

Eiler, B.A., Kallen, R.W., Harrison, S.J. & Richardson, M.J. (2013) Origins of Order in Joint Activity and Social Behavior. *Ecological Psychology*, *25*(3), 316–326.

Fajen, B.R. (2005) Perceiving possibilities for action: on the necessity of calibration and perceptual learning for the visual guidance of action. *Perception*, *34*(6), 717–740.

Fajen, B.R. & Devaney, M.C. (2006) Learning to control collisions: the role of perceptual attunement and action boundaries. *J Exp Psychol Hum Percept Perform*, *32*(2), 300–313.

Fajen, B.R., Riley, M.A. & Turvey, M.T. (2009) Information, affordances, and the control of action in sport. *International Journal of Sport Psychology*, *40*(1), 79–107.

Folgado, H., Lemmink, K.A., Frencken, W. & Sampaio, J. (2014) Length, width and centroid distance as measures of teams tactical performance in youth football. *Eur J Sport Sci*, *14 Suppl 1*, S487–492.

Frencken, W., De Poel, H., Visscher, C. & Lemmink, K. (2012) Variability of inter-team distances associated with match events in elite-standard soccer. *Journal of Sports Sciences*, *30*(12), 1207–1213.

Frencken, W., Lemmink, K., Delleman, N. & Visscher, C. (2011) Oscillations of centroid position and surface area of soccer teams in small-sided games. *European Journal of Sport Science*, *11*, 215–223.

Gibson , J. (1979) *The ecological approach to visual perception*. Lawrence Erlbaum Associates: Hillsdale, NJ.

Haken, H., Kelso, J.A. & Bunz, H. (1985) A theoretical model of phase transitions in human hand movements. *Biol Cybern*, *51*(5), 347–356.

Headrick, J., Davids, K., Renshaw, I., Araújo, D., Passos, P. & Fernandes, O. (2012) Proximity-to-goal as a constraint on patterns of behaviour in attacker-defender dyads in team games. *Journal of Sport Sciences*, *30*(3), 247–253.

Jensen, H. (1998) *Self-Organized Criticality: Emergent Complex Behavior in Physical and Biological Systems*. Cambridge Lecture Notes in Physics: Cambridge.

Kelso, S. (1995) *Dynamic patterns. The self-organization of brain and behavior (complex adaptive systems)*. MIT Press: Cambridge, MA.

Kelso, S. (2009) Coordination dynamics. In R.A. Meyers (Ed.), *Encyclopedia of Complexity and System Science* (pp. 1537–1564). Springer: Heidelberg.

Kugler, P. & Turvey, M.T. (1987) *Information, Natural Law, and the Self-assembly of Rhythmic Movement*. Lawrence Erlbaum Associates: Hillsdale, NJ.

McGarry, T., Anderson, D.I., Wallace, S.A., Hughes, M.D. & Franks, I.M. (2002) Sport competition as a dynamical self-organizing system. *J Sports Sci*, *20*(10), 771–781.

Montagne, G. (2005) Prospective control in sport. *International Journal of Sport Psychology*, *36*(2), 127–150.

Passos, P., Araújo, D. & Davids, K. (2013) Self-organization processes in field-invasion team sports: implications for leadership. *Sports Med*, *43*(1), 1–7.

Passos, P., Araújo, D., Davids, K., Gouveia, L., Milho, J. & Serpa, S. (2008) Information-governing dynamics of attacker-defender interactions in youth rugby union. *Journal of Sports Sciences*, *26*(13), 1421–1429.

Passos, P., Araújo, D., Davids, K., Gouveia, L. & Serpa, S. (2006) Interpersonal dynamics in sport: The role of artificial neural networks and 3-D analysis. *Behavior Research Methods*, *38*(4), 683–691.

Passos, P., Araújo, D., Davids, K., Gouveia, L., Serpa, S., Milho, J. et al. (2009) Interpersonal pattern dynamics and adaptive behavior in multiagent neurobiological systems: conceptual model and data. *J Mot Behav*, *41*(5), 445–459.

Passos, P., Cordovil, R., Fernandes, O. & Barreiros, J. (2012). Perceiving affordances in rugby union. *J Sports Sci*, *30*(11), 1175–1182.

Passos, P. & Davids, K. (2015) Learning design to facilitate interactive behaviours in Team Sports. *Revista Internacional de Ciencias del Deporte*, *11*(39), 18–32.

Passos, P., Milho, J., Fonseca, S., Borges, J., Araújo, D. & Davids, K. (2011) Interpersonal distance regulates functional grouping tendencies of agents in team sports. *J Mot Behav*, *43*(2), 155–163.

Riley, M.A., Richardson, M.J., Shockley, K. & Ramenzoni, V.C. (2011) Interpersonal synergies. *Front Psychol*, *2*, 38.

Rodrigues, M. & Passos, P. (2013) Patterns of Interpersonal Coordination in Rugby Union: Analysis of Collective Behaviours in a Match Situation. *Advances in Physical Education*, *3*(4), 209–214.

Sampaio, J. & Macas, V. (2012) Measuring tactical behaviour in football. *Int J Sports Med*, *33*(5), 395–401.

Schmidt, R.C., Bienvenu, M., Fitzpatrick, P.A. & Amazeen, P.G. (1998) A comparison of intra- and interpersonal interlimb coordination: coordination breakdowns and coupling strength. *J Exp Psychol Hum Percept Perform*, *24*(3), 884–900.

Schmidt, R.C. & O'Brien, B. (1997) Evaluating the Dynamics of Unintended Interpersonal Coordination. *Ecological Psychology*, *9*(3), 189–206.

Scholz, J.P. & Schoner, G. (1999) The uncontrolled manifold concept: identifying control variables for a functional task. *Exp Brain Res*, *126*(3), 289–306.

Silva, P., Travassos, B., Vilar, L., Aguiar, P., Davids, K., Araújo, D. et al. (2014) Numerical Relations and Skill Level Constrain Co-Adaptive Behaviors of Agents in Sports Teams. *PLoS One*, *9*(9).

Travassos, B., Araújo, D., Davids, K., Vilar, L., Esteves, P. & Vanda, C. (2012) Informational constraints shape emergent functional behaviours during performance of interceptive actions in team sports. *Psychology of Sport and Exercise*, *13*(2), 216–223.

Travassos, B., Araújo, D., Vilar, L. & McGarry, T. (2011) Interpersonal coordination and ball dynamics in futsal (indoor football). *Hum Mov Sci*, *6*, 1245–1259.

Travassos, B., Goncalves, B., Marcelino, R., Monteiro, R. & Sampaio, J. (2014) How perceiving additional targets modifies teams' tactical behavior during football small-sided games. *Hum Mov Sci*, *38*, 241–250.

Turvey, M.T. (1990) Coordination. *American Psychologist*, *45*(8), 938–953.

Turvey, M.T. (1992) Affordances and prospective control: An outline of the ontology. *Ecological Psychology*, *4*, 173–187.

Vaz, D.V. (2015) Direct Perception Requires an Animal-Dependent Concept of Specificity and of Information. *Ecological Psychology*, *27*(2), 144–174.

Vilar, L., Duarte, R., Silva, P., Chow, J.Y. & Davids, K. (2014) The influence of pitch dimensions on performance during small-sided and conditioned soccer games. *J Sports Sci*, *32*(19), 1751–1759.

12 Shared affordances guide interpersonal synergies in sport teams

Duarte Araújo, João Paulo Ramos and Rui Jorge Lopes

Introduction

Team sports performance has been described by means of notational techniques with the purpose of inspecting the behaviours of performers during different sub-phases of play in games or matches. The aim of such analysis is to provide accurate, augmented information to practitioners to improve future performance (Vilar et al., 2012). The recorded variables include scoring indicators such as goals, baskets, winners, errors or the ratios of winners to errors and goals to shots, or performance indicators such as turnovers, tackles and passes (see Passos et al., 2016/in press for a recent review). An important criticism of notational analysis research is that it has been somewhat reductionist (Glazier, 2010), typically omitting references to the *why* and *how* of performance that underlie the structure of recorded behaviours, which would define their functional utility (McGarry, 2009). This points towards the need for a sound theoretical rationale of performance behaviours. In fact, there are some theoretical approaches that go beyond analytics, which are relevant *per se* and which present viable hypotheses to test (see Chapter 9 in this book by Araújo and Bourbousson). Ecological dynamics is such a framework for studying behaviours in team games. It has the advantage of recognising the 'degeneracy' of collective biological systems (i.e. teams). Its principles can explain how the same performance outcomes can emerge from different movement or tactical patterns.

Ecological dynamics approach to team behaviour

Ecological dynamics proposes performer-environment relations as the relevant scale of analysis for understanding sport performance (see Chapter 9 of this book by Araújo and Bourbousson; also Davids & Araújo, 2010). The functional patterns of coordinated behaviour emerge through a process of self-organization via performers' interactions with each other under specific task and environmental constraints (Araújo, Davids & Hristovski, 2006). Ecological dynamics analyses of team sports have attempted to explain how the interaction between players and information from the performance environment constrains the emergence of patterns of stability, variability and the transitions in organizational states of such systems. The emergent coordination patterns in team sports are channelled by the surrounding constraints, as they structure the state space of all possible configurations available to the team game as a complex system (Davids, Button, Araújo, Renshaw, & Hristovski, 2006). Constraints are boundaries or features, which interact to shape the emergence of the states of system organization. For example, the surrounding patterned energy distributions that performers can perceive act as important sources of information to support their decisions and actions (e.g. reflected light from the ball) (Araújo & Davids, 2009).

The interaction between constraints of the performance environment and each individual's characteristics allows opportunities for action to emerge (Araújo et al., 2006). For example, an opportunity to score a goal in football may emerge from the interaction between the performer's ability to shoot the ball (individual constraints) and the distance to the goal or to the goalkeeper (task constraints). Moreover, performers are also able to identify relations between other performers and key environmental objects (e.g. the ball and the target zone in team games) that can constrain their behaviours (Richardson, Marsh & Baron, 2007). By perceiving opportunities for others to act, performers make use of environmental information to coordinate their actions with others.

Due to the complex spatial-temporal relations among performers that characterize team sports, performance constraints change on a momentary basis. Opportunities to act (or affordances; Gibson, 1979) may appear and disappear over time (Araújo & Davids, 2009). The concept of affordances presupposes that the environment is perceived directly in terms of what an organism can do with and in the environment (i.e. it is not dependent on a perceiver's expectations, nor on mental representations linked to specific performance solutions) (Richardson et al., 2008). It has been suggested in team sports (Davids, Araújo & Shuttleworth, 2005) that individuals base their movement decisions on locally acquired information sources such as the relative positioning, motion direction, or changing motion direction of significant others operating in a system, making a collective response all the more remarkable. This hypothesis implies that the actions of individuals functioning in a team need to be intimately coordinated. The interactions of agents in sports teams reveal common underlying principles while simultaneously exhibiting their own 'signatures' or idiosyncratic behaviours.

An important understanding of affordances is that they can be perceived by a group of individuals trained to become perceptually attuned to them (Silva et al., 2013). In collective sports, both teams in opposition have the same objective (i.e. to overcome the opposition and win). Hence, the perception of collective affordances acts as a selection pressure (Reed, 1996) for overcoming opponents and achieving successful performance. In this sense, collective affordances are sustained by common goals between players of the same team who act to achieve success for the group. From this perspective, team coordination depends on being collectively attuned to shared affordances founded on joint practice of becoming attuned to specific environmental circumstances (Silva et al., 2013). Through practice, players become perceptually attuned to affordances *of* others and affordances *for* others during competitive performance, and undertake more efficient actions (Fajen et al., 2009) by adjusting their behaviours to functionally adapt to those of other teammates and opponents. This process enables them to act synergistically with respect to specific team task goals (Folgado et al., 2012; Travassos et al., 2012). By means of tracked positional data, recent studies have started to reveal how players and teams *continuously* interact during competition. For example, teams tend to be tightly synchronised in their lateral and longitudinal movements (Vilar et al., 2013) with a counterphase relation regarding their expansion and contraction movement patterns (Yue et al., 2008), commonly caused by changes in ball possession (Bourbousson et al., 2010).

Following these types of compound measures, specific training effects found by Sampaio and Maçãs (2012) indicate that players constantly adjust their positions on the pitch according to the game's ebb and flow, and that more effective team coordination was expressed by the fact that the most powerful variable in distinguishing pre- and post-test conditions was the distance of players from the team's geometric centre (obtained by computing the mean lateral and longitudinal positional coordinates of each performer in a team). As expected, inter-player coordination in pre-test seems to reflect individual affordances and not shared affordances among team players. However, post-test values showed that players became more coordinated with increased expertise. The coordination patterns showed compensatory behaviour within the team, an essential characteristic of synergy (Riley et al., 2011). Thus, the decisions and actions of the players forming a synergy should not be viewed as independent, and the coupling of players' degrees of freedom into interpersonal synergies is based upon a social perception-action system that is supported by the perception of shared affordances.

Specific constraints like the players' individual characteristics, a nation's traditions in a sport, strategy, coaches' instructions etc. may impact on the functional and goal-directed synergies formed by the players to shape a particular performance behaviour. These informational constraints shape shared the affordances that are available for players, viewed as crucial for the assembly of synergies that support the reduction of the number of independent degrees of freedom (Riley et al., 2011). Under this theoretical rationale, the properties of synergies guide the information that could be obtained from the diverse behavioural team measures.

Synergies organize the meaning of measures of team behaviour in performance environments

A synergy is a task-specific organization of elements such that the degrees of freedom of each component are coupled, enabling the degrees of freedom to regulate each other (Bernstein, 1967, Gelfand & Tsetlin, 1966). Latash (2008) identifies the characteristics that should be met for a group of components to be considered a synergy: sharing, error compensation and task-dependence. Sharing means that the components should all contribute to a particular task. A way to quantify the amount of sharing is the matching of the sum of the individual contributions to the task, and the overall measurement of the performance on the task. Error compensation is captured when some components show changes in their contributions to a task, compensating for one component that is not doing its share. Finally, task-dependence is the ability of a synergy to change its functioning in a task-specific way – or, in other words, to form a different synergy for a different purpose based on the same set of components. Therefore, synergies are "task specific devices" (Bingham, 1988).

On the other hand, Riley and colleagues (Riley, Richardson, Shockley & Ramenzoni, 2011) identify two characteristics of a synergy. One is dimensional compression, which means that degrees of freedom that potentially are independent are coupled so that the synergy has fewer degrees of freedom (possesses a lower dimensionality) than the set of components from which it arises. The behaviour of the synergy has even fewer degrees of freedom, a second level of dimensional compression as one moves from structural components to the behaviour that emerged from the interactions among the degrees of freedom. Dimensional compression at both stages results from imposing constraints (environmental, task and individual constraints), which couple components so they that change together rather than independently. The other property of a synergy for Riley et al. (2011) is reciprocal compensation, similar to error compensation as described by Latash (2008).

Here we address three properties of a synergy: 1) dimensional compression (Bingham, 1988; Riley et al., 2011); 2) reciprocal compensation (Latash, 2008; Riley et al, 2011); and 3) degeneracy (Davids et al., 2006; Latash, 2008). We will discuss how these properties organize the existing measures of group behaviour in sport teams.

Dimensional compression

For dealing with the problem of dimensional reduction, there are some useful approaches to data analysis aimed at system identification. For this, experiential knowledge from expert coaches may be a good starting point, by capitalizing on educated guesses about which collective variables are most relevant.

Grouping measurements in sports teams

The joint work of expert coaches and sport scientists has arrived at team measures such as team centre and team dispersion. Coaches often mention the importance of the "centre of gravity" of a team (Grehaigne et al., 2011). An operational approach to this tactical concept is the team's centre (also denominated as the centroid or geometrical centre). This variable can be obtained by computing the mean lateral and longitudinal positional coordinates of each player in a team. It has been used in various ways to evaluate intra- and inter-team coordination in team sports like association football (Frencken, Lemmink, Delleman & Visscher, 2011, and see Clemente, Couceiro, Martins, Mendes & Figueiredo, 2013 for a "weighted" centroid), futsal (Travassos, Araújo, Duarte, & McGarry, 2012) and basketball (Bourbousson, Sève & McGarry, 2010). The teams' centres alone merely represent the relative positioning of both teams in the forward-backward and side-to-side movement displacements, but when analysed in respect to other measures, may provide important descriptions of team tactics.

According to the basic principles of attacking and defending in invasion team sports, the team in possession must create space by stretching and expanding in the field, while the defending team must close down space by contracting and reducing distances between players. Such collective movements may be captured by specific measures of team coordination that quantify the overall spatial dispersion of players. The stretch index (or radius), the team spread and the effective playing space (or surface area) are quantities that have been used to assess such spatial distributions. The stretch index is calculated by computing the average radial distance of all players to their team's centroid. It can also be calculated according to axis expansion, providing distinct measures of dispersion for the longitudinal and lateral directions (e.g. Yue, Broich, Seifriz & Mester, 2008; and see Moura, Martins, Anido, Barros & Cunha, 2012 for team spread – the square root of the sums of the squares of the distances between all pairs of players, excluding the goalkeeper). The effective playing space (or surface area) is defined by the smallest polygonal area delimited by the peripheral players containing all players in the game, and it can also provide information about the surface that is being effectively covered by the two teams. It informs how the occupation of space unfolds throughout the game and how stretched both teams are in the field. This effective playing space may also be computed as a function of attacking and defending, discriminating the surface areas of both teams in confrontation while representing the overall team positioning (Frencken & Lemmink, 2008). As with stretch index and team spread, the relationship between the offensive and defensive surface areas can highlight the balance of the opposition relationship during matches (Gréhaigne & Godbout, 2013). Moreover, Folgado and colleagues (2012) calculated the team's length and width in small-sided games involving by youth football players of different ages. The teams' lengths and widths were calculated by measuring the distance between the players furthest forward and backward and furthest to the left and to the right, respectively. Through these quantities, the authors computed the ratio between length and width, based on the assumption that teams with different tactical approaches would display different length per width ratios. Recently, Silva and colleagues (2014a) developed a "team separateness" metric. It is defined as a measure of the degree of free movement each team has available. In football, it was computed based on a sum of distances (in metres) between each team player and the closest opponent, excluding goalkeepers, and it can be interpreted as the overall radius of action free of opponents. It is different from previous metrics because it accounts for the teams' dispersion differences, which may impact on the players' radius of free movement. A TS value close to zero indicates that all players are closely marked, while a high value indicates more freedom of movement. Interestingly, team separateness increased independently of skill level with the increase in pitch size.

A key idea of invasion team sports assumed to promote effective performance is to outnumber the opposition (creation of numerical overloads) during different performance phases (attack and defense) in spatial regions adjacent to the ball, as expressed by inter-team coordination. Inter-team coordination was recently examined through analysis of the distances separating the teams' horizontal and vertical opposing line-forces in football (Silva et al., 2014b). This measure captures the existence of possible differences in the players' interactive behaviours at specific team locations (e.g. wings and sectors). Each team's horizontal lines are calculated by averaging the longitudinal coordinate values of the two players furthest from and nearest to their own goal line, which corresponded to the forward and back lines respectively. Similarly, the vertical line-forces of each team are computed by averaging the mean lateral coordinates of the players furthest to the left and right on the pitch, corresponding to the left and right lines respectively.

Sharing patterns within teams

A sharing pattern, also known as labour division (Duarte et al., 2012; Araújo et al., 2015), is the specific contribution of each element to a group task (Latash, 2008). The behaviour of each individual in a team is constrained by several factors like his/her position on the field (in relation to the other teammates and opponents), strategic and tactical missions, playing phases (i.e. attacking and defending), game rules etc. Collective behaviour is thus composed of many individual labours (Eccles, 2010) performed by two or more players looking to cooperate together towards common intended goals and linked together by a communication system (Silva, Garganta, Araújo, Davids & Aguiar, 2013). The joint analysis of all these individual behaviours can translate group behaviour as all players constrain and are constrained by the entire dynamic system that they compose (Glazier, 2010). This property could be captured by measures of heat maps, major ranges, player-to-locus distance and Voronoi cells.

Heat maps provide a clear picture of the distribution of each player on the field. Heat maps highlight with warmer colours the zones where each player has lingered for larger periods of time during the match (Araújo et al., 2015). Another approach to assess the division of labour in team sports is by measuring the area covered by each player. Major ranges imply the calculation of an ellipse centred at each player's locus and with semi-axes being the standard deviations in the x- and y-directions respectively (Yue et al., 2008). Through the simple visualization of major ranges it is possible to identify preferred spatial positions, major roles for each player and playing styles (Araújo et al., 2015).

Contrary to the former two measurements, the distance of each player to a private locus on the field, over time, captures the time-evolving nature of their movements' trajectories. The locus represents the player's spatial positional reference around which he/she oscillates (McGarry, Perl, & Lames, 2014).

Individual playing areas attributed to each player on a team delimit the Voronoi cells of players in team ball sports, and offer a time-evolving analysis of the trajectories of these areas (Fonseca, Diniz & Araújo, 2013). A Voronoi cell contains all spatial points that are nearer to the player to whom that cell is allocated than to the other players. By measuring the total area of all Voronoi cells from each team, it is possible to obtain a dominant ratio of one team over the other (Fonseca, Milho, Travassos, & Araújo, 2012).

Order parameters

Dimensional reduction is particularly important because advances in data acquisition have allowed for simultaneous recordings of multiple signals for considerable time spans, resulting in huge data sets. Therefore techniques for *a priori* data reduction, such as principal component analysis (PCA), are very useful (Daffertshofer et al., 2004). Based on the covariance matrix between signals, eigenvectors rank the degree to which a principal component contributes to the entire variance (see Button et al., 2014 for applications in sports). However, an even more principled approach to dimensional reduction is to focus on phase transitions, because theory dictates that very near the critical point the dynamics of the complex system under study are reduced to a small set of collective variables or order parameters. Phase transitions are accompanied by a huge separation of time scales between different system components. In dynamics terms, the transition from one pattern to another implies that the first becomes unstable and the second becomes stable. From the view point of the order parameters, all the subsystems become arbitrarily quick so that they can adapt instantaneously to changes in the order parameters. The system dynamics thus amount to that of the order parameters, implying the ordered states can always be described by a very few variables if they are in the neighbourhood of behavioural transitions. In other words, the state of the originally high-dimensional system can be summarized by a few variables or even a single collective variable – the order parameter (Beek & Daffertshofer, 2014).

However, initially it is important to identify the system (Daffertshofer & Beek, 2014). The identification of a system is nothing more than the characterization of the system's dynamics, in terms of regularity and stability. For this, measures of regularity like sample entropy (see Kuznetsov et al., 2014 for a review) are needed in combination with conventional statistics. Sample entropy quantifies the regularity or repeatability of a signal (i.e. empirical time series). Similarly, Lyapunov exponents are used for estimating the stability properties of dynamical systems involving the emergence and disappearance of behavioural patterns (Stergiou et al., 2004).

Dynamical systems approaches to self-organization have emphasized dimensional compression, where the order parameter "relative phase" (e.g. the difference in the segments' oscillation phases, see Kelso, 1995) captures the low-dimensional behaviour that arises from the high-dimensional neuromuscular system. Relative phase describes the spatiotemporal pattern of rhythmic coordination, and the changes in coordination that occur in response to manipulations of the *control parameters* (e.g. movement frequency). The dynamics of relative phase are understood to reflect the behaviour of a synergy (Kelso, 1995; Turvey and Carello, 1996).

Several coordination variables have been applied in team sports to assess coordination between two oscillatory units (e.g. the coupling of two centroids, or the phase relations of two players' movements in a dyad). For instance, the phase synchronization of two signals has been previously studied through relative phase analysis (e.g. Travassos, Araújo, Vilar, & McGarry, 2011), and through running correlations (e.g. Duarte, Araújo, Freire et al., 2012).

In an attempt to capture group synchrony Duarte and colleagues (Duarte, Araújo, Correia, et al., 2013) shed light on how the players composing a team influence each other to create a collective synergy at the team level. For this, the cluster phase method (Frank & Richardson, 2010), based on the Kuramoto order parameter, was applied to the movements of eleven football players from two teams during a football match to assess whole-team and player-team synchrony. Synergistic relations from the whole team showed superior mean values and high levels of stability in the longitudinal direction when compared with the lateral direction of the field, whereas the player-team synchrony revealed a tendency for a near in-phase mode of coordination. Also, the coupling of the two teams' measures showed that synchronization increased between both teams over time.

Whenever the focus is on phase transitions, mathematical modelling in terms of dynamical systems becomes feasible. Importantly, there were formal demonstrations in sport that did not adhere to the well-known relative phase and related measures, as indicated above. For example, Araújo and colleagues (Araújo, Diniz, Passos & Davids, 2014) conceived behavioural phase transitions as the operational definition of decision-making in sport. They modelled how the order parameter "angle between a vector connecting the participants and the try line" expressed the state of a dyadic system composed of attacker and defender in rugby union. Their model, a potential function with two control parameters (interpersonal distance and relative velocity) and a noise parameter, matches empirical evidence and revealed that this kind of system has three stable attractors.

Reciprocal compensation

Reciprocal compensation indicates that if one element produces more or less than its expected share, other elements should show changes in their contributions such that the task goals are still attained (Latash, 2008).

Contrary to dimensional compression, reciprocal compensation, being intuitively a very important property of a team of players, was only recently operationalized in sport. However, it can be found in motor control with the *uncontrolled manifold (UCM)* approach (Scholz & Schöner, 1999; Latash et al., 2002). This approach assumes that coordinated movement is achieved by stabilizing the value of a performance variable (such as a value for relative phase corresponding to an interlimb coordination pattern). In doing so, a subspace (i.e. manifold) is created within a state space of task-relevant elements (the degrees of freedom that participate in the task), such that within the subspace – called UCM – the value of the performance variable remains constant (Riley et al., 2011).

In sport, Silva and colleagues (Silva et al., in press) created a new metric called readjustment delay (Rd). The football players' co-positioning delay (Rd) in adjusting to teammates' movements (goalkeepers excluded) was computed as a measure of team readiness and synchronization speed during attacking and defending patterns of play. Lower delay values indicate rapid readjustment of movements and faster spatial temporal synchrony between players, whereas a larger readjustment delay might impede the spatial-temporal synchrony of player movements. To analyse Rd, the time series of distances to goal for each dyad were lagged in time, reported through the highest correlation coefficient values. A windowed cross-correlation technique was then used to correlate each player with all his nine teammates, producing a moving estimate of association and lag. The maximum lags were considered to represent the time delay, in seconds, between two players' co-positioning in relation to their own goal. Silva et al. (in press) found that the players' Rd decreased over the fifteen weeks of the study, evidencing faster readjustments of coupled players as a manifestation of how this synergistic property evolved within the team.

Degeneracy

Bernstein (1967) emphasized that motor system degrees of freedom are temporarily coordinated together according to the performance environment and task requirements (aka task dependence; Latash, 2008). It has been well documented that novices typically freeze their motor system degrees of freedom, while experts release the degrees of freedom not useful in task performance (e.g. Vereijken et al., 1992 and Seifert et al., 2013 for a review). Freezing system degrees of freedom corresponds to rigidly fixing the joints to reduce the control problem for a performer. The varying role of these motor system degrees of freedom in assembling actions is essential, and is exemplified by the degenerate networks existing at different levels of human movement systems (Seifert et al., 2013). Degeneracy thus refers to structurally different components performing a similar, but not necessarily identical, function with respect to context (Edelman & Gally, 2001). In this sense, behavioural adaptability reflects the modification of one component of the system and/or a whole modification of the coordination realised by 'redundant' elements (i.e. the presence of isomorphic and isofunctional components) or by 'degenerate' elements (i.e. the presence of heteromorphic variants that are isofunctional) (Mason, 2010). A substantial body of literature has highlighted the functional role of movement variability in a sport performance environment and exemplifies how degeneracy emerges at 'intra-individual' and 'inter-individual' levels in many sports (Davids et al., 2006; Seifert et al., 2013).

A team ball game is sustained by continuous adaptive interactions among players (Araújo et al., 2015). The behaviour of such complex systems emerges from the orchestrated activity of many system components (players) that adaptively interact through local pairwise interactions. A common feature of such complex social networks is that any two nodes or system agents can become interconnected for action through a path of a few links only (Newman, 2003).

Recently, studies with complex networks revealed that certain forms of network growth produce scale-free networks; that is, the distribution of connections per node in the networks is scale invariant (Barabási & Albert, 1999), as happens with phase transitions and critical points. This indicates that degeneracy, as a property of a synergy, might be quantified in the different metrics of social networks.

Passos et al. (2011) showed that social networks could be used to analyse the local structure of communication channels among players, during sub-phases of play in team sports. In these networks, nodes represent players and links are weighted according to the number of passes or positional changes completed between players. Players with major competitive roles (high importance or centrality) may be easily identified through social network analyses, since they display a higher number and thus stronger connections. Additionally, different match networks can be compared to extract the general tactical and strategic features of a team, such as the: i) in-degree that measures the number of players who pass the ball to a focal player; ii) out-degree that measures the number of players to which the focal player passes the ball; and iii) preferential attachments between some team members (Passos et al., 2011; Duch et al, 2010; Grund, 2012).

It is possible to advance the understanding of team sports performance by using other existing metrics that consider more than the links between the focal node and its neighbours. For example, for understanding the playing style of a sports team, Gyarmati et al. (2014) did not use any of the metrics that are "focal node" based, but other metrics that are founded on the identification and quantification of connection patters. In order to include other relevant aspects of the game, notably technical actions, some authors have extended the definition of the network. For example in the analysis of basketball Fewell et al. (2012) included rebounds and steals as nodes in the network. However, other metrics also go beyond the local structure, such as flow centrality which gives a quantification of individual and team performance regarding a specific goal like a goal attempt (shot at the goal), or the clustering coefficient that captures the probability of cooperation between players as a function of their mutual interactions (e.g. Fewell et al., 2012).

Emergent patterns of interaction have also been studied using different representations of the interactions between the different actors. These include hypernetworks, where hyperlinks may connect more than a pair of nodes. This latter approach has been applied to robotic soccer (Johnson & Iravani, 2007) and has proven particularly powerful. Networks are a valuable tool to analyse the structure of such communication channels during sub-phases of play in team sports, since it allows the identification of players engaged in more and less frequent interactions within a team and in particular events.

Conclusions

Specific constraints impact on team synergies formed by players during performance. These constraints shape the perception of shared affordances available for players, which underpin the assembly of interpersonal synergies expressed in collective actions. These group processes form synergies, where their key properties – dimensional compression, reciprocal compensation and degeneracy – guide the meaning of operational variables such as team centre, team dispersion, team synchrony and team communication. Developments in methods of analysis of team coordination and performance can benefit from a theoretical approach that situates and traces relevant team properties as defined by synergies. Here we suggest that shared affordances and synergies embraced by an ecological dynamics perspective present the principles to understand the meaning of existing operational metrics of performance analysis and to guide the search for more meaningful ones.

References

Araújo, D. & Davids, K. (2009) Ecological approaches to cognition and action in sport and exercise: Ask not only what you do, but where you do it. *International Journal of Sport Psychology, 40*(1), 5–37.

Araújo, D., Davids, K. & Hristovski, R. (2006) The ecological dynamics of decision making in sport. *Psychology of Sport and Exercise, 7,* 653–676.

Araújo, D., Diniz, A., Passos, P. & Davids, K. (2014) Decision making in social neurobiological systems modeled as transitions in dynamic pattern formation. *Adaptive Behavior, 22*(1), 21–30.

Araújo, D., Silva, P. & Davids, K. (2015) Capturing group tactical behaviors in expert team players. In J. Baker & D. Farrow (Eds.), *Routledge Handbook of Sport Expertise*. Routledge: London.

Beek, P. & Daffertshofer, A. (2014) Dynamical systems. In R. Eklund & G. Tenenbaum (Eds.), *Encyclopedia of sport and exercise psychology.* (Vol. 4, pp. 224–229). SAGE Publications: Thousand Oaks, CA.

Barabási, A.-L. & Albert, R. (1999) Emergence of Scaling in Random Networks. *Science, 286,* 509–512.

Bernstein, N.A. (1967) *Coordination and Regulation of Movements*. Pergamon Press: New York.

Bingham, G.P. (1988) Task-specific devices and the perceptual bottleneck. *Human Movement Science, 7,* 225–264.

Bourbousson, J., Sève, C. & McGarry, T. (2010) Space-time coordination dynamics in basketball: Part 2. The interaction between the two teams. *Journal of Sports Sciences, 28*(3), 349–358.

Button, C., Wheat, J. & Lamb, P. (2014) Why coordination dynamics is relevant for studying sport performance. In K. Davids, R. Hristovski, D. Araújo, N. Balagué, C. Button & P. Passos (Eds.), *Complex Systems in Sport* (pp. 44–61). Routledge: London.

Clemente, F., Couceiro, M., Martins, F., Mendes, R. & Figueiredo, A. (2013) Measuring collective behaviour in football teams: Inspecting the impact of each half of the match on ball possession. *International Journal of Performance Analysis in Sport, 13,* 678–689.

Daffertshofer, A. & Beek, P. (2014). Modeling. In R. Eklund, & G. Tenenbaum (Eds.), *Encyclopedia of sport and exercise psychology.* (Vol. 12, pp. 445–449). SAGE Publications: Thousand Oaks, CA.

Daffertshofer, A., Lamoth, C.J., Meijer, O.G., & Beek, P.J. (2004) PCA in studying coordination and variability: a tutorial. *Clinical Biomechanics, 19*(4), 415–428.

Davids, K. & Araújo, D. (2010) The concept of 'Organismic Asymmetry' in sport science. *Journal of Science and Medicine in Sport, 13*, 663–640.

Davids, K., Araújo, D. & Shuttleworth, R. (2005) Applications of Dynamical Systems Theory to Football. In T. Reilly, J. Cabri & D. Araújo (Eds.), *Science and Football V* (pp. 547–560). Routledge: London.

Davids, K., Button, C., Araújo, D., Renshaw, I. & Hristovski, R. (2006) Movement Models from Sports Provide Representative Task Constraints for Studying Adaptive Behavior in Human Movement Systems. *Adaptive Behavior, 14*(1), 73–95.

Duarte, R., Araújo, D., Correia, V. & Davids, K. (2012) Sport teams as superorganisms: Implications of sociobiological models of behaviour for research and practice in team sports performance analysis. *Sports Medicine, 42*(8), 633–642.

Duarte, R., Araújo, D., Correia, V., Davids, K., Marques, P. & Richardson, M. (2013) Competing together: Assessing the dynamics of team-team and player-team synchrony in professional association football. *Human Movement Science, 32*, 555–566.

Duarte, R., Araújo, D., Freire, L., Folgado, H., Fernandes, O. & Davids, K. (2012) Intra- and inter-group coordination patterns reveal collective behaviors of football players near the scoring zone. *Human Movement Science, 31*(6), 1639–1651.

Duch, J., Waitzman, J. & Amaral, L. (2010) Quantifying the performance of individual players in a team activity. *PloS One, 5*(6), e10937.

Eccles, D. (2010) The coordination of labour in sports teams. *International Review of Sport and Exercise Psychology, 3*(2), 154–170.

Edelman, G.M. & Gally, J.A. (2001) Degeneracy and complexity in biological systems. *Proc Natl Acad Sci, 98*(24), 13763–8.

Fajen, B.R., Riley, M.A. & Turvey, M. (2009) Information affordances, and the control of action in sport. *Int J Sport Psychol, 40*(1), 79–107.

Fewell, J.H., Armbruster, D., Ingraham, J., Petersen, A. & Waters, J.S. (2012) Basketball Teams as Strategic Networks. *PLoS ONE, 7*(11), e47445.

Folgado, H., Lemmink, K., Frencken, W. & Sampaio, J. (2012) Length, width and centroid distance as measures of teams tactical performance in youth football. *European Journal of Sport Science,* iFirst article, 1–6.

Fonseca, S., Diniz, A. & Araújo, D. (2014) The measurement of space and time in evolving sport phenomena. In K. Davids, R. Hristovski, D. Araújo, N. Balagué, C. Button & P. Passos (Eds.), *Complex Systems in Sport.* Routledge: London.

Fonseca, S., Milho, J., Travassos, B. & Araújo, D. (2012) Spatial dynamics of team sports exposed by Voronoi diagrams. *Human Movement Science, 31*(6), 1652–1659.

Frank, T.D. & Richardson, M.J. (2010) On a test statistic for the Kuramoto order parameter of synchronization: An illustration for group synchronization during rocking chairs. *Physica D, 239*, 2084–2092.

Frencken, W. & Lemmink, K. (2008) Team kinematics of small-sided soccer games: a systematic approach. In T. Reilly & F. Korkusuz (Eds.), *Science and Soccer VI* (pp. 161–6). Routledge: London.

Frencken, W., Lemmink, K., Delleman, N. & Visscher, C. (2011) Oscillations of centroid position and surface area of soccer teams in small-sided games. *European Journal of Sport Science, 11*(4), 215–223.

Gelfand, I.M. & Tsetlin, M.L. (1966/1971) On mathematical modeling of the mechanisms of the central nervous system. In I.M. Gelfand, V.S. Gurfinkel, S.V. Fomin & M.L. Tsetlin (Eds.), *Models of the Structural-Functional Organization of Certain Biological Systems* (pp. 9–26) (1971 English edition). MIT Press: Cambridge, MA.

Gibson, J. (1979) *The ecological approach to visual perception.* Lawrence Erlbaum Associates: Hillsdale, NJ.

Glazier, P.S. (2010) Game, set and match? Substantive issues and future directions in performance analysis. *Sports Medicine, 40*(8), 625–634.

Gréhaigne, J.-F. & Godbout, P. (2013) Collective variables for analysing performance in team sports. In T. McGarry, P. O'Donoghue & J. Sampaio (Eds.), *Routledge handbook of sports performance analysis* (pp. 101–114). Oxon: Routledge: Oxon.

Gréhaigne, J.F., Godbout, P. & Zera, Z. (2011) How the "rapport de forces" evolves in a soccer match: the dynamics of collective decisions in a complex system. *Revista de Psicología del Deporte, 20,* 747–765.

Grund, T.U. (2012) Network structure and team performance: The case of English Premier League soccer teams. *Social Networks, 34*(4), 682–690.

Gyarmati, L., Kwak, H. & Rodriguez, P. (2014) Searching for a Unique Style in Soccer. *Proc. 2014 KDD Workshop on Large-Scale Sports Analytics,* arXiv: 1409.0308.

Johnson, J.H. & Iravani P. (2007) The Multilevel Hypernetwork Dynamics of Complex Systems of Robot Soccer Agents. *ACM Transactions on Autonomous and Adaptive Systems, 2*(2), 1–23.

Kelso, J.A.S. (1995) *Dynamic Patterns: The Self-Organization of Brain and Behavior.* MIT Press: Cambridge, MA.

Kuznetsov, N., Bonnette, S. & Riley, M.A. (2013) Nonlinear time series methods for analyzing behavioral sequences. In K. Davids, R. Hristovski, D. Araújo, N. B. Serre, C. Button & P. Passos (Eds.), *Complex Systems in Sport.* Routledge: London.

Latash, M.L. (2008) *Synergy.* Oxford University Press: Oxford.

Latash, M.L., Scholz, J.P. & Schöner, G. (2002) Motor control strategies revealed in the structure of motor variability. *Exerc. Sport Sci. Rev., 30,* 26–31.

Mason, P.H. (2010) Degeneracy at multiple levels of complexity. *Biol Theory, 5*(3), 277–88.

McGarry, T. (2009) Applied and theoretical perspectives of performance analysis in sport: Scientific issues and challenges. *International Journal of Performance Analysis of Sport, 9,* 128–140.

McGarry, T., Perl, J. & Lames, M. (2014) Team sports as dynamical systems. In K. Davids, R. Hristovski, D. Araujo, N. Balague Serre, C. Button & P. Passos (Eds.), *Complex Systems in Sport* (pp. 208–226). Routledge: London.

Moura, F., Martins, L., Anido, R., Barros, R. & Cunha, S. (2012) Quantitative analysis of Brazilian football players' organisation on the pitch. *Sport Biomechanics, 11*(1), 85–96.

Newman, M. (2003) The Structure and Function of Complex Networks. *SIAM Review, 45*(2), 167–256.

Passos, P., Araújo, D. & Volossovitch, A. (2016/in press) *Performance Analysis in Team Sports.* Routledge: London.

Passos, P., Davids, K., Araújo, D., Paz, N., Minguens, J. & Mendes, J.F.F. (2011) Networks as a novel tool for studying team ball sports as complex social systems. *Journal of Science and Medicine in Sport, 14*(2), 170–176.

Reed, E.S. (1996). *Encountering the World: Toward an Ecological Psychology.* Oxford University Press: Oxford.

Richardson, M.J., Shockley, K., Fajen, B.R., Riley, M.A. & Turvey, M.T. (2008) Ecological psychology: Six principles for an embodied-embedded approach to behavior. In P. Calvo & T. Gomila (Eds.), *Handbook of cognitive science: An embodied approach* (pp. 161–187). Elsevier: New York.

Richardson, M., Marsh, K. & Baron, R. (2007) Judging and Actualizing Intrapersonal and Interpersonal Affordances. *Journal of Experimental Psychology: Human Perception and Performance*, *33*(4), 845–859.

Riley, M., Richardson, M., Shockley, K. & Ramenzoni, V. (2011) Interpersonal synergies. *Frontiers in Psychology*, *2*(38), 1–7.

Sampaio, J. & Maçãs, V. (2012) Measuring tactical behaviour in football. *International Journal of Sports Medicine*, *33*(5), 395–401.

Scholz, J.P. & Schöner, G. (1999) The uncontrolled manifold concept: identifying control variables for a functional task. *Exp. Brain Res.*, *126*, 289–306.

Seifert, L., Button, C. & Davids, K. (2013) Key properties of expert movement systems in sport: An ecological dynamics approach. *Sports Medicine*, *43*, 167–172.

Silva, P., Duarte, R., Sampaio, J., Aguiar, J., Davids, K., Araújo, D. & Garganta, J. (2014a) Field dimension and skill level constrain team tactical behaviours in small-sided and conditioned games in football. *Journal of Sports Sciences*, *32*(20), 1888–1896.

Silva, P., Garganta, J., Araújo, D., Davids, K. & Aguiar, P. (2013) Shared knowledge or shared affordances? Insights from an ecological dynamics approach to team coordination in sports. *Sports Medicine*, *43*, 765–772.

Silva, P., Travassos, B., Vilar, L., Aguiar, P., Davids, K., Araújo, D. & Garganta, J. (2014b) Numerical Relations and Skill Level Constrain Co-Adaptive Behaviors of Agents in Sports Teams. *PLoS ONE*, *9*(9): e107112.

Silva, P., Chung, D., Carvalho, T., Cardoso, T., Aguiar, P., Davids, K., Araújo, D. & Garganta, J. (in press) Practice effects on emergent intra-team synergies in sports teams. *Human Movement Science*.

Stergiou, N., Buzzi, U., Kurtz, M. & Heidel, J. (2004) Nonlinear tools in human movement. In N. Stergiou (Ed.), *Innovative analysis of human movement* (pp. 63–90). Human Kinetics: Champaign, IL.

Travassos, B., Araújo, D., Duarte, R. & McGarry, T. (2012) Spatiotemporal coordination patterns in futsal (indoor football) are guided by informational game constraints. *Human Movement Science*, *31*, 932–945.

Travassos, B., Araújo, Duarte, Vilar, L. & McGarry, T. (2011) Interpersonal coordination and ball dynamics in futsal (indoor football). *Human Movement Science*, *30*(6), 1245–1259.

Turvey, M.T. and Carello, C. (1996) *Dynamics of Bernstein's levels of synergies*. In M.L. Latash and M.T. Turvey (Eds.), *Dexterity and Its Development* (pp. 339–377). Erlbaum: Mahwah, NJ.

Vereijken, B., Van Emmerik, R.E., Whiting, H. & Newell, K.M. (1992) Free(z)ing degrees of freedom in skill acquisition. *J Motor Behav.*, *24*, 133–142.

Vilar, L., Araújo, D., Davids, K. & Bar-Yam, Y. (2013) Science of winning soccer: Emergent pattern-forming dynamics in association football. *Journal of Systems Science and Complexity*, *26*(1), 73–84.

Vilar, L., Araújo, D., Davids, K. & Button, C. (2012) The role of ecological dynamics in analysing performance in team sports. *Sports Med*, *42*(1), 1–10.

Yue, Z., Broich, H., Seifriz, F. & Mester, J. (2008) Mathematical analysis of a soccer game. Part i: Individual and collective behaviours. *Studies in Applied Mathematics*, *121*, 223–243.

13 Interpersonal coordination in competitive sports contexts

Martial arts

Yuji Yamamoto, Motoki Okumura,
Keiko Yokoyama and Akifumi Kijima

Introduction

Kendo as martial arts

Kendo, or Japanese fencing, a modern Japanese martial art that uses a bamboo sword (*shinai*) and protective armor in a square court with sides measuring nine to eleven metres, is said to have been originally practiced as swordsmanship (*kenjutsu*) with traditional Samurai spirits. Today it is widely practiced across Japan and many other nations. Generally, performance constraints require the kendo player to make instantaneous decisions and execute appropriate motor behaviours in response to diverse environments. A player must simultaneously strike an opponent and avoid an opponent's counterstrike. Scoring a point (*ippon*) in a kendo competition requires an accurate strike on the opponent with the uppermost third, or the top 0.3–0.4m, of the total length of the *shinai* (which is 1.2 metres in total). Generally, a maximum of two to three points are scored during a five-minute match between experts, and the strike movement, from the start of the upward swing of the *shinai* or the start of the forward movement of

Figure 13.1 Kendo match. © Gekkan Kendo Nippon.

the right foot to the end of the downward swing of the *shinai*, requires less than 0.4 seconds and is executed from an average interpersonal distance of 2.37 metres (Okumura et al., 2012). Thus, a split-second offensive or defensive move may decide the outcome of a match. A player must carefully and constantly monitor, maintain and change interpersonal distances to balance the gain/loss of offensive and defensive behaviour. Thus, according to offensive-defensive trade-offs, if the interpersonal distance decreases or increases, the potential offensive gain or defensive loss simultaneously increases or decreases, respectively, as a function of reaction and movement times.

Theoretical rationale

Oscillatory dynamics

Interpersonal coordination has been examined from the perspective of two coupled oscillators. This idea originated with the Haken-Kelso-Bunz model (Haken et al., 1985) of interlimb coordination and was then extended to interpersonal coordination (Schmidt et al., 1990; Schmidt & Turvey, 1994). Schmidt and O'Brien (1997) introduced the relative phase region analysis to our understanding of interpersonal coordination. This analysis can reveal the dominant mode of synchronisation between two oscillators, from in-phase (0 degree) to anti-phase (180 degree) synchronization during a certain time period (Coey et al., 2011; Fine & Amazeen, 2011; Miles et al., 2011; Richardson et al., 2005).

This model also has been applied to one-on-one games such as tennis (Carvalho et al., 2013; Lames, 2006; Palut & Zanone, 2005) and squash (McGarry et al., 1999); these studies found that players' movements while taking turns hitting a ball frequently switched between in- and anti-phase synchronization in the direction of the short axis of the court. Chow et al. (2014) also proposed treating the speed scalar product as a collective variable to describe different patterns in a badminton game. Dietrich et al. (2010) modelled variations in the interpersonal distance between two players in kendo as a linear mass spring pendulum. In that study, it was assumed that the two players' movements were mechanically coupled and simulated a linear coupled harmonic oscillator that integrated the intensity of the players' strategy involving stepping backward from and forward toward opponents.

In court-net sports, the ball is considered a physical linkage that constrains the opponent's movements. These systems of two or more players can be regarded as strongly coupled oscillators. Although the player needs to co-adapt to the opponent's movements in kendo, the two players do not join together physically and can move freely around each other. Consequently, more accurately, the system of two kendo players can be considered as weakly coupled nonlinear oscillators.

Return map analysis

Of additional concern in understanding performance dynamics in martial arts is capturing the abrupt switching among distinct movement patterns (such as sudden blocking or attacking manoeuvres) that punctuate competitive match play. Generally, the movements we observe during sporting activities appear to be complex continuous phenomena. However, the time series of these movements is not one of a harmonic oscillation but of continuous abrupt switching among several movement patterns. Moreover, the observed timing of events in these systems is aperiodic. The timing of events evolves from one unstable periodicity to another. Furthermore, these unstable periodicities are characterized by recurring patterns that can be quantitatively understood by examining the relationship between the timing of sequential events. This can be visualised using a plot, which is a type of a return map used in various fields (Garfinkel et al., 1992; Schiff et al., 1994). The most famous return map is a Lorentz map, which is derived from a three-dimensional Lorentz equation using inter-peak intervals. The Lorentz map provides short-term predictions about chaotic systems, and a return map analysis can reveal the regularity underlying a complex continuous time series.

Such a map plots the present peak X_n versus the next peak X_{n+1}. Periodicities are revealed on such a plot as intersections with the line of identity $X_n = X_{n+1}$. These intersections are known as an attractive fixed point and repellers or saddle points. These attractive fixed points are deterministically approached from a direction called the stable direction or manifold, and the repellers diverge from these attractive fixed points along the unstable direction or manifold as a linear function. Theoretically, we postulated the linear function $X_{n+1} = aX_n + b$. The intersections can be classified into two properties depending on the absolute value of a. When a is less than 1, $|a| < 1$, then the intersection is considered to be an attractive fixed point (i.e. an "attractor"). When the absolute value of a is more than 1, $|a| > 1$, then the intersection is referred to as a repellent fixed point (i.e. a "repeller"). An attractor can be further classified into two types. When $0 < a < 1$, the trajectories are asymptotically close to the attractor. When $-1 < a < 0$, the trajectories are rotationally close to the attractor. A repeller also has two types of trajectories: $1 < a$, and $a < -1$, corresponding to asymptotical and rotational trajectories respectively. Trajectories also approach and diverge from points that do not cross the line $X_n = X_{n+1}$. We can postulate that these functions, an exponential function, $X_{n+1} = b \ exp(aX_n)$, and a logarithmic function, $X_{n+1} = a \log X_n + b$, represent intermittency.

In following sections two experiments are summarized that measured competitive kendo match play. The first experiment was primarily concerned with determining if interpersonal dynamics of expert kendo practitioners can be understood when considered as a system of weakly coupled nonlinear oscillators. The second experiment applied a return map analysis for identifying switching amongst distinct movement patterns during kendo match play. Of additional interest in this experiment was to consider the skill effects on measures of the aforementioned parameters.

Research results

Perceptions of interpersonal distance and abrupt mode switches of synchronization

We observed twelve kendo matches among six expert members of a university kendo team; all matches followed official kendo rules. The movements of players were recorded during matches using an optical motion capture system (100Hz, OQUS300, Qualysis Inc. Sweden). Additionally, measurements were taken after all matches to analyse the players' ability and the physical constraints of a strike movement. The measurement involved a player holding a *shinai* and stepping toward and away from an opponent who was standing and also holding a *shinai*. The player would then determine and stop at his own striking distance, which was considered a possible striking distance. The trajectories of the two players' head positions were expressed as a time-dependent vector $A : [x(t), y(t)]$ and $B : [u(t), v(t)]$ (hereafter, players A and B). These time-series vectors were calculated using software (Qualysis Track Manager, Qualysis, Inc.) and flattened using a fourth-order Butterworth filter with a cut-off frequency of 6Hz. The time series for the Euclidean distance $D(t)$ between two players was calculated using the following equation:

$$D(t) = \sqrt{\left(x(t) - u(t)\right)^2 + \left(y(t) - v(t)\right)^2}$$ (Eq. 1)

$D(t)$ was calculated for the entire duration of each of the twelve matches.

Step toward-away velocity (SV) was calculated to estimate the movement of each player (Kijima et al., 2012). To calculate these variables, the displacement vector from time t−1 to time t+1 (the sampling frequency was 100Hz and the time lag in real time was 0.02s) was defined as $A_{t-1\rightarrow t+1} = [x(t+1) - x(t-1), y(t+1) - y(t-1)]$ for player A and as $B_{t-1\rightarrow t+1} = [u(t+1) - u(t-1), v(t+1) - v(t-1)]$ for player B. When we postulated that the movement of A was a uniform linear motion, the scalar, $|P_A(t)|$ was calculated as the half of the projection of vector of $A_{t-1\rightarrow t+1}$ to a linked vector, $L_A(t) = [u(t) - x(t), v(t) - y(t)]$ with the direction from the position of A to that of the position of B at time t. The scalar $|PB(t)|$ was also obtained as half of the projection of the vector of $B_{t-1\rightarrow t+1}$ to a linked vector, $L_B(t) = [x(t) - u(t), y(t) - v(t)]$ with equations (Eq. 2) and (Eq. 3):

$$|P_A(t)| = \frac{A_{t-1\rightarrow t+1} \cdot L_A(t)}{2|L_A(t)|}$$ (Eq. 2)

$$|P_B(t)| = \frac{B_{t-1\rightarrow t+1} \cdot L_B(t)}{2|L_B(t)|}$$ (Eq. 3)

The scalars $|P_A(t)|$ and $|P_B(t)|$ were then defined as SV for players A and B (i.e. SV_A and SV_B respectively). These values represent each player's velocity while stepping toward ($SV < 0$) or away from ($0 < SV$) the opponent.

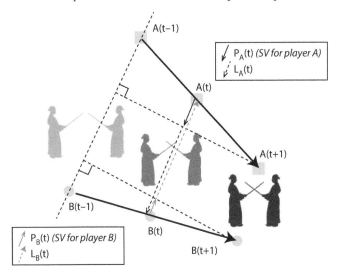

Figure 13.2 Schematic representation of step toward-away velocity (SV). In this example, player A moved from position $A(t-1)$ to $A(t+1)$ during t time from $t-1$ to $t+1$. The *SV* for player A is depicted as $|P_A(t)|$ and was defined as the projection of the vector $A(t-1) \rightarrow A(t+1)$ to vector $L_A(t)$.

To calculate the lag between both players' phases [$\Delta\phi_{AB}(t)$], we deconstructed each player's *SV* time series into real and imaginary parts using the Hilbert transform, formulated as equations (Eq. 4) and (Eq. 5):

$$SV_A(t) = s_A(t) + is_{H_A}(t) \qquad\qquad\qquad \text{(Eq. 4)}$$

$$SV_B(t) = s_B(t) + is_{H_B}(t) \qquad\qquad\qquad \text{(Eq. 5)}$$

Finally, $\Delta\phi_{AB}(t)$ was obtained by equation (Eq. 6) (Boashash, 1992a, 1992b):

$$\Delta\phi_{AB}(t) = \left| \tan^{-1} \frac{s_{H_A}(t)\, s_B(t) - s_{H_B}(t)\, s_A(t)}{s_A(t)\, s_B(t) + s_{H_A}(t)\, s_{H_B}(t)} \right| \qquad \text{(Eq. 6)}$$

Figure 13.3 shows the average frequency of interpersonal distance in twelve matches. We observed two peaks, at 1.0–1.1 metres and at 2.7–2.8 metres. The one-way repeated-measures analysis of variance (ANOVA) of distance regions revealed a significant main effect [$F(2.84, 31.23) = 58.03$, $p < 0.01$], and the results of the multiple-comparisons analysis are shown in Figure 13.3. The frequencies around these peaks were 11.6% at 0.8–1.3 m and 56.5% at 2.4–3.1 m, which accounted for 68.1% of the total. The data clearly demonstrate that the players tended to stay within their preferred interpersonal distances and executed their offensive and defensive movements during these matches based on these distances.

The interpersonal distance around 1.0–1.1 metres represented a situation in which the distances between two opponents were at their closest and where they

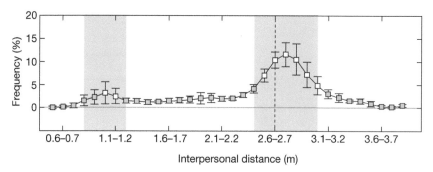

Figure 13.3 Frequencies of interpersonal distances. The grey boxes indicate distances that were significantly less frequent than were those within the range of 2.5–2.8 metres. The dotted line indicates the average of the possible striking distances. Hatched areas show preferred interpersonal distances.

continued contacting each other (called "*tsubazeriai*"), which is similar to clinches in boxing matches. In these situations, the player with a long *shinai* had to move backward to strike the opponent. Our data showed 1.0 ±1.29 instances of striking and 0 points per match at this distance. In fact, the players executed few striking movements in this context, and they primarily separated and returned at around 2.7–2.8 metres. Thus, it may be that executing offensive techniques and scoring a point by stepping away are less likely at around 1.0–1.1 metres than at around 2.7–3.0 metres.

The interpersonal distance between two players separated by about 2.7–2.8 metres represented a neutral posture that allowed a balance between offensive and defensive techniques while maintaining of a ready stance ("*kamae*"). To efficiently strike the opponent, the player needed to move forward from approximately 2.7–2.8 metres. On the defensive end, the player trying to prevent his opponent from striking needed to maintain a safety space with respect to the interpersonal distance. In general, the smaller the interpersonal distance, the less time needed for the striking movement. With this rule in mind, the task difficulty for the offensive players was gradually reduced as they approached from a far distance. Conversely, as the interpersonal distance gradually decreased, the task difficulty for the defensive players gradually increased owing to the reduction in the available reaction and movement times. In this situation, the gain and loss of offensive and defensive movements resulted in a disturbance of the normal balance. In general, the players were able to continuously maintain the interpersonal distance to balance gains and losses.

The players' possible striking distance averaged 2.65 metres, which is indicated by the dotted line in Figure 13.3. A striking distance of 2.65 metres was almost consistent with the preferred interpersonal distance. However, the 0.05–0.15 metre difference between the peak value of the preferred distance and the possible striking distance indicates that players tended to engage in subtle manoeuvres while maintaining a safety margin for defensive movements (Higuchi et al., 2011; Rand et al., 2007). In this study, an interpersonal distance of around 2.7–2.8 metres was optimal for subtle manoeuvres in preparation for the next striking or defensive movement. Thus, this distance constitutes the preferred interpersonal distance for maintaining a balance between the gain and loss of offensive and defensive movements. Hence, the remaining distance, 1.1–2.7 metres, may be the distance at which players are forced to quickly return to the distances that they prefer and that are appropriate for the task constraint.

The average frequencies of the relative phase during the twelve matches are presented in Figure 13.4a. We observed a slight tendency for players' movements to be synchronized in the anti-phase as a whole. A one-way repeated-measures ANOVA of phase ranges revealed a significant main effect [$F(1.60, 17.57) = 23.35, p < .01$]. The results of multiple comparisons are presented in Figure 13.4a.

Relative phase was altered by a change in interpersonal distance (Figure 13.4b to k). A two-way repeated-measures ANOVA of interpersonal distance × relative phase regions revealed a significant interaction [$F(6.86, 75.43) = 8.31, p < .01$]. The results of the simple main-effect tests are given in Figure 13.4b to k. These results indicate that the relative phase showed clear trends toward anti-phase synchronization at near distances (i.e. those closer than 2.7 metres) and toward in-phase synchronisation at far distances (i.e. those farther than 3.0 metres). This in- and anti-phase transition was characterized by a regular pattern in that distance elicited the phase transition and corresponded to the preferred interpersonal distance at around 2.7–2.8 metres.

We then focused on distances of 2.7–3.0 metres to fully analyse phase transitions. The frequency of the relative phase in distance per 0.1-metre range is indicated in Figure 13.5. A two-way repeated-measures ANOVA of interpersonal distance × relative phase regions revealed a significant interaction [$F(4.85, 52.97) = 9.05, p < .01$]. The results of the simple main-effect tests are given in Figure 13.5. Surprisingly, the relative phase at this important distance transformed anti-phase synchronisation into in-phase synchronisation based on a minimal 0.1-metre difference in distance. These results clearly show that a higher anti-phase synchronisation occurred at 2.7–2.8 metres than at other distances, and this switched to in-phase synchronization at 2.9–3.0 metres, setting a boundary at 2.8–2.9 metres. Thus, 2.8–2.9 metres was the critical interpersonal distance needed to change the kendo players' relative phase, on a slight 0.1-metre scale.

Our analysis of these data revealed three different relationships between interpersonal distances and coordination modes. 1) At distances closer than 2.8 metres, including a possible striking distance of 2.65 metres, one player would try to execute a strike attempt and step forward while the other player would try to avoid the strike by stepping backward. This resulted in the players' coordinating their interpersonal distances in an anti-phase synchronization. 2) At distances farther than 2.9 metres the two players could not strike each other, and as a result defensive movements were not necessary. Consequently, the players would approach each other with in-phase synchronization to make it possible to strike the opponent. 3) At distances of 2.8–2.9 metres (i.e. 0.15–0.25 metres from a

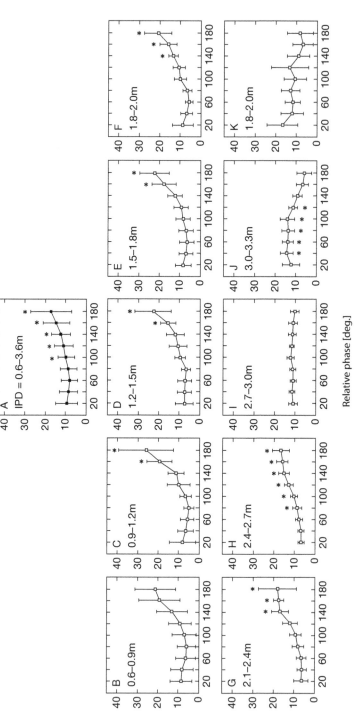

Figure 13.4 Frequency of relative phases at each interpersonal distance.

Notes: Relative phases were divided into nine ranges (20: 0–20 deg, 40: 20–40 deg, ... 180: 160–180 deg). (**A**) Total frequency at all distances (0.6–3.6 m). (**B–K**) Partial frequencies at each distance (0.6–0.9 m, 0.9–1.2 m, ... 3.3–3.6 m). The asterisks (* $p < .05$) above the line indicate distances that were significantly more frequent than were those at 160 deg and/or 180 deg, and the asterisks below the line indicate distances that were significantly more frequent than were those at 20 deg and/or 40 deg.

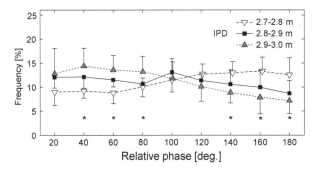

Figure 13.5 Frequency of relative phases at an interpersonal distance of 2.7–3.0 m. The asterisks below the lines indicate significant differences between 2.7–2.8 m and 2.9–3.0 m in each relative phase range. IPD = interpersonal distance.

possible striking distance), interpersonal distance was critical. At these distances, players had to coordinate their movements using both anti-phase synchronization for distances closer than 2.8 metres and in-phase synchronization for distances farther than 2.9 metres. This indicates that the players performed a variety of movements to execute subtle active and passive offensive and defensive manoeuvres at these distances. As a result, these manoeuvres seem to increase IPD frequency and to be the clearest peak of preferred interpersonal distance, around 2.7–2.8 metres.

The phase transition phenomenon demonstrated in other studies investigating behavioural dynamics in cooperative tasks can also be observed in competitive kendo matches. The coordination between two players' movements reflect open and regular switching between the in- and anti-phase synchronizations as an order parameter by changing the interpersonal distance as the control parameter. Surprisingly, the synchronization mode switched abruptly at interpersonal distances of only 0.10 metres during rapid movement in kendo. This means these players perceived and understood, in behavioural terms, the need for minute changes on a 0.10-metre scale, and consequently they regularly switched their movements to appropriate directions or synchronizations. Thus, examinations of the interpersonal coordination in martial arts such as kendo should include interpersonal distance as a key variable. However, little is known about the dynamics underlying the continuous abrupt switching behaviour observed in martial arts that require rapid decision-making and execution.

Only several basic patterns and switching dynamics

To determine how skill level affects intentional switching dynamics during interpersonal competition, we observed a time series of the interpersonal distances (IPDs) between two players based on their movement trajectories during twenty-four kendo matches, from the perspective of a hybrid dynamical system consisting of discrete or symbolic and continuous dynamical system (Yamamoto et al., 2013). Each of the six expert and six less skilled (substitute) players were matched against four different opponents of the same skill level. Following official kendo rules, each match lasted five minutes, and players' movement trajectories were recorded using an optical motion capture system at 100 Hz.

We eliminated unrelated scenes and then divided each of the twelve matches into 67 and 54 sequences involving expert and intermediate competitors, respectively. Furthermore, we divided the time-series data into discrete episodes of trials so as to include only a single striking action per trial. This was because interpersonal competition in kendo is interrupted by striking actions, and movements after striking actions were considered transitions to a new confrontation. Each trial started at the farthest interpersonal distance and ended at the nearest distance, for striking, or at the middle distance, for slow detachment (Figure 13.6c). Trials with fewer than four positive peaks of IPDs were excluded from further analysis. As a result, 184 trials involving experts and 162 trials involving intermediate players remained.

We defined displacement, $X_{IPD}(t)$, and velocity, $V_{IPD}(t)$, as the state variables of the behaviour of the system using the following equations, respectively:

$$X_{IPD}(t) = X_B(t) - X_A(t)$$
$$= \sqrt{\left(x_B(t) - x_A(t)\right)^2 + \left(y_B(t) - y_A(t)\right)^2} \qquad \text{(Eq. 7)}$$

$$V_{IPD}(t) = \left(X_{IPD}(t+1) - X_{IPD}(t-1)\right)/2 \qquad \text{(Eq. 8)}$$

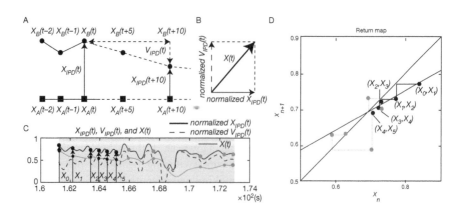

Figure 13.6 Schematic representation of state variables ($X(t)$).

Notes: (**A**) Schematic representation of $X_{IPD}(t)$ and $V_{IPD}(t)$. (**B**) Schematic representation of the state variables. (**C**) Grey, broken, and black lines show a time series of normalized $X_{IPD}(t)$, normalized $V_{IPD}(t)$, and $X(t)$, respectively, for a 12s trial with more than five peaks. The black and grey circles show the corresponding values of $X(t)$ for the peaks of $X_{IPD}(t)$. (**D**) Return map of the time series of the observed data, X_n versus X_{n+1} using the amplitude of $X(t)$ at the peaks of $X_{IPD}(t)$ corresponding to the series of points (black and grey circles) in panel shown in **C**.

The trajectory of the player's head position was expressed as time-dependent vectors $X_A(t) = [x_A(t), y_A(t)]$ for player A and $X_B(t) = [x_B(t), y_B(t)]$ for player B. Both $X_{IPD}(t)$ and $V_{IPD}(t)$ were normalised between 0 and 1, and state variables $X(t)$ were calculated as composite vectors of two time-dependent vectors using the following equation (Figure 13.6b):

$$X(t) = \sqrt{\left(normalized\ X_{IPD}(t)\right)^2 + \left(normalized\ V_{IPD}(t)\right)^2} \qquad \text{(Eq. 9)}$$

A total of 346 trials with more than five peaks of $X(t)$ were fitted to three types of function: $X_{n+1} = aX_n + b$, $X_{n+1} = b\exp(aX_n)$, and $X_{n+1} = a\log X_n + b$ (Figure 13.7). The number of fitted points was altered from three to six on the return map using moving windows from the beginning of the data to the end of each trial.

We found 291 trials that could be fitted by the candidate functions; 162 of these trials included expert competitors, and 129 trials included intermediate competitors. In total, 485 series of points in these 291 trials were well fitted to the functions: 284 trajectories were revealed as attractors, 146 trajectories were fitted as repellers, and 55 trajectories were identified as intermittency. All six types of candidate functions could be found using three-, four-, and five-point fitting; 16 trajectories were fitted using six points for four types of function. Examples of well-fitted series of points for each function are shown in Figure 13.7a"–f".

We found that 121 trials switched among two to nine different functions in each trial; 80 trials switched between two functions, 22 trials switched among three functions, 11 trials switched among four functions, six trials switched among five functions, and one trial switched among seven and nine functions (Figure 13.8).

These results suggest that complex movements occurring during the interpersonal competition of a kendo match may be generated by simple rules that involve attraction toward or repulsion from fixed attractive and/or repellent points. We identified two discrete states in each histogram of the return maps using the following thresholds as the minimum value for each match: the "farthest apart" high-velocity state (F), and the "nearest (closest) together" low-velocity state (N) (Figure 13.9a–d). Thus, we identified four trajectories, $\{X_n = F,\ X_{n+1} = F\}$, $\{X_n = N,\ X_{n+1} = N\}$, $\{X_n = F,\ X_{n+1} = N\}$ and $\{X_n = N,\ X_{n+1} = F\}$, as second-order transitions. The state transition diagrams for experts and intermediate players are shown in Figures 13.9e and 13.9f, respectively.

The conditional probabilities for second-order state transitions were calculated for each skill level. The differences in transition probabilities between expert and intermediate players for each discrete state were significant according to Fisher's exact test (F: $p < 1.45 \times 10^{-12}$, N: $p < 1.08 \times 10^{-5}$). The offensive and defensive manoeuvres of the experts were more often in the "farthest apart" high-velocity F-state. In contrast, those of intermediate players were more likely to be found in the "nearest (closest) together" low-velocity N-state. Two peaks can be observed in each histogram in the "farthest apart" high-velocity state (Fig. 13.9c–d). This indicates that the current discrete state has two second-order states that depend on the previous state: $\{X_{n-1} = F,\ X_n = F\}$ (FF) and $\{X_{n-1} = N,\ X_n = F\}$ (NF). The

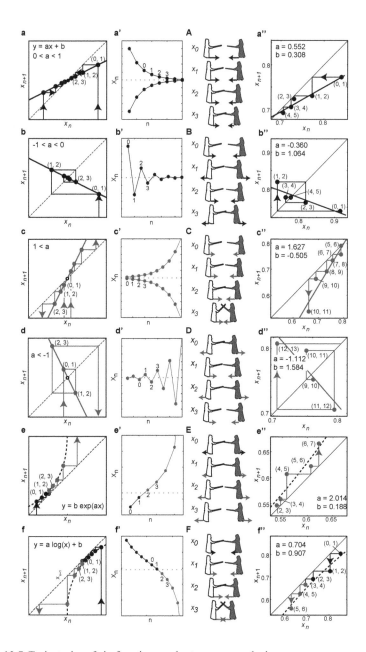

Figure 13.7 Trajectories of six functions and return map analysis.

Notes: (**a–d**) Linear functions, $X_{n+1} = a X_n + b$, with four different slopes for $0 < a < 1$, $-1 < a < 0$, $1 < a$, and $a < -1$, respectively. (**e**) Exponential function, $X_{n+1} = b \exp(a X_n)$. (**f**) Logarithmic function, $Xn+1 = a \log(Xn) + b$. (**a'–f'**) Corresponding time series for each function (**a–f**), respectively. (**A–F**) Corresponding movements for each function (**a–f**). (**a"–f"**) Examples of well–fitted series of points by each function (**a–f**).

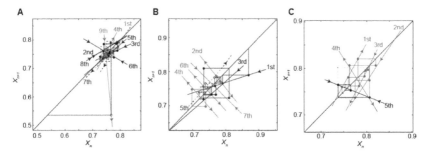

Figure 13.8 Examples of return maps and switching functions from a series of points in a trial. Black lines show attractors, and grey lines show repellers. Dotted lines show intermittency in all panels.

Notes: (**A**) Nine functions, (**B**) seven functions, and (**C**) five functions switched sequentially.

probabilities of third-order trajectories between four second-order states were calculated for each skill level (Figure 13.9g–h) for experts and intermediates. Experts exhibited higher probabilities for the third-order transitions to the state of the "farthest apart" high velocity in all sub-states. On the other hand, intermediate players showed higher transition probabilities to the state of the "nearest together" low velocity in all sub–states (FF: $p < 1.97 \times 10^{-9}$, NF: $p = .16$, NN: $p < .04$, FN: $p < 2.3 \times 10^{-4}$). These results reveal that the second-order trajectories between two discrete states, that is, the "farthest apart" high velocity and the "nearest together" low velocity, and the third-order trajectories among four discrete states, depend not only on the current state but also on the previous state. This suggests that these state transitions of offensive and defensive manoeuvres have a hierarchical structure in kendo.

The return map analysis revealed that continuous interpersonal competition, which may appear to be quite complex, could be expressed in terms of six discrete dynamics represented by simple functions. The state transition analysis revealed second-order transition probabilities between two states: the "farthest apart" high-velocity state (F) and the "nearest (close) together" low-velocity state (N). These two states have a hierarchical structure that depends on the previous state. Third-order transition probabilities also revealed differences between expert and intermediate competitors. This result suggests that intentional switching dynamics is embedded in complex continuous interpersonal competitions (such as a martial arts competition) and is thus better described as a hybrid dynamical system consisting of higher discrete and lower continuous modules connected via a feedback loop (Nishikawa & Gohara, 2008a, 2008b).

Applications

The behaviour of two individuals engaged in interpersonal coordination in martial arts can be depicted as a synchronization phenomenon involving two weakly coupled oscillators using the interpersonal distance as a control parameter.

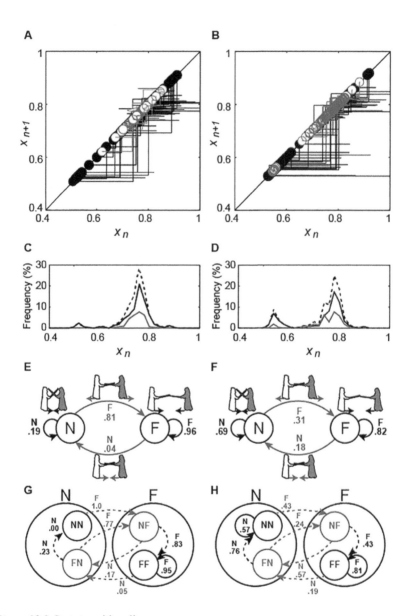

Figure 13.9 State transition diagrams.

Notes: (**A, B**) Return maps were plotted using observed points as four different linear functions of an attractor (black) and a repeller (grey) for expert and intermediate competitors respectively. The circles show crossing points with the line of identity, $X_n = X_{n+1}$. (**C, D**) Black, grey and dotted lines show histograms of crossing points for an attractor, a repeller and the sum of these respectively. (**E, F**) Second-order state transition diagrams with the conditional probabilities consisting of the "farthest apart" high-velocity states (F) and the "nearest together" low-velocity state (N) for expert and intermediate competitors respectively. (**G, H**) The third-order state transition diagrams comprised four second-order sub-states.

The velocity for stepping toward/away from an opponent is an important variable that describes individual behaviour in martial arts, because both offensive and defensive movements must be executed simultaneously. This analysis revealed that experts show amazing perception and action involving minute changes in interpersonal distance. However, the behaviour of individuals and interpersonal coordination do not constitute a harmonic oscillation such as rhythmic finger movements (Haken et al., 1985). Other approaches are required to understand the underlying abrupt switching dynamics in martial arts. One possible approach is a return map analysis using interpersonal distance as state variables; this enables us to extract several simple rules from complex phenomena and to understand that these phenomena may emerge from continuous switching among several patterns.

In martial arts, the goal is to beat the opponent. The dynamics in a competitive coordination task such as martial arts differs from that in a coordination task, and we need to focus on the behaviour that emerges from competition. However, we understand stable states in only the context of interpersonal coordination, the coordination phase (Dietrich et al., 2010). To achieve their goal, players must transition from a stable to an unstable state. The nature of this abrupt transition or bifurcation remains unclear, and the hysteresis and bifurcation parameter should be clarified in the near future.

Interpersonal distance is regarded as a control parameter that describes a partial system for each individual. In addition, this variable can be regarded as a state variable or an order parameter that describes interpersonal coordination as a whole system. Interpersonal distance is a key variable for both a partial system and a whole system in efforts to understand the dynamics of martial arts, suggesting that research regarding interpersonal coordination is also relevant to part-whole relations.

References

Boashash, B. (1992a) Estimating and interpreting the instantaneous frequency of a signal. Part 1: fundamentals. *Proceedings of the IEEE*, *80*, 520–538.

Boashash, B. (1992b) Estimating and interpreting the instantaneous frequency of a signal. Part 2: algorithms and applications. *Proceedings of the IEEE*, *80*, 540–568.

Carvalho, J., Araújo, D., Travassos, B., Esteves, P., Pessanha, L., Pereira, F. & Davids, K. (2013) Dynamics of players relative positioning during baseline rallies in tennis. *Journal of Sports Sciences*, *31*, 1596–1605.

Chow, J.Y., Seifert, L., Hérault, R., Chia, S.J.Y. & Lee, M.C.Y. (2014) A dynamical system perspective to understanding badminton singles game play. *Human Movement Science*, *33*, 70–87.

Coey, C., Varlet, M., Schmidt, R.C. & Richardson, M.J. (2011) Effects of movement stability and congruency on the emergence of spontaneous interpersonal coordination. *Experimental Brain Research*, *211*, 483–493.

Dietrich, G., Bredin, J. & Kerlirzin, Y. (2010) Interpersonal distance modeling during fighting activities. *Motor Control*, *14*, 509–527.

Fine, J.M. & Amazeen, E.L. (2011) Interpersonal Fitts' law: when two perform as one. *Experimental Brain Research, 211,* 459–469.

Garfinkel, A., Spano, M.L., Ditto, W.L. & Weiss, J.N. (1992) Controlling cardiac chaos. *Science, 257,* 1230–1235.

Haken, H., Kelso, J.A.S. & Bunz, H. (1985) A theoretical model of phase transitions in human hand movements. *Biological Cybernetics, 51*(5), 347–356.

Higuchi, T., Murai, G., Kijima, A., Seya, Y., Wagmane, J.B. & Imanaka, K. (2011) Athletic experience influences shoulder rotations when running through apertures. *Human Movement Science, 30,* 534–549.

Kijima, A., Kadota, K., Yokoyama, K., Okumura, M., Suzuki, H., Schmidt, R.C. & Yamamoto, Y. (2012) Switching dynamics in an interpersonal competition brings about 'Deadlock' synchronization of players. *PLoS ONE, 7,* e47911.

Lames, M. (2006) Modelling the interaction in game sports – relative phase and moving correlations. *Journal of Sports Science and Medicine, 5,* 556–560.

McGarry, T., Khan, M.A. & Franks, I.M. (1999) On the presence and absence of behavioural traits in sports: an example from championship squash match-play. *Journal of Sports Sciences, 17,* 297–311.

Miles, L.K., Lumsden, J., Richardson, M.J. & Macrae, C.N. (2011) Do birds of a feather move together? Group membership and behaviour al synchrony. *Experimental Brain Research, 211,* 495–503.

Nishikawa, J. & Gohara, K. (2008a) Anomaly of fractal dimensions observed in stochastically switched systems. *Physical Review E, 77,* 036210.

Nishikawa, J. & Gohara, K. (2008b) Automata on fractal sets observed in hybrid dynamical systems. *International Journal of Bifurcation and Chaos, 18,* 3665–3678.

Okumura, M., Kijima, A., Kadota, K., Yokoyama, K., Suzuki, H. & Yamamoto, Y. (2012) A critical interpersonal distance switches between two coordination modes in kendo matches. *PLoS ONE, 7,* e51877.

Palut, Y. & Zanone, P.-G. (2005) A dynamical analysis of tennis: concepts and data. *Journal of Sports Sciences, 23,* 1021–1032.

Rand, M.K., Lemay, M., Squire, L.M., Shimansky, Y.P. & Stelmach, G.E. (2007) Role of vision in aperture closure control during reach-to-grasp movements. *Experimental Brain Research, 181,* 447–460.

Richardson, M.J., Marsh, K.L. & Schmidt, R.C. (2005) Effects of visual and verbal interaction on unintentional interpersonal coordination. *Journal of Experimental Psychology: Human Perception and Performance, 31,* 62–79.

Schiff, S.J., Jerger, K., Duong, D.H., Chang, T., Spano, M.L. & Ditto, W.L. (1994) Controlling chaos in the brain. *Nature, 370,* 615–620.

Schmidt, R.C., Carello, C. & Turvey, M.T. (1990) Phase transitions and critical fluctuations in the visual coordination of rhythmic movements between people. *Journal of Experimental Psychology: Human Perception and Performance, 16,* 227–247.

Schmidt, R.C. & O'Brien, B. (1997) Evaluating the dynamics of unintended interpersonal coordination. *Ecological Psychology, 9,* 189–206.

Schmidt, R.C. & Turvey, M.T. (1994) Phase-entrainment dynamics of visually coupled rhythmic movements. *Biological Cybernetics, 70,* 369–376.

Yamamoto, Y., Yokoyama, K., Okumura, M., Kijima, A., Kadota, K. & Gohara, K. (2013) Joint action syntax in Japanese martial arts. *PLoS ONE, 8,* e72436.

14 Interpersonal coordination in competitive sports contests

Racket sports

Tim McGarry and Harjo J. de Poel

Introduction

Visual inspection suggests structure in sports contests, but identifying behavioural patterns in sports contests using a formal level of description remains challenging. Put simply, the patterns that we think we observe in sports contests have proved difficult to find in the recorded data that are supposed to describe them. That is, the reliable detection of patterned behaviours in sports contests remains equivocal. In this article, these challenges lead us to consider evidence from other research enquiries that may inform on our ongoing search for behavioural structure in sports contests. First, we trace research developments leading to the present consideration of the kinematic patterns of racket sports players as a self-organizing dynamical system, one in which the interactions of the two players produce unique yet patterned behaviours on the basis of interpersonal coordination.

In the first instance, emergent structured behaviours in sports contests are constrained by the game rules. In this regard, sports contests may be generally divided into two separate categories: score-dependent sports and time-dependent sports. Score dependent sports such as squash, tennis and badminton finish when a given final score is reached. However, time-dependent sports such as football (soccer), rugby football and basketball finish on the expiry of some fixed amount of time, thus allowing for the possibility of a tied result after regular time. Score-dependent sports are more likely to demonstrate structured sequences of action behaviours, given that possession of the sports article is usually traded between opponents in equal fashion by virtue of the game rules. Time-dependent sports invariably exhibit looser structured behaviours because of the unequal exchanges that occur as possession is won and lost between players or teams at times unspecified in the game rules. In this article, we focus on the racket sports which fall under the rubric of score-dependent sports.

Investigating action behaviours and outcomes as stochastic (random) processes

The game rules influence the behavioural structure of different sports contests by means of external constraints, thus yielding recognized behaviour patterns that characterize a given sport. However, the game rules do not prescribe the specific action behaviours produced, as the unique behaviours in any given sports contest attests. Assuming that sports players demonstrate unique playing behaviours on the basis of various factors, including past experiences producing habitual preferences, game strategies and the opposing player, it was suggested that individual players demonstrate behavioural signatures (McGarry & Franks, 1994). Indeed, the sports practice of scouting opponents suggests as much, in that information observed from a past contest is held to apply to a future one for the purpose of advance preparation. On this supposition, we undertook research in squash contests, looking to describe sports behaviours on the basis of Markov (probability) chain sequences (McGarry and Franks, 1994; 1996). Here, the term sports behaviours refer to the squash shot types (i.e. serve, drive, drop, volley etc.) and any associated outcomes (i.e. winners, errors and lets).

As noted above, while the behavioural patterns in sports contests are produced within the constraints of the game rules, these same patterns are not produced by the game (system) constraints. Instead, the patterns emerge by way of interactions among the players, with the game rules and other constraints only serving to limit the array of behavioural patterns possible at any instant. Thus, the playing dyad self-organizes within the constraints under which it competes. Other than game rules, playing behaviours are likely to be constrained by some hierarchy of customs and habits that result in a set of strategies and tactics for both players. In squash contests, for instance, each player seeks control of the T-position (approximate centre-court) while awaiting the opponent's shot. Other likely constraints include cognitive, psychological, physiological, physical and technical factors, each of which will influence the decisions and actions of players.

If the action behaviours of sports contests at any instant are uncertain, as it appears, then they may appropriately be described on the basis of chance (or probability). Thus, the action and outcome sequences comprising a squash contest can be analysed using probability structures, where the next action in sequence is linked to the former action on the basis of probability. Investigating sports behaviours in terms of probability sequences allows for managing chance so as to maximize the upsides (rewards) and minimize the downsides (risks). On this thinking, we analysed the shot selection patterns in squash contests and their consequent outcomes as a probability structure (McGarry and Franks, 1994; 1996). In this way, the frequencies of shot behaviours were quantified as probabilities, thereby establishing a "playing profile" or "playing signature" for each player. These playing profiles were then subjected to statistical analysis for evidence of signature behaviours unique to individual players.

Signature behaviours by definition require similar playing profiles for any given player observed in different contests (cf. scouting), meaning that both shot type and associated outcome probabilities should be invariant, within reason. Using χ^2 analysis, we examined the playing profiles of different squash players for these expected invariant properties, with mixed results (McGarry & Franks, 1996). In general, the playing profiles of squash players changed as the opponent changed, although the playing profiles tended to demonstrate increased invariance as the context detail for the analysis increased. Regardless, in summary, these results highlighted the importance and challenges of finding invariant features in sports behaviour, a necessary requirement if the data from sports contests are to offer useful information for future application. These findings indicate that a system description of sports contest behaviours using transition probabilities is not as straightforward as might otherwise be imagined.

In sum, sports analysts have tended to record all available data from a sports contest and looked to describe sports behaviour using various measures as indicators of sports performance. This approach assumes that each datum item has an equal weighting in terms of its information value, an assumption that appears inconsistent with the research findings, prompting the suggestion that some behavioural actions in sports contests might be more important than others (McGarry & Franks, 1996). The shot that forces an opening for a successful attack might be considered as more important, say, than a shot that trades possession in some type of prolonged stable rally exchange. If some behavioural actions in a given sequence are indeed more important than others, then some rethinking for an appropriate system description is required. In consequence, we introduced the notion of behavioural perturbations as part of a new system description for squash contests and, by extension, racket sports in general (McGarry, Kahn & Franks, 1996; 1999).

On the presence of behavioural perturbations

In his keynote address to the First World Congress of Science and Racket Sports (July 1993), Jake Downey introduced a dance analogy to describe the behavioural interactions of two players in a badminton rally. In this analogy, it is held that behavioural synchrony within the playing dyad is maintained as both players search for an opportune time to break rhythmic stability with the aim of gaining an advantage in the rally by seizing control of the behavioural exchange. Following on from this observation, we introduced the general concept of behavioural perturbations for sports contests in more formal terms, whereupon behavioural stability within a squash dyad, say, is maintained until the synchrony gets broken by way of some behavioural perturbation. (Note: the idea of behavioural perturbations is proposed as a general concept for describing sports behaviours in sports contests; it is not limited only to the racket sports.) Following perturbation, behavioural instability in the squash dyad is observed as one player looks to press the advantage while the other player seeks to recover the disadvantage by returning the dyad to behavioural stability. Thus, the squash rally transits from behavioural stability to behavioural instability and back again in time (Figure 14.1B); otherwise the behavioural instability marks the onset of a winning, losing or neutral outcome (Figures 14.1A and 14.1B). Evidence supporting this concept of behavioural perturbations in squash rallies, and more specifically the demarcations between behavioural stabilities and behavioural instabilities, was demonstrated by independent observers who produced good agreement regarding their detection (see Figures 14.1A and 14.1B, reproduced

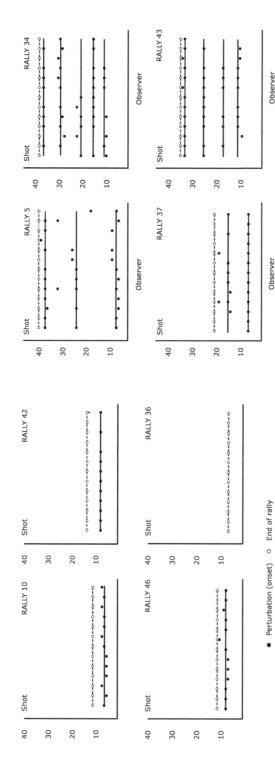

Figure 14.1 Behavioural perturbations identified by independent observers (N=13).

Notes: Each solid circle in a single panel represents an individual shot in the rally sequence identified by the independent observer. The horizontal lines represent agreement on behavioural perturbations as identified by the independent observers as indicated. Each solid circle for each independent observer represents the final shot in that rally sequence. For a given rally, single behavioural perturbation (A) and multiple behavioural perturbations (B). (Note. Figure A, Rally 36 indicates no behavioural perturbation detected by any independent observer for that rally.)

from McGarry et al., 1999). The notion that behavioural perturbations mark the boundaries between behavioural stability and behavioural instability, together with their subsequent demonstrations, led us towards considering sports contest behaviours as dynamical self-organizing systems (see McGarry et al., 1999, 2002), whereupon the spatial-temporal kinematic relations in the squash dyad are a key descriptor for system behaviour.

The idea that tendencies towards behavioural synchrony in squash dyads, with both players looking for opportunities to break synchrony to their advantage, might be restated in terms of coordinated behaviours. Here, the playing dyad produces coordinated behaviours, for the most part at least, even though squash contests are necessarily competitive and not cooperative. This said, as McGarry et al. (1999) noted, a common feature of squash contests are controlled rally exchanges to length, deep to the back of court, while waiting for an opening to attack, perhaps by means of a loose shot from the opponent. These rally exchanges might be considered cooperative with the aim of both players achieving and maintaining a manageable physiological state before and/or after interspersed anaerobic efforts following behavioural perturbations. This argument for cooperation in sport contests is consistent with a general commentary in a foreword by Yates (in Kugler & Turvey, 1987) who reflected on "the problem of coordination, competition and cooperativity present at all levels in the hierarchies within and among living organisms".

Self-organizing coordinated behaviour

In 1664, Huygens reported sympathetic behaviour between coupled pendulums swung separately but suspended from a common frame, meaning that in time the two pendulums coordinate themselves and produce in-phase or anti-phase coordination (Meijer, 2001). In-phase and anti-phase coordination constitute attractors, with the coupled pendulums drawn to one or the other attractor given the initial conditions. Thus, the coordinated behaviours of coupled pendulums emerge from within the system by virtue of common information (energy) flows. That is, the coupled pendulums self-organize by means of shared information exchange.

Relative phase reports the relative position of two points in their given cycles at any instant. In-phase (zero or unity) represents the same positions in the given cycles at any instant, whereas anti-phase (half) represents opposite positions, with other phase relations expressed within the limits of zero through unity. (Alternatively, relative phase may be expressed in degrees with in-phase and anti-phase represented as 0° or 360° and 180°, respectively.) For example, in-phase represents two pendulums at the same points in their respective cycles with both pendulums reaching the same zeniths at the same time, whereas anti-phase represents the anti-symmetric relation with the two pendulums reaching opposing zeniths at the same time.

The coordination problem: from intrapersonal to interpersonal coordination

Relative phase is a common measure used in research investigations on human rhythmic coordinated actions, as reported for example in the well-known "finger waggling" studies of Kelso and colleagues (Kelso, Holt, Rubin & Kugler, 1981; Kelso, 1984; Haken, Kelso and Bunz, 1985). Here, the experimental task was to flex and extend the index fingers of both hands in the transverse plane when paced at different oscillating frequencies. Starting with anti-phase, increasing the cycling frequencies led to increasing anti-phase coordination instabilities with spontaneous phase transitions to in-phase observed when the cycling frequency reached some critical value. Starting with in-phase, however, increasing the cycling frequencies produced no increasing in-phase coordination instabilities and hence no phase transitions. These dynamics were modelled as two coupled oscillators (pendulums) representing the two index fingers (see Haken et al., 1985), leading to a theory of human rhythmic coordinated action based on dynamical self-organizing principles. This dynamical systems theory for coordinated action behaviour was subsequently applied to other rhythmic movements including coordination between different limbs (Kelso & Jeka, 1992), multi-limb movements (Kelso, Buchanan & Wallace, 1991) and other coordination patterns (de Guzman & Kelso, 1991), including coordination processes between the swinging legs of different persons (Schmidt et al., 1990). Thus, the same system description predicated on coupled oscillator dynamics explains the coordinated behaviours for different rhythmic actions. Importantly, this dynamical description for coordinated behaviour extends from coordination within persons to coordination between persons (Schmidt et al., 1990). This latter finding prompted us to consider coupled oscillator dynamics between players in squash contests, and beyond as the theoretical underpinning for the unique but nonetheless patterned game behaviours that typify different sports and sports contests (McGarry et al., 2002).

Self-organizing dynamical systems: racket sports

The underlying premise for dynamical systems theory is that of coupled oscillators whose behaviours self-organize, and perhaps reorganize under changing system constraints, as a result of mutual influences by means of common information exchanges. Furthermore, as noted, this premise extends from intrapersonal coordination dynamics to interpersonal coordination dynamics. In considering the squash dyad behaviours in terms of stability and instability, demarcated by behavioural perturbations, we also demonstrated from single squash rallies that squash players indeed exhibit alternating oscillatory movement kinematics, as proposed, as they travel to and from the T-position to await and retrieve shots respectively (McGarry et al. 1999). Thus, given the game constraints, both players in the squash dyad produce oscillatory movements about the T-position in alternating sequence, thus yielding attraction toanti-phase as expected. From this

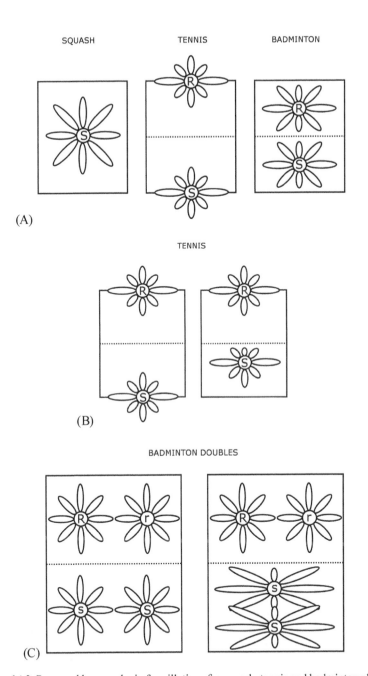

Figure 14.2 Proposed locus or loci of oscillations for squash, tennis and badminton players (A). The flower-like schematic simply demonstrated oscillatory movements about a given locus. Possible change in locus of oscillation in tennis during a rally (B). Proposed loci of oscillations for doubles badminton players, including a possible change in loci for a double pair during a rally (C).

Notes: S = Server, R = Receiver. s = Server double. r = Receiver doubles partner.

result, we proposed a common dynamical system description for racket sports – squash, tennis and badminton – and, possibly, beyond to also encompass teams sports (McGarry et al., 2002).

As reported above, preliminary analysis of squash dyad kinematic data from single rallies provided a good indication of anti-phase attraction, as hypothesized. In general terms, as one player leaves some neutral position to make a shot, the other player waits for the pending shot in the same or equivalent neutral position. This description applies equally well to the other racket sports when accounting for a common locus or separate loci of oscillations as appropriate. Here, it is acknowledged that players in squash, tennis and badminton oscillate around separate and different loci given the nature of the sports contest: for squash, the locus of oscillation is the T-position, a common locus that both players vie for control of; for tennis, the locus of oscillation is the mid-point of the baseline for each player (or thereabouts); and for badminton, the centre half-court (or thereabouts). Thus, for squash players the locus might be considered public (or shared) whereas for tennis and badminton the locus is private to each player. This same reasoning of coupled oscillator dynamics for the squash, tennis and badminton dyads was extended to include doubles play in these same sports. Now it is suggested that doubles players couple with their double (teammate) as well as with their opponents, thereby introducing the concept of multiple couplings (S-s, R-r, S-R, s-r, S-r and s-R; Figure 14.2C) as well as that of layered, or nested, couplings – a coupling of couplings, if you will (Ss-Rr). While outside the mandate of this article, this same reasoning was speculatively applied to other team sports, for example basketball and football.

Dynamical analysis of squash game behaviour

The suggestion that the playing dyads in racket sports subscribe to dynamical self-organizing principles prompted research investigations in both tennis (Palut & Zanone, 2005; Lames, 2006) and squash (McGarry, 2006). In research findings reported in McGarry (2006), kinematic data from both players comprising the squash dyad were obtained from forty-seven rallies selected at random from a squash competition comprising of quarter-finals, semi-finals and a final game. Using the Hilbert transform (Palut & Zanone, 2005), the lateral (side-to-side), longitudinal (forward-backward) and radial displacement data were subjected to relative phase analysis. To recap, relative phase for a squash dyad expresses where one player is in his (her) movement cycle in respect of the other player. For instance, if the instant that a player leaves and returns to the T-position marks the start and end of a cycle, then anti-phase (180°) indicates that as one player is at the start of a cycle the other player is half-way through their cycle, as evidenced when one player is at the T-position and the other player at the point of shot retrieval and thus furthest from the T-position. Similarly, in-phase (0° or 360°) indicates that both players are at the same point in their respective cycles, with other values spanning 0°–360° assuming all other possible phase relations for the playing dyad. Given the nature of squash contests, anti-phase attraction in the squash dyad is expected for the most part.

Figure 14.3 (upper panels) presents the results of individual squash rallies (N = 47) collapsed into relative phase bins as a frequency histogram for the radial (Figure 14.3A), lateral (Figure 14.3B) and longitudinal (Figure 14.3C) displacements. Figure 14.3 (lower panels) presents the instantaneous phase relations for radial (A), lateral (B), and longitudinal (C) displacements from a single squash rally, together with the individual displacement data in respect of the T-position. The upper part of each figure represents the phase relations (left y-axis) for the squash dyad and the lower part the displacements (right y-axis). In Figure 14.3A (lower panel), the anti-phase relation is represented using dotted lines for ease of data interpretation. In Figures 14.3B and 14.3C (lower panels), the dotted lines are likewise used to indicate the quarter- and three-quarter phases, respectively. (Note: Adding or subtracting 360°, or multiples thereof, to a phase relation does not affect it because of circular statistics.) In the first example (Figure 14.3A, lower panel), anti-phase is represented in each of the 180°, 540° and 900° dashed lines with different expressions of anti-phase relation attributed to skipped, or extra, cycles in one set of displacement data when compared to the other – or, more frequently, a failure of the data sets to meet a condition of the Hilbert analysis in some instances.

Figure 14.3A (upper panel) demonstrates strong attraction to anti-phase (180°), as expected, suggesting a complementary relation between the two players, presumably by virtue of the alternating shot sequences. Figure 14.3A (lower panel) from a single rally likewise demonstrates that the squash dyad spends most of its time approximating anti-phase. Some wandering from anti-phase, indicated in the bi-modal peaks in the frequency histogram, suggests phase variability in the squash dyad albeit with underlying anti-phase attraction. These findings may lend support to the previous interpretation of McGarry et al. (1999) who noted from visual inspection, albeit from only a few single rallies, that the radial displacements of squash players demonstrate strong anti-phase attraction with which brief fluctuations, or perturbations, might be occasionally witnessed. Since perturbations serve to disturb the system, the short-lasting fluctuations from anti-phase might possibly be a result of system perturbation.

Since the travel of squash players on the playing surface takes place in two dimensions, ignoring vertical movements, we reasoned that radial data could be considered the principal measure for describing squash behaviour using relative phase analysis. This said, separate analyses of the displacements in the lateral and longitudinal directions offer additional important information on the space-time dynamics of squash game behaviour. Interestingly, the lateral data (Figure 14.3B) demonstrate phase attractions (upper panel) and phase transitions (lower panel) between the quarter and three-quarter phases, with times spent in the respective phases indicating strong attraction for both coordination patterns. The longitudinal data (Figure 14.3C) likewise exhibit similar features of phase attractions in keeping with expectations from dynamical system descriptions. These frequent switches in the squash dyad between quarter phase and three-quarter phase for both lateral and longitudinal data indicate that the two most stable patterns for the squash dyad are when one player is a quarter-cycle

(A)

RADIAL displacement

N = 47 trials

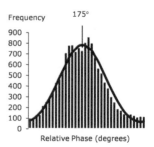

Single phase attraction

Individual trial

········· 180°

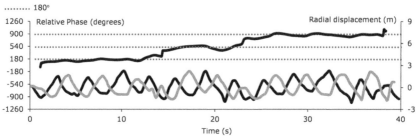

(B)

LATERAL displacement

N = 47 trials

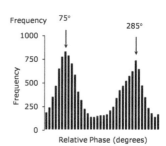

Dual phase attraction

Individual trial

········· 90°   ········· 270°

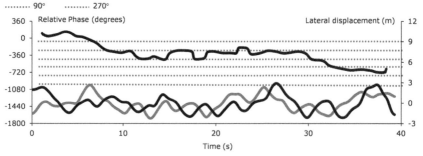

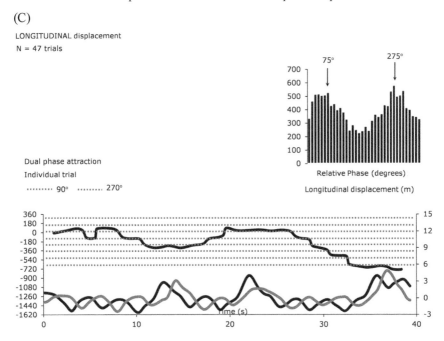

Figure 14.3 Relative phase frequency histograms (upper panels) and instantaneous relative phase and displacement data for a single rally (lower panels) for radial (A), lateral (B) and longitudinal (C) directions. See text for further details.

displaced from the other. Switches between the quarter and three-quarter phases are most likely a result of shot exchanges, as would be seen if, say, a player leaves the T-position a short time before or after the other player begins to return to it. Interestingly, the step features observed as a staircase effect in both lateral and longitudinal data, particularly the latter, demonstrate that system transitions may sometimes occur on a shot-by-shot basis, though not necessarily so as indicated in longer durations of these same phase relations in other instances.

The lateral and longitudinal data demonstrated similar phase attractions towards quarter phase and three-quarter phase, albeit with lesser phase attraction for the longitudinal direction. Thus, squash players exhibit marginally stronger phase couplings in the lateral direction than the longitudinal one, which presumably then makes symmetry breaking for the longitudinal direction easier than for the lateral one. This finding might be important if an advantage should be awarded to the player who manages to break symmetry and gain control of the rally exchange. This said, the consequences of breaking symmetry in the squash dyad – resulting in the winning or losing of a squash rally, for example – remain unclear and require determination in future research.

In sum, the results indicate that the space-time kinematics of squash players seemingly subscribe to dynamical principles of self-organizing systems. Importantly, these dynamical features are observed from relative phase using different measures of displacement – radial, lateral and longitudinal – demonstrating that the dynamical patterns evidenced are not specific to the selected direction. Instead, the dynamical tendencies are seemingly the result of space-time interactions within the squash dyad, a product of shared information within the squash coupling as each player vies for control of the squash rally.

Quarter and three-quarter phase continued: competition and cooperation revisited

Other than research investigations of squash contests, the lateral movements of two players in a singles tennis rally were also examined as coupled oscillations (Palut & Zanone 2005). As such, tennis matches have also been analysed in terms of relative phasing between the lateral displacements of the two players (Lames, 2006; Palut & Zanone 2005; Walter, Lames & McGarry, 2007). In line with related research in other sports settings, these studies showed mutual tuning of the players' movements leading to stable periods approximating in-phase and anti-phase coordination, and demonstrated how such analysis can reveal perturbations to the otherwise stable balance of a tennis rally (as before, in the context of squash contests; see McGarry et al., 1999).

These studies were inspired by experimental work on between-person coordination in cyclical movement actions (e.g. Schmidt et al., 1990; Schmidt et al. 1998) which was later discussed regarding potential applications for sports (Schmidt, O'Brien & Sysko, 1999). Indeed, many studies have confirmed that between-person coordination abides by similar coordinative phenomena as within-person coordination (for reviews, see Schmidt et al., 2011; Schmidt & Richardson, 2008), as formalized by a model of two coupled oscillators known as the HKB model (Haken et al., 1985), and that this also holds in cyclical interpersonal behaviour in real-life tasks (e.g. Schmidt et al., 2011). An important notion for the purposes of examining competitive sport is that, mathematically, in the HKB model the in-phase and anti-phase patterns emerge merely because the HKB coupling functions are formulated in such a way that they cause attraction of both oscillators towards each other's behaviour. Hence, the model oscillators are coupled in a '*cooperative*' manner. However, particularly in sports, next to cooperative interactions between teammates, direct competition between opponents is ubiquitous and fundamental. The question begs then as to what degree the cooperative HKB model is applicable to competitive interaction between players as in squash or tennis singles play.

Interestingly, a study by Kelso, de Guzman, Reveley & Tognoli (2009) provided inspiration for addressing this question. In their laboratory experiment, these authors examined coordination between the oscillatory hand movements of a subject with that of a computer controlled "avatar" presented on a computer monitor. Importantly, the avatar hand movement was made dependent on the movements of the subject, which was programmed based on HKB coupling (for specific details see Dumas et al., 2014; Kelso et al., 2009). As such, it was also possible to couple the movements of an avatar in an inverse manner with those of the subject, meaning that the avatar wanted to achieve anti-phase with the subject while the subject was instructed to perform in-phase with the avatar. Thus, the inverse coupling reflects a "movement intention conflict", which is highly related to attacker-defender coupling in sports: a defender wants to achieve in-phase with the attacker, while the attacker wants to do exactly the opposite and achieve anti-phase. Most interestingly, and perhaps somewhat unexpectedly, numerical simulations of this type of HKB model with inverted (or antagonistic) coupling between oscillators revealed novel predictions, namely that coordination would not be attracted towards in-phase and/or anti-phase, but instead to quarter-phase (90°) (Kelso et al., 2009). As such, in truly competitive stages of player-opponent interaction, as in some instances of a racket sport rally, one might then expect to see phase attractions towards the quarter and three-quarter phases, rather than in-phase and anti-phase.

Indeed, in the study on squash contest behaviours reported by McGarry (2006) higher occurrences of phase relations approximating quarter and three-quarter phase was observed in both the lateral and longitudinal displacements, but these findings were not associated with possible inverse couplings at the time. To further explore whether the predictions of the inverse coupling model could be supported by competitive dyadic interaction in sports, we analysed forty long baseline rallies of equal or greater than ten strokes in length taken from official tennis matches (obtained from Association of Tennis Professionals tournaments). (For recent studies regarding inverse coupling, see De Poel, Van de Laarschot & Noorbergen, 2014; De Poel, McGarry, Noorbergen & Van de Laarschot, in preparation.) As in previous studies, relative phase was calculated from the lateral positions of both players on the tennis field. Preliminary analysis of this data revealed that relative phase values near to -90° (three-quarter phase), 0° (in-phase), and 90° (quarter phase) were observed more often than other phase relations. Notably, anti-phase patterns were not observed in these data. Taken together, these results were not in line with previous findings and the interpretations of Palut and Zanone (2005), but they are nonetheless consistent with predictions from a model of coupled oscillators with inverse coupling – that is, where there was a 'conflict of intention' (or competition) in the interaction (Kelso et al., 2009). (Note: In the Palut and Zanone study, in an attempt to extend rally length the tennis dyad was instructed to trade baseline shots with only "indirect winning points" as a result of errors being permitted until the seventh exchange, after which "direct winning points" were permitted. This aspect of cooperation between players stands in contrast to the tennis study of De Poel et al. and the squash study of McGarry, which both analysed high level sports performance in tournament competition.) Stable periods in a rally approximating in-phase likely indicate that players were both trying to maintain balance within the rally – for example, both playing in defensive mode while waiting or searching for possibilities to perturb the steady state of the rally. Those periods observed within the rally approximating quarter (90°) and three-quarter (-90°) phase may possibly reflect the attacker-defender dynamics after a perturbation was made by one of the players, considered the attacking player in this instance. Furthermore, whether the interpersonal pattern is located around quarter (90°) or three-quarter (-90°) phase indicates which player phase led the other. In other words, in periods of quarter (90°) phase there is one attacking player that is a quarter cycle ahead and forces the defending player to follow. Of course, the coordination does not remain steady but changes within a rally. Within a long rally, then, the odds change back and forth, reflected by interpersonal pattern switches between -90° (three-quarter phase), 0° (in-phase), and 90° (quarter phase) states. In fact, who is leading may possibly be indicative of who is most likely to eventually win the rally. Further study of the competitive interpersonal dynamics in sports, and beyond, alongside the model of inversely coupled oscillators is an obvious worthwhile undertaking.

Summary

n this chapter we have traced some research leading to developing considerations of the rackets sports as dynamical self-organizing systems based on properties of interpersonal coordination. The notion of behavioural perturbations was introduced to demarcate the onset of behavioural instability from behavioural stability in squash rallies, a notion that received support from the reported ability of independent observers to recognize their presence with a good level of inter-observer agreement. Moreover, relative phase analysis of squash dyad kinematics measured in both lateral and longitudinal directions reported attractions to quarter and three-quarter phase. Recent preliminary results from analysis of tennis dyad kinematics measured in the lateral direction report similar findings, from which an interpretation of inverse coupling dynamics has been proposed. Inverse coupling dynamics speak to competition within the dyad with one agent looking to break synchrony and the other looking to maintain it. Further research in terms of additional data gathering and subsequent theoretical grounding, if appropriate, is recommended with the ultimate aim of developing a mathematical description for the racket sports akin to the HKB model developed for dual limb rhythmic action.

References

De Guzman, G.C. & Kelso, J.A.S. (1991) Multi-frequency behavioral patterns and the phase attractive circle map. *Bioloical Cybernetics, 64*, 485–495.

De Poel, H.J., McGarry, T., Noorbergen, O.S. & Van de Laarschot, A. (in preparation).

De Poel, H.J., Van de Laarschot, A. & Noorbergen, O.S. (2014) Competitively coupled oscillator dynamics revealed in ATP top 20 tennis rallies. In H.J. de Poel, C.J.C. Lamoth & F.T.J.M. Zall (Eds.), *International Congress on Complex Systems in Sport and Healthy Ageing, Book of Abstracts.* Groningen, The Netherlands

Dumas, G., de Guzman, G.C., Tognoli, E. & Kelso, J.A.S. (2014) The human dynamic clamp as a paradigm for social interaction. *Proceedings of the National Academy of Sciences, 111*(35), E3726–E3734.

Haken, H., Kelso, J.A.S. & Bunz, H.A. (1985) Theoretical model of phase transitions in human hand movements. *Biological Cybernetics, 51*, 347–356.

Kelso, J.A.S. (1984) Phase transitions and critical behavior in human bimanual coordination. *American Journal of Physiology: Regulatory, integrative and Comparative Physiology, 15*, R1000–1004.

Kelso, J.A.S., Buchanan, J.J. & Wallace, S.A. (1991) Order parameters for the neural organization of single, multijoint limb movement patterns. *Experimental Brain research, 85*, 432–444.

Kelso, J.A.S., de Guzman, G.C., Reveley, C. & Tognoli, E. (2009) Virtual partner interaction (VPI): Exploring novel behaviors via coordination dynamics. *PloS One, 4*(6), e5749.

Kelso, J.A.S., Holt, K.G., Rubin, P. & Kugler, P.N. (1981) Patterns of human interlimb coordination emerge from the properties of non-linear, limit cycle oscillatory processes: Theory and data. *Journal of motor Behaviour, 13*, 226–261.

Kelos, J.A.S. & Jeka, J.J. (1992) Symmetry breaking dynamics of human multilimb coordination. *Journal of Experimental Psychology: Human Perception and Performance, 18*, 645–668.

Kugler, P.N. & Turvey, M.T. (1987) Information, natural law, and the self-assembly of rhythmic movement. Hillsdale, NJ: Lawrence Erlbaum Associates.

Lames, M. (2006) Modelling the interaction in game sports – relative phase and moving correlations. *Journal of Sports Science and Medicine, 4*, 189–190.

McGarry, T. (2006): Identifying patterns in squash contests using dynamical analysis and human perception. *International Journal of Performance Analysis in Sport, 6*(2), 134–147.

McGarry, T., Anderson, D.I., Wallace, S.A., Hughes, M. & Franks, I.M. (2002) Sport competition as a dynamical self-organizing system. *Journal of Sports Sciences, 20*, 771–781.

McGarry, T. & Franks, I.M. (1996) In search of invariant athletic behaviour in competitive sport systems: An example from championship squash match-play. *Journal of Sports Sciences, 14*, 445–456.

McGarry, T. & Franks, I.M. (1994) A stochastic approach to predicting competition squash match-play. *Journal of Sports Sciences, 12*, 573–584.

McGarry, T., Khan, M.A. & Franks, I.M. (1996) Analyzing championship squash match-play: In search of a system description. In S. Haake (Ed.), *The Engineering of Sport* (pp.263–269). Balkema: Rotterdam.

McGarry, T., Khan, M.A. & Franks, I.M. (1999) On the presence and absence of behavioural traits in sport: An example from championship squash match-play. *Journal of Sports Sciences, 17*, 297–311.

Meijer, O.G. (2001) An introduction to the history of movement science. In M.L. Latash & V. Zatsiorsky (eds.), *Classics in Movement Science* (pp. 1-57). Champaign, IL: Human Kinetics.

Palut, Y. & Zanone, P.S. (2005) A dynamical analysis of tennis players' motion: Concepts and data. *Journal of Sports Science, 23*, 1021–1032.

Schmidt, R.C., Bienvenu, M., Fitzpatrick, P.A. & Amazeen, P.G. (1998) A comparison of intra- and interpersonal interlimb coordination: Coordination breakdowns and coupling strength. *Journal of Experimental Psychology: Human Perception and Performance, 24*(3), 884–900.

Schmidt, R.C., Carello, C. & Turvey, M.T. (1990) Phase transitions and critical fluctuations in the visual coordination of rhythmic movements between people. *Journal of Experimental Psychology: Human Perception and Performance, 16*(2), 227–247.

Schmidt, R.C., Fitzpatrick, P., Caron, R. & Mergeche, J. (2011) Understanding social motor coordination. *Human Movement Science*, *30*(5), 834–845.

Schmidt, R.C., O'Brien, B. & Sysko, R. (1999) Self-organization of between-person cooperative trials and possible applications to sport. *International Journal of Sport Psychology*, *30*, 558–579.

Schmidt, R.C. & Richardson, M.J. (2008) Dynamics of interpersonal coordination. In *Coordination: Neural, behavioral and social dynamics* (pp. 281–308). Springer: Berlin, Heidelberg.

Walter, F., Lames, M. & McGarry, T. (2007) Analysis of sports performance as a dynamical system by means of the relative phase. *International Journal of Computer Science in Sport*, *6*(2), 35–41.

15 Impact of mental disorders on social motor behaviours

Jonathan Del-Monte, Stéphane Raffard,
Delphine Capdevielle, Richard C. Schmidt,
Manuel Varlet, Robin N. Salesse,
Benoît G. Bardy, Jean Philippe Boulenger,
Marie Christine Gély-Nargeot and
Ludovic Marin

Introduction

Impact of mental disorders on social motor behaviours

The goal of this chapter is to present the impact of mental disorders on interpersonal coordination. We will present two mental disorders (schizophrenia and social anxiety disorder) characterized by social functional deficits, and discuss their consequences for nonverbal social behaviours in general and social motor coordination in particular. Finally, we will discuss the importance of considering social motor behaviours as a potential target of psychosocial treatment in psychiatric disorders.

Social deficits and mental disorders: the role of social motor behaviours

Schizophrenia is a serious psychiatric condition with unknown etiologic origin, characterized by positive symptoms (such as hallucinations and delusions), negative symptoms (such as avolition, social isolation and psychomotor poverty) and cognitive deficits. The onset of symptoms typically occurs in young adulthood. There are wide variations in reported incidence, psychopathology and course of the illness. A systematic review of epidemiological data indicates that schizophrenia affects about 1% of the world population (McGrath et al., 2008). This disease is also characterized by social withdrawal with an important impact on social relationships and professional activities (Thornicroft et al., 2009). Although the precise societal burden of schizophrenia is difficult to estimate, because of the wide diversity of accumulated data and methods employed, cost-of-illness indications uniformly point to disquieting human and financial costs (Rajagopalan et al., 2013). The costs to society include direct costs, such as treatment, as well as indirect costs, such as loss of manpower at work due to the illness or due to caring by relatives (Mangalore & Knapp, 2007; Konnopka et al., 2009). Current pharmacological treatments of schizophrenia are successful at reducing psychotic symptoms but do not provide a cure (Carpenter & Koenig, 2008), especially given that neurocognitive deficits and negative symptoms appear less amenable to treatment than psychotic phases of the disorder. Schizophrenia does not just affect mental health; people with a diagnosis of schizophrenia die twelve to fifteen years before the average population, with this mortality difference increasing in recent decades (Saha et al., 2007).

Social anxiety disorder (SAD), also called social phobia, is a chronic and disabling condition. It is assumed to be the most common anxiety disorder, with a lifetime prevalence of 12% (Kessler et al., 2005). SAD is defined by a persistent fear of embarrassment or negative evaluation while engaged in social interaction or public performance and tends to be followed by avoidant behaviour (Lecrubier et al., 2000). SAD is associated with significant functional impairment in daily activities (Aderka et al., 2012), social relationships (Wittchen et al., 2000), reduced quality of life (Safren et al., 1996), as well as increased risks of comorbid disorders such as depression, other anxiety disorders and alcohol abuse (Merikangas & Angst, 1995; Stein et al., 2001).

Classically, social interaction disorders are explained through social cognition deficits in both schizophrenia (Green et al., 2008) and social anxiety disorder (Tibi-Elhanany & Shamay-Tsoory, 2011). Recently, several studies provided new ways to understand social functioning deficits in both mental disorders by examining social motor behaviours, which yield data more objective with deficits less heterogeneous than cognitive deficits (Savla et al., 2013).

Past research has shown that schizophrenia patients show a decrease in facial expressive behaviours (Trémeau et al., 2005), and more particularly, display fewer spontaneous smiles in funny situations compared to healthy controls (Henry et al., 2007). Facial expressive behaviour reductions were accompanied by a reduction of body movements during social interactions. Lavelle et al. (2012) showed that schizophrenia patients speak less and use less hand gestures when speaking compared to healthy controls during social interaction. Furthermore, results indicated that reductions in expressiveness in schizophrenia seem to be modulated by the type and intensity of symptomatology.

Concerning social anxiety disorders, few studies have assessed motor behaviours. Past research suggests that adults with social phobia exhibit expressive behaviour deficits (Schneier et al., 2011). Moukheiber et al. (2010) has observed gaze avoidance during social interaction. Also, Melfsen et al. (2000) assessed facial expressions in socially anxious children aged from eight to twelve years. Children's spontaneous facial expressions were covertly recorded when they were trying to solve a puzzle or watching a funny movie. Additionally, they were asked to produce voluntary facial expressions. Results showed that children with social anxiety exhibited fewer spontaneous facial expressions and had poorer performance in voluntary expression of happiness, surprise and fear than healthy children. However, the lack of studies documenting expressive behaviours in social phobia suggests a need for further investigation of this domain. In order to verify the hypothesis of a link between nonverbal social behaviours and social functioning deficits, Del-Monte et al. (2013a) have investigated and compared nonverbal social behaviours in schizophrenia and social anxiety disorder using the Motor-Affective-Social Scale (MASS, Trémeau et al., 2008). During a structured interview, the Motor-Affective-Social Scale is used to assess fundamental expressive deficits through evaluating spontaneous smiles, spontaneous hand gestures and voluntary facial expression. Each occurrence of these behaviours is recorded on a pre-defined rating sheet. A high score on the MASS indicates less impairment of expressiveness. Results of Del-Monte et al.'s (2013a) study showed that schizophrenia and social phobia patients both exhibited significantly fewer expressive behaviours compared to healthy controls. Moreover, results showed specific differences between pathologies. Schizophrenia patients performed fewer spontaneous gestures (hand gestures and smiles) whereas social phobia patients had an impaired ability to produce voluntary facial expression in comparison to healthy controls. Finally, poor social functioning was significantly correlated with a decrease of nonverbal social behaviours for schizophrenia patients. This study shows that two disorders with social interaction deficits lead to different and sometimes opposite nonverbal social behaviour sub-deficits. In social motor behaviours, social motor coordination plays a fundamental role in the quality of social interaction with others (Schmidt et al., 2011). Few studies have evaluated the dynamics of social motor coordination in people with schizophrenia or social anxiety disorder in order to further characterize their abnormal interpersonal movements and understand the processes affected by these pathologies. Varlet et al. (2012a) attempted to answer these questions in schizophrenia. The authors examined the unintentional and intentional social motor coordination of individuals with schizophrenia using a task in which they coordinated the swinging of a pendulum with someone else (Schmidt et al., 1997; Richardson et al., 2005). Participants sat on chairs approximately 1 metre from one another facing in the same direction and swung wrist pendulums attached to a structure that allowed only movements from front to back. The swinging of the pendulums made no noise that could be used as a cue for coordination. The length of the two pendulums was 60 cm and a mass of 150 g was attached either at the bottom or at the middle to give the pendulums different inertias, which would manipulate the self-selected comfort tempo (preferred frequencies) of participants. For those pendulums that had greater inertia, and hence slower comfort tempos, past research found that they would lag their partner's pendulum throughout the cycle, whereas pendulums with less inertia were found to lead their partner's. Consequently the individual pendulums will be referred to as FOLLOWER and LEADER, respectively. Two computer-monitored potentiometers measured the angular displacements of the pendulums during the trials at a sampling rate of 50 Hz (for more information see Varlet et al., 2012a). Results demonstrated that social motor coordination is impaired in schizophrenia, but only intentional coordination is impaired while spontaneous, unintentional coordination remains unaffected. This result is in line with previous research on cognitive and emotional processes that have demonstrated that explicit processing mechanisms are generally more affected than implicit processing mechanisms in schizophrenia (Danion et al., 2001; Roux et al., 2010). Additionally, such a dissociation has also been reported for social processes impairments (Frith, 2004; Schwartz et al., 2010). In the intentional coordination task, the analysis of the relative phasing of the pendulum oscillations showed that the dyads in which a patient was present had a lower stability, and also showed that the patient never led in the coordination. For Varlet et al. (2012a), these results were not related to preferred movement frequencies of schizophrenia patients, and a decrease of the coupling strength, which corresponds to a lower sensitivity to the movements of the other, was not sufficient in explaining the pattern of results observed for the phase shift measure. To explain this result, authors have assumed that patients with schizophrenia have a delay in processing information for visuomotor control. In addition to their contribution in decreasing coupling strength, attention and visual perception deficits may also impair information transmission and be understood as an increase of the time delay in the coupling function (Banerjee & Jirsa, 2007). This hypothesis is supported by previous research showing slower reaction times in schizophrenia (Vinogradov et al., 1998) and more specifically by anatomical correlates such as a degradation of the degree of myelination (Lim et al., 1999). To further explore these hypotheses and determine how a decrease of the strength or/and a delay in information transmission in patients could explain their impairments of intentional social motor coordination, Varlet et al. (2012a) have compared phase shift results in schizophrenia with those obtained during a simulation of a non-linear coupling of two limit-cycle oscillators used in the past for understanding social motor coordination (Haken et al., 1985; Schöner & Kelso, 1988). Simulated data showed that both a decrease in coupling strength and an increase in time delay explained the patient's results.

This study shows that social motor coordination was impaired in schizophrenia. In intentional coordination, the schizophrenia group displayed coordination patterns that had lower stability and in which the patient never led the coordination. A coupled oscillator model suggests that the schizophrenia group coordination pattern was due to a decrease in the amount of available information together with a delay in information transmission. According to authors, social motor coordination assessment seems to open a new perspective for understanding social disorders in mental illness. However, two new questions emerge from these results. First, are the motor coordination abnormalities specific to schizophrenia? Second, could they be a potential trait-marker for this mental disease?

To answer the first question, Varlet et al. (2014) assessed the unintentional and intentional social motor coordination of individuals with social anxiety disorder (SAD). Authors used the same pendulum paradigm and the same methodology as for the schizophrenia study, and they assessed psychological variables such as anxiety level. Results showed that unintentional social motor coordination was preserved with social anxiety disorder while intentional coordination was impaired. But unlike schizophrenic patients, who show an overall reduction in intentional coordination and for whom the patient never led in the coordination, social anxiety patients had a specific coordination impairment. In particular, intentional coordination became impaired when patients with SAD had to lead the coordination. This result might be explained by the fact that intended coordination requires a social goal, and consequently stronger social involvement. Important behavioural adaptations are needed to adequately maintain in-phase and anti-phase patterns of coordination, and people with SAD might have been afraid to fail and to be judged (Lecrubier et al., 2000). Of particular interest with regard to this suggestion is that the SAD group demonstrated the increased variability of social motor coordination only for some pendulum combinations. Stability became problematic when the leader and follower positions were not mechanically constrained (there was a significant correlation between the anxiety level and the standard deviations of relative phase), and even more problematic when patients were constrained to lead the coordination (there was a significant difference between the groups in the relative phase standard deviation ANOVA, and a significant correlation between anxiety level and standard deviation of relative phase). However, the social coordination of the SAD group was not more variable than the control group when the SAD participant was constrained to follow in the coordination, a condition in which patients might have been less afraid of being judged. This result supports past results that have found poorer leadership skills in patients with SAD, and confirms the finding that anxiety can modulate motor performance (Lecrubier et al., 2000; Bernstein et al., 2008; Hatfield et al., 2013). Together, these studies (Varlet et al., 2012a, Varlet et al., 2014) suggest that social disorders might preferentially affect intentional rather than spontaneous unintended interpersonal coordination. However, it is clear that two disorders with social interaction deficits lead to specific social motor coordination impairments – but could they be a potential trait-marker for mental diseases, especially for schizophrenia?

To try to answer this second question, Del-Monte et al. (2013b) have considered that social motor coordination impairment is a motor trait-marker of this pathology – in other words, a potential intermediate endophenotype of schizophrenia. For this, the trait-marker of the pathology must be present in schizophrenia patients but also in their unaffected first-degree relatives (mothers and fathers). In order to verify these intermediate endophenotype criteria, the authors used the same pendulum paradigm and the same methodology as Varlet et al. (2012a) to evaluate social motor coordination in schizophrenia patients and in unaffected first-degree relatives of schizophrenia patients. These results showed that unaffected relatives of schizophrenia patients had impaired intentional interpersonal coordination compared to healthy controls while unintentional interpersonal coordination was preserved. More specifically, in intentional coordination, the unaffected relatives of schizophrenia patients exhibited coordination patterns that had greater variability and in which relatives did not lead the coordination. Thus, the results of this study show that, similar to schizophrenia patients, their relatives did not lead the coordination even in the conditions in which they should have. Therefore, social motor coordination impairments seem to be a potential endophenotype of schizophrenia.

A possible explanation of this result is that unaffected relatives not only have a weaker perceptual coupling but also have a time delay in processing information for visual motor coupling (Banerjee & Jirsa, 2007; Varlet et al., 2012b). Research on neuroanatomical connectivity might explain such a time delay. For example, in a bimanual task there is a contribution from the contralateral and ispsilateral motor cortices (Rokni et al., 2003). Thus, to move the right hand, information must be transmitted between both cortices. Callosal fibres allow interhemispheric crosstalk between the left and right hemispheres, and several studies have elucidated their important role in creating temporal couplings (Kennerley et al., 2002). An anomaly in the callosal fibres could result in a slowdown in the transmission of interhemispheric information. Recently, Knöchel et al. (2012) have suggested that the volume of callosal fibres was significantly reduced in schizophrenia patients and in their relatives compared to healthy controls. Thus, in unaffected relatives, as in schizophrenia, the reduction of volume of callosal fibres would lead to a reduction in the transmission of interhemispheric crosstalk and thus to the occurrence of an offset in temporal coupling. Moreover, results in the attentional processes task showed that relatives have reduced capacity for sustained attention compared to the control group. This observation strengthens the time delay explanation of the unaffected relatives' phase shift. Thus, this offset in temporal coupling could result in the unaffected relatives not leading the coordination, even in the conditions in which they should have.

All these impairments might have contributed to an impaired social motor coordination in unaffected relatives; A number of results suggest that social motor coordination impairment matches the intermediate phenotype criteria for schizophrenia (Gottesman & Gould, 2003). First, social motor coordination impairment is associated with schizophrenia (Varlet et al., 2012b). Secondly, Del-Monte et al.'s study (2013b) indicates that this impairment is heritable to some degree because we find it among unaffected relatives of schizophrenia patients. Third, Del-Monte et al.'s study (2013b) shows that social motor coordination impairment is observed in unaffected family members to a greater extent than among the general population. Unfortunately, the cross-sectional design of Del-Monte et al.'s study (2013b) does not allow us to conclude for the fourth criterion, "state independent in affected individuals" (i.e. manifests in an individual whether or not illness is active). Nevertheless, several studies suggest that adolescents who later develop schizophrenia exhibit motor behaviour disorders, particularly motor dyscoordination, well before overt manifestations of the illness (Leask et al., 2002; Mittal et al., 2011). Consequently, this study has provided evidence to support the claim that social motor coordination could be considered as a trait-marker that might be associated with the risk for schizophrenia. This study showed the necessity to assess the motor dimension as an embodiment of mental disorders and to determine the impact of such impairments in daily living on patients and their relatives. The clinical significance of this finding lies in early diagnostics and in the field of preventative interventions, which therapeutically might target the motor aspect of social abilities in non-clinical individuals with a genetic risk of schizophrenia.

Social deficits, social motor behaviour impairments and therapeutically interventions in schizophrenia

Social deficits are classically explained through cognitive function deficits in schizophrenia (Green et al., 2004, 2008; Lepage et al., 2014). Moreover, traditional neurocognitive remediation effects (Kurtz et al., 2001) and psychotic symptom management alone (Heydebrand et al., 2004) show limited generalizability to functional improvements. Recent years have been marked by increased efforts to identify causes of poor social functioning in schizophrenia and develop novel treatments targeting these domains to optimize the outcome. One promising area that has garnered a great deal of attention is social cognition (Horan et al., 2008; Fiszdon & Reddy, 2012). Social cognition is a multidimensional construct defined as the way we understand, perceive and interpret our social world (Penn et al., 1997), including emotion processing, Theory of Mind, social perception, social knowledge, and attributional style (Green et al., 2005). As reviewed by Horan et al. (2008), there is substantive evidence of social cognitive impairments across domains in schizophrenia, with a general consensus that these deficits are distinct from other hallmark neurocognitive and clinical features of schizophrenia (Penn et al., 1997; Green et al., 2005). Social cognition appears to maintain a unique (and stronger) relationship to functional outcome in schizophrenia, above and beyond that of basic cognition (Penn et al., 1996; Pinkham & Penn, 2006). Based on this observation, several social cognition remediation programmes (Penn et al., 2005; Horan et al., 2009) were developed to promote social cognition in schizophrenia. However, these remediation programmes only train the identification and analysis of facial expressions in other people; they do not promote nonverbal social behaviour production in the schizophrenia patients themselves. Brüne et al. (2009) have explored links between social cognition and nonverbal expressiveness in schizophrenia spectrum disorders, showing that schizophrenia patients had fewer prosocial behaviours compared to healthy controls, and a partial correlation existed between reduced prosocial behaviour and social competence for the schizophrenia group. This study shows the necessity of targeting both social cognition and nonverbal social behaviours to promote the quality of social functioning in schizophrenia patients.

Despite this evidence, no study has focused on nonverbal social behaviours to promote social functioning in schizophrenia. However, the literature shows that it is possible to increase the production of motor behaviours with a priming task, for example. Priming classically refers to the process by which a given stimulus activates mental pathways, thereby enhancing the ability to process subsequent stimuli related to the priming stimulus. For a long time semantic priming has been the standard paradigm for evaluating the priming phenomenon (Maxfield, 1997). There is now an accumulation of evidence that priming tasks can facilitate cooperative behaviours and induce a feeling of social attitude/affiliation between two individuals (Schröder & Thagard, 2013). Over and Carpenter (2009) have shown that children primed with photographs evoking affiliation helped a person in need more often and more spontaneously than when primed with photographs evoking individuality. Lakin and Chartrand (2003) also showed that individuals primed by subliminal words related to the concept of social affiliation (e.g. *together* or *friend*) increased mimicry in a subsequent social interaction. According to the authors, the social priming task leads to an increase of the attentional capacity on partner's social body movements, which would directly promote behavioural mimicry between individuals.

Although their results indicated that enhanced psychological (implicit) attitudes using priming could lead to more automatic mimicry in healthy individuals, it remains largely unknown whether such effects can be generalized to a social illness such as schizophrenia. To our knowledge, only two experimental studies (Del-Monte et al., 2014; Raffard & Salesse et al., 2015) have investigated the influence of semantic social priming on nonverbal social behaviours in schizophrenia. First, Del-Monte et al. (2014) assessed the impact of semantic social priming on nonverbal social behaviour production (spontaneous hand gestures, spontaneous smile and voluntary facial expression) in schizophrenia patients. Second, Raffard and Salesse et al. (2015) evaluated the influence of semantic social priming on the stability of social motor coordination and its effect on the feeling of connectedness between a schizophrenia patient and a healthy control. Both studies used the same implicit priming task inspired by Leighton et al. (2010).

Three versions of the Scrambled-Sentence Task were constructed: one was intended to prime a pro-social attitude (affiliation), another an anti-social attitude (non-affiliation) and a third was intended to prime no attitude, the non-social priming condition (for more information see Del-Monte et al., 2014). The results of these studies confirmed that schizophrenia patients are sensitive to priming effects (Rossell & David, 2006; Pomarol-Clotet et al., 2008; Mathalon et al., 2010). Del-Monte et al. (2014) revealed a significant main effect for priming conditions. Schizophrenia patients primed with pro-social words showed significantly more nonverbal behaviours than schizophrenia patients primed with anti-social priming words, and than schizophrenia patients primed with non-social priming words. Schizophrenia patients in the non-social and anti-social priming conditions did not differ in their performance. However, pro-social priming affected different nonverbal behaviours in schizophrenia patients. Compared to anti-social priming, pro-social priming significantly increased spontaneous hand gestures and marginally enhanced voluntary facial expression. Schizophrenia patients in pro-social priming also had a higher score on voluntary facial expression than patients with non-social priming.

Raffard and Salesse et al. (2015) assessed the influence of implicit social priming on the stability of social motor coordination and its effect on the feeling of connectedness between a schizophrenia patient and a healthy control. This study used the same experimental pendulum paradigm as Varlet et al. (2012a) and assessed the subjective feeling of connectedness between participants through two questions: "Did you find that your partner was likeable?" and "Would you have enjoyed spending more time with your partner?" In Raffard and Salesse et al.'s study only patients were primed in the dyad, in one of three conditions: pro-social, anti-social or non-social. Three important findings were observed in this study: first, the results of Varlet et al.'s study were confirmed. Second, the pro-social priming significantly increased the coupling strength between schizophrenia patients and healthy participants, whereas anti-social or non-social priming did not change the stability of the coordination. Third, Raffard and Salesse et al. (2015) found that whatever the social valence of the priming, being primed did not affect the affiliation judgement score of patients toward their partners, but it did change the affiliation judgment score of partners toward the patients. More specifically, synchronization partners reported a better feeling of connectedness towards pro-socially primed schizophrenia patients than towards non-socially or anti-socially primed schizophrenia patients. In addition, affiliation judgment scores significantly correlated with an increase in the stability of the coordination. Therefore, the authors inferred that increasing motor coordination stability by virtue of a pro-social priming procedure increased the feeling of connectedness towards schizophrenia patients from non-primed participants (synchronization partners). According to the authors, this better stability among pro-socially primed schizophrenia patients was perceived by their synchronization partners, leading to a "normalization" of their affiliation towards them in a similar way to healthy controls. In other words, the perception of good stability in the social motor coordination of the patient leads healthy individuals to judge the patient more positively.

For the first time, both studies demonstrate that it is possible to promote nonverbal social behaviours and increase the feeling of connectedness in the healthy partner interacting with an individual with schizophrenia using an experimental semantic priming task. These results may have several implications for social clinical perspective therapies in this mental pathology. First, training patients to improve interpersonal motor coordination through a sensorimotor approach (Varlet et al., 2012a) may improve social attunement. Second, pro-social priming may facilitate such training. It has recently been shown that a priming procedure administered twice daily as an adjunct of cognitive behavioural therapy increased cognitive modifications in social phobic patients (Borgeat et al., 2013). Taken together, such interventions could make a significant contribution and constitute an augmentation strategy to the existing behavioural therapies dedicated to promote social skills in schizophrenia patients.

Classical psychiatry has long believed that mental disorders were mainly characterized by cognitive and psychological impairments. Today, some studies show the presence and the central role of reductions of social motor behaviour in social functioning in schizophrenia and social anxiety disorder (Varlet et al., 2012a; Del-Monte et al., 2013a; Varlet et al., 2014). In addition, the consideration of motor dimensions allows a better understanding of pathologies by highlighting new and original phenotype motor markers for early diagnosis (Del-Monte et al., 2014) and thus more prompt management of difficulties encountered by patients. Finally, two studies (Del-Monte et al., 2014; Raffard & Salesse et al., 2015) have shown that it is possible, with an experimental priming task, to promote the stability of social motor coordination and to change the social perception of healthy controls towards the schizophrenia patients. Thus, it seems unquestionable that the social motor behaviour dimension must be considered as an important therapeutic target to increase the quality of social functioning of schizophrenia patients.

References

Aderka, I.M., Hofmann, S.G., Nickerson, A., Hermesh, H., Gilboa-Schechtman. E. & Marom, S. (2012) Functional impairment in social anxiety disorder. *Journal of Anxiety Disorders*, *26*(3), 393–400. doi: 10.1016/j.janxdis.2012.01.003.

Banerjee, A. & Jirsa, V. (2007) How do neural connectivity and time delays influence bimanual coordination? *Biological Cybernetics*, *96*, 265–278.

Bernstein, G.A., Bernat, D.H., Davis, A.A. & Layne, A.E. (2008) Symptom presentation and classroom functioning in a nonclinical sample of children with social phobia. *Depression and Anxiety*, *25*, 752–760. doi: 10.1002/da.20315

Borgeat, F., O'Connor, K., Amado, D. & St-Pierre-Delorme, M.E. (2013) Psychotherapy Augmentation through Preconscious Priming. *Frontiers in Psychiatry*, 4-15.

Brüne, M., Abdel-Hamid, M., Sonntag, C., Lehmkämper, C. & Langdon, R. (2009) Linking social cognition with social interaction: non-verbal expressivity, social compentence and "mentalising" in patients with schizophrenia spectrum disorders. *Behavioral and Brain Functions*, *5*, 6.

Carpenter, W.T. & Koenig, J.I. (2008) The evolution of drug development in schizophrenia: past issues and future opportunities. *Neuropsychopharmacology*, *33*(9), 2061–79.

Danion, J., Meulemans, T., Kauffmann-Muller, F. & Vermaat, H. (2001) Intact implicit learning in schizophrenia. *American Journal of Psychiatry*, *158*, 944–948.

Del-Monte, J., Raffard, S., Salesse, R.N., Marin, L., Schmidt, R.C., Varlet, M., Bardy, B.G., Boulenger, J.P., Gély-Nargeot, M.C & Capdevielle, D. (2013a) Nonverbal expressive behaviour in schizophrenia and social phobia. *Psychiatry Research*, *210*, 29–35.

Del-Monte, J., Capdevielle, D., Varlet, M., Marin, L., Schmidt, R.C., Salesse, R.N., Bardy, B.G., Boulenger, J.P., Gély-Nargeot, M.C., Attal, J. & Raffard, S. (2013b) Social motor coordination in unaffected relatives of schizophrenia patients: a potential intermediate phenotype. *Frontiers in Behavioral Neuroscience*, *7*, 137.

Del-Monte, J., Raffard, S., Capdevielle, D., Salesse, R.N., Schmidt, R.C., Varlet, M., Bardy, B.G., Boulenger, J.P., Gély-Nargeot M.C. & Marin, L. (2014) Social priming increases nonverbal expressive behaviors in schizophrenia. *PlosOne*.

Ekman, P. & Friesen, W. (1969) The repertoire of nonverbal behavior. Categories, origins, usage, and coding. *Semiotica*, *1*, 49–98.

Fiszdon, J.M. & Reddy, L.F. (2012) Review of social cognitive treatments for psychosis. *Clinical Psychology Review*, *32*(8), 724–40. doi: 10.1016/j.cpr.2012.09.003.

Frith, C. (2004) Schizophrenia and theory of mind. *Psychological Medicine*, *34*, 385–389.

Goldin-Meadow, S. (1999) The role of gesture in communication and thinking. *Trends in Cognitive Sciences*, *3*(11), 419–429.

Gottesman, I.I., & Gould, T.D. (2003) The endophenotype concept in psychiatry: etymology and strategic intentions. *The American Journal of Psychiatry*, *160*, 636–645. doi:10.1176/appi. ajp.160.4.636

Green, M.F., Kern, R.S. & Heaton, R.K. (2004) Longitudinal studies of cognition and functional outcome in schizophrenia: implications for MATRICS. *Schizophrenia Research*, *72*(1), 41–51.

Green, M.F., Olivier, B., Crawley, J.N., Penn, D.L. & Silverstein, S. (2005) Social cognition in schizophrenia: recommendations from the measurement and treatment research to improve cognition in schizophrenia new approaches conference. *Schizophrenia Bulletin*, *31*(4), 882–7.

Green, M.F., Penn, D.L., Bentall, R., Carpenter, W.T., Gaebel, W., Gur, R.C., Kring, A.M., Park, S., Silverstein, S.M. & Heinssen, R. (2008) Social cognition in schizophrenia: an nimh workshop on definitions, assessment, and research opportunities. *Schizophrenia Bulletin*, *34*, 1211–1220.

Haken, H., Kelso, J. & Bunz, H. (1985) A theoretical model of phase transitions in human hand movements. *Biological Cybernetics*, *51*, 347–356.

Hatfield, B.D., Costanzo, M.E., Goodman, R.N., Lo, L.C., Oh, H., Rietschel, J.C., Saffer, M., Bradberry, T., Contreras-Vidal, J. & Haufler, A. (2013) The influence of social evaluation on cerebral cortical activity and motor performance: a study of "real-life" competition. *International Journal of Psychophysiology: Official Journal of the International Organization of Psychophysiology*, *90*, 240–249. doi: 10.1016/j.ijpsycho.2013.08.002.

Henry, J.D., Green, M.J., de Lucia, A., Restuccia, C., McDonald, S. & O'Donnell, M. (2007) Emotion dysregulation in schizophrenia: reduced amplification of emotional expression is associated with emotional blunting. *Schizophrenia Research*, *95*, 197–204.

Heydebrand, G., Weiser, M., Rabinowitz, J., Hoff, A.L., DeLisi, L.E. & Csernansky, J.G. (2004) Correlates of cognitive deficits in first episode schizophrenia. *Schizophrenia Research*, *68*(1), 1–9.

Horan, W.P., Kern, R.S., Shokat-Fadai, K., Sergi, M.J., Wynn, J.K. & Green, M.F. (2008) Social cognitive skills training in schizophrenia: an initial efficacy study of stabilized outpatients. *Schizophrenia Research*, *107*(1), 47–54. doi: 10.1016/j.schres.2008.09.006.

Horan, W.P., Kern, R.S., Green, M.F. & Penn, D.L. (2008) Social cognition training for individuals with schizophrenia: emerging evidence. *The American Journal of Psychiatric Rehabilitation*, *11*, 205–252.

Horan, W.P., Kern, R.S., Shokat-Fadai, K., Sergi, M.J., Wynn, J.K. & Green, M.F. (2009) Social cognitive skills training in schizophrenia: an initial efficacy study of stabilized outpatients. *Schizophrenia Research*, *107*, 47–54.

Kennerley, S.W., Diedrichsen, J., Hazeltine, E., Semjen, A. & Ivry, R.B. (2002) Callosotomy patients exhibit temporal uncoupling during continuous bimanual movements. *Nature Neuroscience*, *5*, 376–381. doi:10.1038/ nn822.

Kessler, R.C., Chiu, W.T., Demler, O., Merikangas, K.R. & Walters, E.E. (2005) Prevalence, severity, and comorbidity of 12-month DSM-IV disorders in the National Comorbidity Survey Replication. *Archive General in Psychiatry*, *62*(6), 617–27. Erratum in: *Archive General in Psychiatry*, *62*(7), 709.

Knöchel, C., Oertel-Knöchel, V., Schönmeyer, R., Rotarska-Jagiela, A., van de Ven, V., Prvulovic, D., Haenschel, C., Uhlhaas, P., Pantel, J., Hampel, H. & Linden, D.E. (2012) Interhemispheric hypoconnectivity in schizophrenia: fiber integrity and volume differences of the corpus callosum in patients and unaffected relatives. *Neuroimage*, *59*, 926–934. doi: 10.1016/j.neuroimage.2011.07.088.

Konnopka, A., Klingberg, S., Wittorf, A. & König, H.H. (2009) The cost of schizophrenia in Germany: a systematic review of the literature. *Psychiatrische Praxis*, *36*(5), 211–8. doi: 10.1055/s-0028-1090234.

Kurtz, M.M., Moberg, P.J., Gur, R.C. & Gur, R.E. (2001) Approaches to cognitive remediation of neuropsychological deficits in schizophrenia: a review and meta-analysis. *Neuropsychology Review*, *11*(4), 197–210.

Lakin, J.L. & Chartrand, T.L. (2003) Using nonconscious behavioral mimicry to create affiliation and rapport. *Psychological Science*, *14*, 334–339.

Lavelle, M., Healey, P.G. & McCabe, R. (2012) Is nonverbal communication disrupted in interaction involving patients with schizophrenia. *Schizophrenia Bulletin*, http://dx.doi.org/10.1093/schbul/sbs091.

Leask, S.J., Done, D.J. & Crow, T.J. (2002) Adult psychosis, common childhood infections and neu- rological soft signs in a national birth cohort. *The British Journal of Psychiatry: The Journal of Mental Science*, *181*, 387-392. doi:10.1192/bjp.181.5.387.

Lecrubier, Y., Wittchen, H.U., Faravelli, C., Bobes, J., Patel, A. & Knapp, M. (2000) A European perspective on social anxiety disorder. *European Psychiatry*, *15*(1), 5–16.

Leighton, J., Bird, G., Orsini, C. & Heyes, C.M. (2010) Social attitudes modulate automatic imitation. *Journal of Experimental Social Psychology*, *46*, 905–910.

Lepage, M., Bodnar, M. & Bowie, C.R. (2014) Neurocognition: clinical and functional outcomes in schizophrenia. *The Canadian Journal of Psychiatry*, *59*(1), 5–12.

Lim, K.O., Hedehus, M., Moseley, M., de Crespigny, A., Sullivan, E.V. & Pfefferbaum, A. (1999) Compromised white matter tract integrity in schizophrenia inferred from diffusion tensor imaging. *Archive of General Psychiatry*, *56*, 367–374.

Mangalore, R. & Knapp, M. (2007) Cost of schizophrenia in England. *The Journal of Mental Health Policy and Economics*, *10*(1), 23–41.

Mathalon, D.H., Roach, B.J. & Ford, J.M., (2010) Automatic semantic priming abnormalities in schizophrenia. *International Journal of Psychophysiology*, *75*(2), 157–66. doi: 10.1016/j.ijpsycho.2009.12.003.

Maxfield, L. (1997) Attention and semantic priming: a review of prime task effects. *Consciousness and Cognition*, *6*, 204–218.

McGrath, J., Saha, S., Chant, D. & Welham, J. (2008) Schizophrenia: a concise overview of incidence, prevalence, and mortality. *Epidemiologic Reviews*, *30*, 67–76. doi: 10.1093/epirev/mxn001. Epub 2008 May 14.

Melfsen, S., Osterlow, J. & Florin, I. (2000) Deliberate emotional expressions of socially anxious children and their mothers. *Journal of Anxiety Disorders*, *14*, 249–261.

Merikangas, K.R. & Angst, J. (1995) Comorbidity and social phobia: evidence from clinical, epidemiologic, and genetic studies. *European Archives of Psychiatry and Clinical Neuroscience*, *244*(6), 297–303.

Mittal, A.V., Jalbrzikowski, M., Daley, M., Roman, C., Bearden, C.E. & Cannon, T.D. (2011) Abnormal movements are associated with poor psychosocial functioning in adolescents at high risk for psychosis. *Schizophrenia Research*, *130*, 164–169. doi:10. 1016/j.schres.2011.05.007.

Moukheiber, A., Rautureau, G., Perez-Diaz, F., Soussignan, R., Dubal, S., Jouvent, R. & Pelissolo, A. (2010) Gaze avoidance in social phobia: objective measure and correlates. *Behaviour Research and Therapy*, *48*, 147–151.

Over, H. & Carpenter, M. (2009) Eighteen-month-old infants show increased helping following priming with affiliation. *Psychological Science*, *20*, 1189–1193.

Penn, D.L., Corrigan, P.W., Bentall, R.P., Racenstein, J.M. & Newman, L. (1997) Social cognition in schizophrenia. *Psychological Bulletin*, *121*(1), 114–32.

Penn, D.L., Spaulding, W., Reed, D. & Sullivan, M. (1996) The relationship of social cognition toward behavior in chronic schizophrenia. *Schizophrenia Research, 20*(3), 327–35.

Penn, D., Roberts, D.L., Munt, E.D., Silverstein, E., Jones, N. & Sheitman, B. (2005) A pilot study of social cognition and interaction training (SCIT) for schizophrenia. *Schizophrenia Research, 80*, 357–359.

Pinkham, A.E. & Penn, D.L. (2006) Neurocognitive and social cognitive predictors of interpersonal skill in schizophrenia. *Psychiatry Research, 143*(2–3), 167–78.

Pomarol-Clotet, E., Oh, T.M., Laws, K.R. & McKenna, P.J. (2008) Semantic priming in schizophrenia: systematic review and meta-analysis. *The British Journal of Psychiatry, 192*(2), 92–7. doi: 10.1192/bjp.bp.106.032102.

Raffard, S., Salesse, R.N., Marin, L., Del-Monte, J., Schmidt, R.C., Varlet, M., Bardy, B.G., Boulenger, J.P & Capdevielle, D. (2015) Social priming enhances interpersonal syncronization and feeling of connectedness towards schizophrenia patients. *Scientific Reports, 2*(5), 8156. doi: 10.1038/srep08156.

Rajagopalan, K., O'Day, K., Meyer, K., Pikalov, A. & Loebel, A. (2013) Annual cost of relapses and relapse-related hospitalizations in adults with schizophrenia: results from a 12-month, double blind, comparative study of lurasidone vs quetiapine extended-release. *Journal of Medical Economics, 16*(8), 987–96. doi: 10.3111/ 13696998.2013.809353.

Richardson, M., Marsh, K. & Schmidt, R.C. (2005) Effects of visual and verbal interaction on unintentional interpersonal coordination. *Journal of Experimental Psychology: Human Perception and Performance, 31*, 62–79.

Richardson, M.J., Marsh, K., Isenhower, R., Goodman, J. & Schmidt, R.C. (2007) Rocking together: dynamics of intentional and unintentional interpersonal coordination. *Human Movement Science, 26*, 867–891. doi:10.1016/ j.humov.2007.07.002.

Rokni, U., Steinberg, O., Vaadia, E. & Sompolinsky, H. (2003) Cortical representation of bimanual movements. *The Journal of Neuroscience: The Official Journal of the Society for Neuroscience, 23*, 11577–11586.

Rossell, S.L. & David, A.S. (2006) Are semantic deficits in schizophrenia due to problems with access or storage? *Schizophrenia Research, 82*(2–3), 121–34.

Roux, P., Christophe, A. & Passerieux, C. (2010) The emotional paradox: dissociation between explicit and implicit processing of emotional prosody in schizophrenia. *Neuropsychologia, 48*, 3642–3649.

Safren, S.A., Heimberg, R.G., Brown, E.J. & Holle, C. (1996–1997) Quality of life in social phobia. *Depression and Anxiety, 4*(3), 126–33.

Saha, S., Chant, D. & McGrath, J. (2007) A systematic review of mortality in schizophrenia: is the differential mortality gap worsening over time? *Archive of General Psychiatry, 64*(10), 1123–31.

Savla, G.N., Vella, L., Armstrong, C.C., Penn, D.L. & Twamley, E.W. (2013) Deficits in domains of social cognition in schizophrenia: a meta-analysis of the empirical evidence. *Schizophrenia Bulletin, 39*(5), 979–92. doi: 10.1093/schbul/sbs080.

Schmidt, R.C. & O'Brien, B. (1997) Evaluating the dynamics of unintended interpersonal coordination. *Ecological Psychology: a publication of the International Society for Ecological Psychology, 9*, 189–206. doi:10. 1207/s15326969eco0903_2.

Schmidt, R.C., Bienvenu, M., Fitzpatrick, P.A. & Amazeen, P.G. (1998) A comparison of intra- and interpersonal coordination: Coordination breakdowns and coupling strength. *Journal of Experimental Psychology: Human Perception and Performance, 24*, 884–900.

Schmidt, R.C., Carello, C. & Turvey, M.T. (1990) Phase transitions and critical fluctuations in the visual coordination of rhythmic movements between people. *Journal of Experimental Psychology. Human Perception and Performance, 16,* 227. doi:10.1037/0096- 1523.16.2. 227.

Schmidt, R.C., Fitzpatrick, P., Caron, R. & Mergeche, J. (2011) Understanding social motor coordination. *Human Movement Science, 30,* 834–845.

Schmidt, R.C. & Richardson, M.J. (2008) Dynamics of interpersonal coordination. In A. Fuchs & V. Jirsa (Eds.), *Coordination: Neural, Behavioral and Social Dynamics* (pp. 281–308). Springer-Verlag: Heidelberg.

Schneier, F.R., Rodebaugh, T.L., Blanco, C., Lewin, H. & Liebowitz, M.R. (2011) Fear and avoidance of eye contact in social anxiety disorder. *Comprehensive Psychiatry, 52,* 81–87.

Schöner, G. & Kelso, J.A.S. (1988) Dynamic pattern generation in behavioral and neural systems. *Science, 239,* 1513–1520.

Schröder, T. & Thagard, P. (2013) The affective meanings of automatic social behaviors: three mechanisms that explain priming. *Psychological Review,* 120, 255–280.

Schwartz, B.L., Vaidya, C.J., Howard, J.H. & Deutsch, S.I. (2010) Attention to gaze and emotion in schizophrenia. *Neuropsychology, 24,* 711–720. doi: 10.1037/a0019562.

Stein, M.B., Fuetsch, M., Müller, N., Höfler, M., Lieb, R. & Wittchen, H.U. (2001) Social anxiety disorder and the risk of depression: a prospective community study of adolescents and young adults. *Archive of General Psychiatry, 58*(3), 251–6.

Thornicroft, G., Brohan, E., Rose, D., Sartorius, N. & Leese, M. (2009) I.N.D.I.G.O. Study Group. Global pattern of experienced and anticipated discrimination against people with schizophrenia: a cross-sectional survey. *The Lancet, 373*(9661), 408–15. doi: 10.1016/S0140- 6736(08)61817-6.

Tibi-Elhanany, Y. & Shamay-Tsoory, S.G. (2011) Social cognition in social anxiety: first evidence for increased empathic abilities. *The Israel Journal of Psychiatry and Related Sciences, 48,* 98–106.

Trémeau, F., Goggin, M., Antonius, D., Czobor, P., Hill, V. & Citrome, L. (2008) A new rating scale for negative symptoms: the Motor-Affective-Social Scale. *Psychiatry Research, 160,* 346–355.

Trémeau, F., Malaspina, D., Duval, F., Corrêa, H., Hager-Budny, M., Coin-Bariou, L., Macher, J.P. & Gorman, J.M. (2005) Facial expressiveness in patients with schizophrenia compared to depressed patients and nonpatient comparison subjects. *The American Journal of Psychiatry, 162,* 92–101.

Varlet, M., Marin, L., Capdevielle, D., Del-Monte, J., Schmidt, R.C., Salesse, R.N., Boulenger, J.P., Bardy, B.G. & Raffard S. (2014) Difficulty leading interpersonal coordination: towards an embodied signature of social anxiety disorder. *Frontiers in Behavioral Neuroscience, 8,* 29. doi: 10.3389/fnbeh.2014.00029.

Varlet, M., Coey, C.A., Schmidt, R.C. & Richardson, M.J. (2012b) Influence of stimulus amplitude on unintended visuomotor entrainment. *Human Movement Science, 31,* 541–552. doi:10.1016/j.humov.2011.08.002.

Varlet, M., Marin, L., Raffard, S., Schmidt, R.C., Capdevielle, D., Boulenger, J.P., Del-Monte, J. & Bardy, B.G. (2012a) Impairments of social motor coordination in schizophrenia. *PLoS ONE, 7,* e29772.

Vinogradov, S., Poole, J., Willis-Shore, J., Ober, B. & Shenaut, G. (1998) Slower and more variable reaction times in schizophrenia: what do they signify? *Schizophrenia Research, 32,* 183–190.

Wittchen, H.U., Carter, R.M., Pfister, H., Montgomery, S.A. & Kessler, R.C. (2000) Disabilities and quality of life in pure and comorbid generalized anxiety disorder and major depression in a national survey. *International Clinical Psychopharmacology,* *15*(6), 319–28.

16 Horse-rider interactions in endurance racing

An example of interpersonal coordination

Rita Sleimen-Malkoun,
Jean-Jacques Temprado, Sylvain Viry,
Eric Berton, Michel Laurent and
Caroline Nicol

Theoretical rationale behind the study of horse-rider interactions

Horse-rider functional coordination

The understanding of horse-human interaction became critical when humans acquired the ability to tame horses and train them to carry individuals and merchandise (Robinson, 1999). Independently of the context (horseback fighting, hunting, sports and leisure), the dimension of movement control and synchronization is seen as a prominent aspect of the functional interactions between horses and humans. More recently, it has led to the introduction of synchronization/coordination between the horse and the rider as an important issue in equine research (cf. Lagarde et al., 2005; Pfau et al., 2009; Wolframm et al., 2013). The term *coordination* is employed to describe the spatiotemporal relationship between two or more components or agents that are coupled, by virtue of informational exchanges and/or mechanical entrainment. Accordingly, in equestrian practices, the horse-rider (H-R) coordination is an assembly of spatiotemporal patterns emerging as a result of an informational (haptic, visual) coupling that governs their ongoing interactions.

For a long time, the equine research literature has been predominantly addressing the horse and the rider as separate systems by focusing either on the ongoing inter-limb coordination of the horse (gaits or energy cost, Wickler et al., 2001) or the rider (riding techniques), and overemphasizing the influence of the rider on the horse's behaviour. Consequently, in spite of a recent growing interest (Lagarde et al., 2005; Schöllhorn et al., 2006; Visser et al., 2008; Pfau et al., 2009; Witte et al., 2009; Wolframm et al., 2013; Munz et al., 2014), knowledge of how the intra-individual coordination patterns of the horse and the rider are orchestrated to give rise to a stable and (more or less) synchronized global dynamics has only taken its first steps. In this respect, we have shown the major contribution that can be offered by combining the framework of

Social Coordination Dynamics and sophisticated portable motion recording technology (Viry et al., 2013).

Horse-rider interactions as a specific case of social coordination

Recently, it became obvious that H-R interactions are not solely biomechanically driven or purely controlled by the rider, but also, and more importantly, mediated by a mutual informational exchange (Visser et al., 2003; Visser et al., 2008; Munsters et al., 2012; Wolframm & Meulenbroek, 2012). Thus, to the extent that H-R interactions comply with the definition that *"two a priori independent living systems must mutually interact and (intentionally or not) coordinate their motions on the basis of information exchange (haptic, visual) in order to achieve specific goals"* (Bernieri & Rosenthal, 1991; Marsh et al., 2009), it can be considered as social coordination.[1] Such an interplay of forces and informational exchanges entails a mutual coupling between the respective intra-individual dynamics, and results in establishing stable H-R coordination patterns.

The framework of Social Coordination Dynamics applied to the study of horse-rider interactions

The basic assumption of Coordination Dynamics (CD) is that whatever the nonlinear complex system under consideration, coordination patterns arise spontaneously as the result of self-organization from mutual coupling among interacting subsystems (e.g. neural, muscular, mechanical, energetic, environmental) (Kelso et al., 1981; Haken, 1983; Kelso, 1984; Haken et al., 1985; Kelso, 1995; Kelso, 2009a). The complex dynamics can thus be studied through the changes over time of low dimensional collective variables (so-called order parameters, Haken, 1983). This framework has been shown to be relevant to the study of inter-personal coordination (see Oullier & Kelso, 2009 for an overview). In the same vein, synchronization between the horse and the rider can be viewed as a special case of inter-personal coordination that belongs to a family of processes generic to the organization of complex brain-behaviour systems (Kelso, 1995; Rosenblum et al., 2001). Indeed, in horse riding, both horse and rider move in 3D space trying to manage mechanical constraints associated with the horse's gait, but the rider must also continuously adapt his/her movement (in particular through riding techniques) to assemble a common coordination pattern that: i) optimizes efficiency (Pfau et al., 2009), ii) preserves horse integrity (Viry et al., 2014), and iii) facilitates informational exchanges (using rider-saddle contact, for instance; Lagarde et al., 2005).

Supporting evidence for information coupling in the (complex) horse-rider system has been provided in the literature. For instance, Lagarde et al. (2005) showed that synchronization between rider and horse motions is more stable in expert riders than in novices, suggesting a stronger and more efficient horse-rider coupling (HRC). This view can also be used to account for Peham and Schobesberger's (2004) findings showing that skilled riders were able to further stabilize horse trotting (on a treadmill) relative to free gait (without rider). To account for the dynamics resulting from intentional forcing of HRC, Lagarde et al. (2005) suggested that the contact between the horse and the rider (through the saddle) conveys haptic information that is functional for driving horse gait. These different findings inspired us to take part in the development of a firm framework to study horse-rider dynamics through macroscopic variables capturing HRC in natural riding conditions, namely during an endurance race, which presents a rich context for studying HRC without any experimental constraints on either horse gait or riding techniques. To meet the challenge of the race while preserving horse integrity, the horse-rider couple must assemble adequate patterns of coordination allowing the accommodation of transient external and internal constraints (related to either or both horse and rider). The nature, the distribution and the stability of the different H-R coordination patterns can be studied a posteriori from blind analysis of raw kinematic data. In this respect, we showed that H-R coordination patterns emerging in an endurance race can be consistently characterized by analysing the vertical displacements of the rider (craniocaudal) and the horse (dorsoventral) (Viry et al., 2013; Viry et al., 2014). In the following section, we present the methodology for extracting coordination patterns resulting from the functional combination of horse gaits (trot[2], canter[3]) and riding techniques (sitting, rising and two-point[4]). We then discuss the outcome of this methodology and review the main findings of previous research, before concluding with a discussion of more general applications and future perspectives.

Developing a method for extracting horse-rider coordination patterns

Data acquisition and pre-processing

The dynamical system derived method to characterize H-R coordination pattern can be applied to kinematic data recorded via two tri-axial accelerometer data loggers (see Viry et al., 2013 for a detailed description). Effectively, an acceleration time series of the horse and the rider can be recorded by two accelerometers positioned to record data along the 'vertical' (dorsoventral for the horse and craniocaudal for the rider), anteroposterior (fore-aft), and mediolateral (side to side) axes.

Acceleration signals between horse and rider should be precisely synchronized and then analysed within stride segments, with a stride being defined as a full cycle of limb motion. The respective displacement data are obtained by double integration. The different following steps allowing the identification of H-R coordination patterns are applied to the displacement data along the vertical axis, i.e. dorsoventral (DV) for the horse and craniocaudal (CC) for the rider, which appear to be the most reliable across studies (Barrey et al., 1994; Barrey et al., 1995; Peham et al., 2004; Lagarde et al., 2005; Viry et al., 2013a). An illustration of how horse locomotion patterns relate to changes in acceleration and displacement profiles can be seen in Figure 16.1.

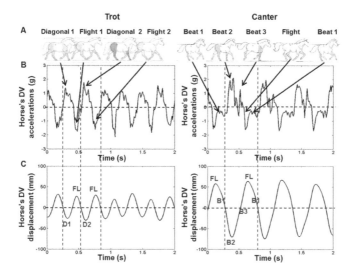

Figure 16.1 Acceleration and displacement profiles of the horse along the vertical axis at the trot and canter.

Notes: A: Representative coordination patterns between horse's limbs for each gait. B: Recorded dorsoventral (DV) acceleration profile. C: Calculated DV displacement profile. As shown by the vertical dashed lines, the DV displacement presents two repeated oscillations per stride at the trot and only one at the canter. The corresponding DV frequency at the trot is thus higher than at the canter.

Step 1. Identification of horse gaits

Horse gait patterns are differentiated according to their range of stride frequency (Minetti, 1998): the walk consists of a four-beat sequence with no suspension phase, which is characterized by a frequency of 55 strides/min; the trot is a two-beat gait with two stride-stance phases and two suspension phases, currently ridden at about 75 strides/min; and the canter is a three-beat pattern, which is performed at a higher frequency (99 strides/min) relative to the other gait patterns. Even when stride frequencies are very close to each other as in trot and canter (slow gallop), the distinction can still be made through the DV displacement profiles presenting two repeated oscillations per stride in trot (two contact and flight phases) and only one in canter. Barrey and collaborators (Barrey et al., 1994; Barrey et al., 1995; Barrey, 1999; Barrey et al., 2001) have shown that the DV displacement frequency corresponds to the half stride in trot and to the stride in canter. The differentiation of the horse's gait patterns can thus be inferred from the period of the DV oscillation, which allows counting the number of strides.

Step 2. Identification of riding techniques

In natural race conditions, a few riding techniques are actually used at each horse gait (e.g. Liesens, 2010). For instance, in endurance races the rider may use the so-called "two-point technique" at trot (*two-point trot*) and canter (*two-point canter*). In both cases, the rider remains off the saddle in equilibrium in his two stirrups, with only his calves in contact with the horse. The second technique is the so-called *rising trot*, in which the rider moves up and down at each stride, rising out from the saddle for one beat (half stride) and going down to touch the saddle lightly on the second one. Finally, the *sitting canter* technique requires that the rider remains mostly in contact with the saddle, making a sweeping motion with the hips. The riding technique being used can be identified based on the distribution of the amplitude of the rider's craniocaudal displacement. A method of minimization of error threshold can be used to determine the optimal threshold to differentiate the distributions obtained in each gait (a detailed description can be found in Kittler and Illingworth (1986)). These distributions allow the distinction of two groups of strides in each gait, which correspond to different riding techniques. The strides associated with the smaller amplitude of the rider's CC displacement correspond to the *two-point* technique (Pfau et al., 2009). Then, it is possible to unambiguously distinguish the *two-point trot* and *two-point canter* from the other techniques. For the two remaining groups of strides, those corresponding to the *sitting canter* technique are identified as being the ones mechanically associated with the largest CC displacements of the rider due to the prolonged contact with the saddle (see Wolframm et al., 2013 for a full description). The group of strides identified at the trot necessarily corresponds to the *rising* technique, which can be confirmed by the specificity of the rider's DV displacement.

Step 3. Representation of the H-R coordination patterns with Lissajous plots

Simple Lissajous plots can be used to represent the horse-rider coordination patterns. They provide a simplified picture of the spatiotemporal relationship between limb motions corresponding to different coordination patterns (e.g. Lee et al., 1995; Swinnen et al., 1997; Summers, 2002). These plots allowed us to capture the horse-rider spatiotemporal relationship along the vertical axis over stride cycles through a horse-displacement/rider-displacement representation, thereby revealing the existence of different patterns that emerge under unspecific constraints (Figure 16.2). It is noticeable in this respect that, before plotting, the vertical displacements of both the horse and the rider per stride should be normalized over time, but not for amplitude to allow subsequent differentiation of riding techniques within each gait.

Changes in horse-rider interactions are analysed through Lissajous plots under the hypothesis that, for a given stride, each horse-rider coordination pattern would correspond to a specific Lissajous profile. For instance, a Lissajous profile close to an oblique line (45° to the right) reflects an almost perfect in-phase vertical horse and rider displacements along the stride (Figure 16.2). Deviations from a perfect line will be observed in cases where there is a constant time delay in the H-R spatiotemporal relationship, thereby leading to a somewhat flattened ellipsoidal (Figure 16.2). In addition to global profiles of Lissajous plots, the H-R

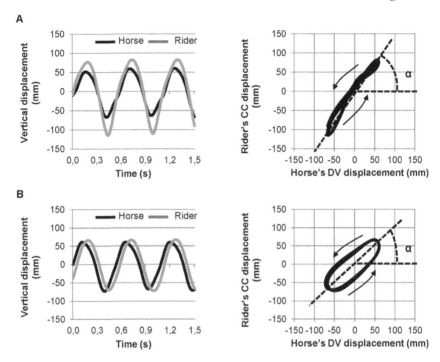

Figure 16.2 Lissajous plots obtained from the vertical displacements of the horse and
rider.

Notes: The angle reflects the ratio between the amplitude of the horse's dorsoventral (DV) and rider's
craniocaudal (CC) displacements. A reduction of the amplitude of the rider's displacement while that
of the horse is kept constant (B, left side plot compared to the one in A), reduces α angle (B vs. A, right
side plots). The global form of the Lissajous plot illustrates the spatiotemporal relationship (relative
phase) between the horse and the rider along the stride cycle. The sign of the relative phase is related
to the direction of rotation of the plot as indicated by black arrows.

relationship can be quantified by calculating the relative phase (RP) – that is, the
phase difference between the rider and the horse in their cycle. However, since
displacement profiles are not perfectly harmonic, the RP should be calculated at
the lower reversal point of the displacement using the discrete point estimate
method (see Zanone and Kelso, 1997; Hamill et al., 2000 for an overview of the
calculation methods). The time delay that separated the rider from the horse at
their respective lower point of vertical displacement is measured for each stride
and transformed into RP by applying the following formula:

$$RP=((Tmin(rider) – Tmin(horse))/Stride\ duration)*360$$

RP is thus based on the time difference at the lowest point of the vertical
displacements of the rider and the horse, and is expressed in degrees relative to the
period of the horse stride. Whenever the horse's and the rider's displacements are

in phase at this point, the RP value is close to zero (Figure 16.2, A). A negatively signed RP is obtained when the rider is ahead of the horse. Conversely, a positive RP (the horse is leading) is associated with a counter-clock wise sense of rotation of the Lissajous plot (Figure 16.2, B). The H-R coordination patterns corresponding to each of the two riding techniques per horse's gait can thus be characterized by their mean RPs and standard deviations.

The Lissajous profiles can also be characterized by their orientation (i.e. α angle, see Figure 16.2). For a similar DV displacement of the horse, α angle decreases with the reduction in the amplitude of the rider's CC displacement (Figure 16.2, B *vs.* A). Hence, this variable can be used to investigate the difference in coordination patterns between the two riding techniques within each type of gait.

Patterns of horse-rider coordination during endurance race

The study of H-R interactions during an endurance race is an interesting challenge since this type of event has specific constraints. First of all, it covers up to 160 km in a one-day competition, with various tracks and altitude conditions in one race. Second, in an endurance race efficiency is critical since the ultimate objective is to beat the opponents while crossing the finish line after having passed all veterinary inspections (vet gates) occurring before and after the race as well as between the different loops.[5] Accelerometers make it possible to acquire kinematic data over long durations and in natural conditions, while not imposing strong temporal and spatial constraints or compromising the precision and the quality of the recording.

Validating the Lissajous method to study horse-rider coordination: example from one dyad

In the following, we provide a practical example of the use of the aforementioned method, through representative data extracted from an expert female rider (27 years; 50 kg; 1.63 m) and a gelding horse (10 years; 320 kg; 1.50 m), who won a 130 km race by riding at a mean speed of 18.1 km/h and passing all the vet gates.

Step 1. Identification of horse gaits

The mean DV oscillation frequencies for a full stride were found to be 1.85 ± 0.17 Hz for canter and 1.60 ± 0.07 Hz for trot, with a more frequent use of trot than canter (L2: 63 *vs.* 37% and L3: 71 *vs.* 29% respectively).

Step 2. Identification of riding techniques

The analysis of the maximal amplitude of the rider's CC displacement per stride showed a bimodal distribution for both trot and canter (Figure 16.3), revealing the strides in which the *two-point technique* was used – that is, the ones with the lower range of displacement (63.7 ± 13.0 mm for *two-point trot*, and 134.7 ± 17.3 mm in *two-point canter*).

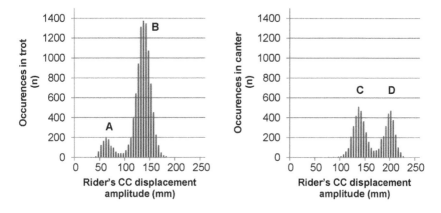

Figure 16.3 Distribution of the amplitude of the rider's craniocaudal displacement per horse's gait.

Notes: The rider's craniocaudal (CC) displacement has lower amplitude in the two-point technique (A and C) as compared to both sitting canter (D) and rising trot (B).

In trot, the distribution around the highest displacement amplitude (131.4 ± 14.2 mm) gathers strides, where the *rising trot* technique was used as demonstrated by the specific shape of the horse-rider coordination pattern for this technique. In canter, the distribution around the highest displacement amplitude (194.6 ± 11.2 mm) shows strides in which the *sitting canter* technique was used. Mean amplitude was the largest for the *sitting canter* technique.

In terms of percentage of use (Figure 16.3), in trot, the rider used predominantly the *rising trot* technique (mode labelled as B in Figure 16.4, occurring in 89 % of the strides), while in canter the *two-point* (*C*) and *sitting canter* (*D*) techniques were almost equally used (53 and 47 %, respectively).

Step 3. Representation of the H-R coordination patterns with Lissajous plots

The Lissajous plots show a distinct profile for each gait-technique combination (Figure 16.4). H-R coordination patterns can be actually differentiated on the basis of the Lissajous profiles.

The ellipsoidal profiles observed during canter for both the two-point and the sitting techniques indicated that the rider and the horse were globally coupled in-phase during the whole cycle. The thicker profile observed for the *two-point* technique roughly reflects a positive time-delay between the rider and horse displacements, which was confirmed by the RP analysis. Indeed, it revealed that the *two-point* riding technique was associated with a larger RP value (29.9 ± 8.2 deg) than the *sitting* technique (8.0 ± 6.3 deg) (Figure 16.4). Thus, the coordination patterns adopted in the *two-point* and the *sitting* techniques differed mostly with respect to the time delay between the displacements of the horse and the rider at the lowest point. The analysis of the α angle shows a lower value of inclination for

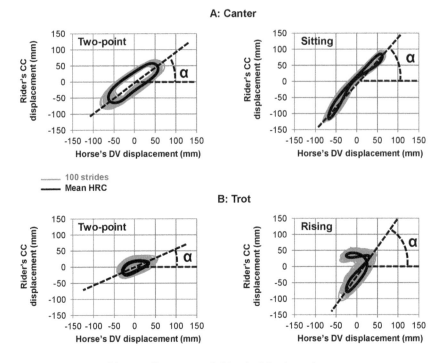

Figure 16.4 Horse-rider coupling as revealed by the Lissajous plots.

Notes: In each panel the rider's craniocaudal (CC) displacement is plotted against the horse's dorsoventral (DV) displacement. The shape of the Lissajous plot is clearly gait specific: the trot (B) is characterized by two successive coordination patterns versus one for canter (A). In (B) the two successive patterns either overlap (two-point technique) or are different (rising trot). For both trot and canter the riding technique can be differentiated based on the inclination of the Lissajous plot (α angle). The black curves represent the mean plot across strides and the grey curves correspond to 100 successive strides.

the *two-point* technique relative to the *sitting* one (47.2 ± 2.8 deg and 57.0 ± 1.5 deg, respectively) consistent with smaller CC displacements of the rider.

Profiles of Lissajous plots observed during trot were qualitatively different from those observed during canter, and they differed also between the two adopted techniques (Figure 16.4). Specifically, the *rising trot* was found to be characterized by two successive, but different, coordination patterns (De Cocq et al., 2013), whereas the *two-point trot* has two similar (overlapping) ones (Figure 16.4). In *rising-trot* two ellipses were observed, one with an inclination (α = 66.8 ± 4.1 deg) and the other flat. The flat part reflects a freezing-like behaviour of the rider during one horse DV oscillation. So, for the inclined portion, the two oscillators (horse and rider) were coupled in-phase (RP = 13.3 ± 5.8 deg), in a similar fashion as in canter. However, during the flat portion, the rider's dynamics were decoupled from those of the horse. In the *two-point trot* the coordination pattern presented two overlapping ellipses. In this technique, the inclination of

Lissajous plot was smaller (α = 37.9 ± 8.7 deg) than in *rising* technique. As in canter, the time delay between horse and rider's displacement was found to be larger in the *two-point* technique (24.0 ± 10.0 deg) than in the *rising* technique (13.3 ± 5.8 deg at the *rising trot*). This former technique appeared to be dynamically less consistent than the *rising trot* (more variable RP, Figure 16.5). It was actually the least used, and should certainly correspond to specific situations of infrequent occurrence.

Generalization to expert horse-rider dyads in endurance race

In Viry et al. (2014), it was shown that this method applies nicely to endurance race data and that it shows consistent results across different horse-rider dyads. Mainly it was shown that the riders used two riding techniques per horse gait, which are clearly distinguishable in the Lissajous profiles that capture the HRC during the stride period. Overall, it seems that in endurance races the spatiotemporal patterns of coordination between the horse and the rider are robust and reproducible over multiple dyads despite changes in environmental conditions within and between races (see Figure 16.6 for mean plots taken from Viry et al., 2013). The group analysis offered additional insights on how race conditions and racers may affect the adopted coordination patterns. For instance, the mean frequencies of the different combinations observed in the different dyads – that is, during the six different races – showed large variability (see Viry et al., 2013). Such a finding was thought to suggest that the occurrence of the different patterns strongly depended on race conditions (length of the loops, nature of the terrain etc.).

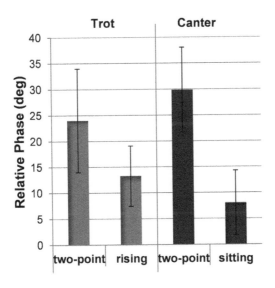

Figure 16.5 Relative phase at lower reversal point.

Notes: Mean and standard deviations of the relative phase at the lowest point of the horse's and the rider's vertical displacements for all technique-gait combinations.

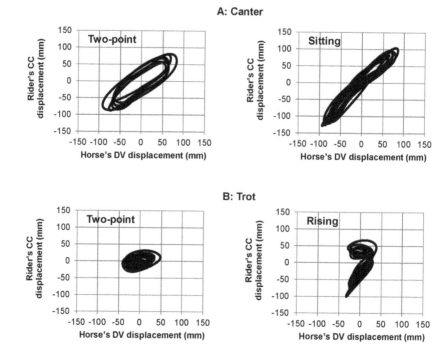

Figure 16.6 Lissajous plots of the six dyads.

Notes: The curves represent the six individual Lissajous plots (averaged over 100 strides) for each combination of horse's gait (A and B) and riding technique (left and right panels).

The statistical analysis of the α angles confirmed the robustness of the differences in-between the Lissajous profiles (Figure 16.6), with the *rising trot* having the greatest inclination ($\alpha = 66.9 \pm 1.8$ deg) followed by the *sitting canter* ($\alpha = 55.6 \pm 6.9$ deg). The *two-point canter* and *trot*, differentiated a priori based on stride frequency, had no significantly different α angles ($\alpha = 47.6 \pm 5.8$ and 44.8 ± 1.5 deg). Observations with the RP analysis were also quite robust: the *sitting canter* showing the lowest RP value (-2.3 $\pm$ 9.4 deg), followed by the inclined portion of *rising trot* (7.9 $\pm$ 7.3 deg), and finally the *two-point* technique that showed similar values in both canter and trot (15.6 $\pm$ 7.8 and 19.8 $\pm$ 5.8 deg).

It can be concluded that in *canter* the riders' displacements were in phase with those of their horses. The difference in values of the α angle between the *two-point* (α close to 45 deg) and the *sitting* techniques ($\alpha > 45$ deg) actually results from the fact that in *sitting canter*, the riders do not remain seated in the saddle. Effectively, they only maintain contact during a brief portion of the cycle – that is, at the minimal amplitude of the horse's CC oscillation – which helps their propulsion above the saddle. Thereafter, they move down in synchrony with the horse till the reversal point of the downward oscillation. The RP values calculated for the *sitting canter* were significantly lower and less variable than those observed in the *two-point* technique, showing that at lower reversal points the riders maintain a close contact with the saddle (source of haptic information).

During *trot*, different spatiotemporal relations (and hence coordination patterns) are observed where, at least for a portion of the cycle, the riders relatively froze their displacements and decoupled from the horse. This can be seen, although it is less marked, through the rather flat ellipsoidal Lissajous profile of the *two-point trot*, showing the weakness of HRC. More obviously, in the *rising trot*, the 7-like profile captures how for half a stride the riders remained in close contact with the saddle (oblique portion), whereas during the subsequent half-stride they were in equilibrium on the stirrups (flat portion). This profile suggests that the rider accomplished one oscillation per stride while the horse oscillated twice as often. The larger α angles indicate that for the portion during which the horse and the rider were coupled in-phase, the displacements of the rider were larger than those of the horse.

The stability of the different patterns was attested by the variability analysis, which showed lower values of SD of relative phase when compared to those observed in classic laboratory studies on interlimb coordination (e.g. Salesse et al., 2005) or even on interpersonal coordination (Temprado and Laurent, 2004). Thus, even when measured in natural situations and over a very large number of strides, constraints led to very stable horse-rider coordination patterns, at least at lower reversal points of the gait cycle. It was noticeable that the RP's variability at lower reversal points was smaller for *sitting canter* and *rising trot* patterns than for the two others. These results suggest that the rider's technique was more constraining on the type and stability of the adopted coordination pattern than was the horse gait. This gives support to the existence of a specific coordination within the horse-rider dyad that would not depend on the horse's gait. The stability of the coordination pattern observed at *sitting canter* and *rising trot* is presumed to result from the anchoring provided by haptic contact information of the rider with the saddle. This observation points out the importance of analyzing discrete relative phase at the lower reversal point of the cycle, which is more functional relative to the upper portion (reflecting the rider's off-the-saddle freezing-like behaviour in *rising trot*, for instance). The larger RP values found in the *two-point trot* and *canter* techniques were attributed to the fact that the coordination pattern is less stable in the *two-point* technique due to the lack of informational contact with the saddle than in the *rising trot* and *sitting canter*.

Discussion, applications and perspectives

The concepts, methods and tools of dynamical systems analysis can be considered as heuristics to describe the nature of coordination patterns assembled within H-R systems (Viry et al., 2013) and to identify the multiple constraints at the origin of their occurrence (Viry et al., 2014).

Based on the kinematics obtained from accelerometric data, the patterns of H-R coordination can be extracted and the evolution of H-R coupling can be then characterized. The different results that are presented in this chapter illustrate the application of the framework of Social Coordination Dynamics to a quite challenging natural equestrian context, namely an endurance race. This constitutes a significant addition to what has been previously shown in dressage (e.g. Lagarde et al., 2005; De Cocq et al., 2013).

H-R synchronization is presented here as a special case of inter-personal coordination. The shapes and angles of Lissajous plots, together with the values of relative phase between horse and rider displacements at lower reversal points, allow an unambiguous characterization of the different coordination patterns. The only pre-requisite is the correct identification of horse gait patterns and riding techniques, which can be done through specific kinematic signatures. What is noticeable in endurance races is that the frequency at which different gait patterns are used is variable and non-equivalent (cf. Viry et al., 2013). This is not surprising since, unlike in dressage, horses are not strictly constrained to adopt a single pattern. Instead, the adoption of different gait patterns reflects the online adaptation to constraints of various origins, including those related to the strategies of the rider. Nevertheless, with a sufficient number of observations (strides), the combinations between horse gaits and riding techniques can be identified, and subsequently. so can the specific H-R coordination patterns. In endurance races four patterns, resulting from the use of two riding techniques per horse's gait (*trot* and *canter*), appear to be consistently present (Viry et al., 2013; Viry et al., 2014). In addition to the observed general tendencies, the presence of a certain variance between races also offers a considerable validation to the introduced method of identification of H-R coordination patterns. Indeed, in addition to being robust in detecting invariances, it preserved individual and environmental specificities.

Our different studies showed that H-R coordination patterns are highly reproducible, easily identifiable and clearly distinguishable on the basis of relative phase, its variability and α angle analyses. For instance, the *sitting canter* was shown to be characterized by the largest intra-dyad variability in α angle, and by the lowest mean RP value. These observations suggested that the riders, at least the expert ones, were able to adjust and synchronize their vertical displacement with the horse's motion in order to be in phase at the lowest point of their respective movement reversal. This is consistent with the hypothesis that the contact between the horse and the rider (through the saddle) might haptically convey effective communication between them (Lagarde et al., 2005). Conversely, the patterns observed during canter (e.g. *sitting canter*) were quite different from those observed during trot. As an illustration, the pattern characterizing the *rising trot* technique corresponded to a 7-like profile, which is presumably due to the freezing of the rider's displacement during half a cycle, requiring an active spring system to account for such displacement (De Cocq et al., 2013). This freezing was also observed, although it was less marked, in the *two-point* technique. Thus, at a more abstract level, the adopted representation of H-R coordination patterns allowed us to capture the outcome of the coalition of constraints acting on the different systems (e.g. fatigue) and on their coupling (e.g. frequency detuning between horse and rider oscillations). The analysis of RP and its variability provide complementary information about coordination patterns. For instance, it was noteworthy that RP variability was lower in such a natural situation than in most experimental studies on inter-limb and inter-personal coordination. Differences of stability between the coordination patterns were also observed, which shows informational exchanges between the horse and the rider at specific points in their oscillation cycle. For instance, in *rising trot* and *sitting canter*, low reversal points of the cycles appeared to be very functional relative to upper portions. This presumably resulted from the anchoring provided by the haptic contact of the rider with the saddle. As a consequence, in the *two-point* technique (where this contact was absent or less marked), larger values of relative phase and variability were observed.

The emergence of different coordination patterns revealed by Lissajous plots is consistent with the interplay of metabolic and biomechanical factors related to riding techniques and gait patterns (see Viry et al., 2013; Viry et al., 2014 for detailed explanations). This is easily understandable since, in this type of race, the rider's energy expenditure is affected by their horse's gait (Devienne and Guezennec, 2000), their riding technique (Westerling, 1983; Trowbridge et al., 1995) and their expertise (Lagarde et al., 2005). For instance, the *two-point* technique has been reported as beneficial for the horse's back loading and speed, although metabolically and mechanically costly for the rider. Therefore, during endurance races, the proportional use of different combinations of gaits and riding techniques could be expected to depend on the racing strategy as well as on the fatigue status of the horse and rider.

Overall, H-R coordination patterns appear to obey general principles of self-organization in dynamical systems. Thus, the methods introduced by Viry et al. (2013) could serve as a basis for investigating the factors and constraints that influence the use and stability of the different coordination patterns: horses' and riders' respective fatigue, their mental and physical states, the temperament of horses, typologies of riders, the terrain profile etc. However, beyond the application of the framework of Social Coordination Dynamics to H-R interactions, our studies might also have practical consequences for equestrians (riders, trainers, veterinarians etc.). Indeed, in the future, the association of accelerometer measurement devices with sophisticated technologies and software that allow recording additional information (positioning, physiological parameters etc.) might lead to a better understanding of the H-R system.

Notes

1   It is noticeable that this definition applies even if it is not clear whether this exchange is true the other way around – that is, if the horse adapts to its rider's personality, and not only to constraints imposed through changes in body positions, stirrups actions, or whip (see Visser et al., 2008).
2   The trot is a two-beat diagonal (the forefoot of one side and the hind food of the opposite side move together) gait, with an even cadence of footfalls in a 1-2 pattern and a suspension period (flight) in between beats.

3   The canter is a rhythmical three-beat gait with an even cadence of footfalls in a 1-2-3 pattern. The independent moving foreleg is the lead.
4   The main difference between these riding techniques is the contact of the rider with the saddle: in the sitting technique the rider remains mostly in contact, in the rising technique the rider moves away from the saddle on some beats, and finally in two-points the rider remains off the saddle.
5   FEI rules for endurance riding can be found online: http://www.fei.org/disciplines/endurance/rules.

References

Barrey, E. (1999) Methods, applications and limitations of gait analysis in horses. *Vet J*, *157*, 7–22.

Barrey, E., Auvinet, B. & Couroucé, A. (1995) Gait evaluation of race trotters using an accelerometric device. *Equine Vet J*, *27*, 156–160.

Barrey, E., Evans, S.E., Evans, D.L., Curtis, R.A., Quinton, R. & Rose, R.J. (2001) Locomotion evaluation for racing in thoroughbreds. *Equine Vet J Suppl*, 99–103.

Barrey, E., Hermelin, M., Vaudelin, J.L., Poirel, D. & Valette, J.P. (1994) Utilisation of an accelerometric device in equine gait analysis. *Equine Vet J*, 7–12.

Bernieri, F. & Rosenthal, R. (1991) Interpersonal coordination: Behavior matching and interactional synchrony. In R. Feldman & B. Rime (Eds.), *Studies in emotion and social interaction* (pp. 401–432). Cambridge University Press): New York.

De Cocq P., Muller M., Clayton H.M. & van Leeuwen J.L. (2013) Modelling biomechanical requirements of a rider for different horse-riding techniques at trot. *J Exp Biol*, *216*, 1850–1861.

Devienne, M.F. & Guezennec, C.Y. (2000) Energy expenditure of horse riding. *Eur J Appl Physiol*, *82*, 499–503.

Haken, H. (1983) *Synergetics: An introduction*. Springer-Verlag: Berlin.

Haken, H., Kelso, J.A.S. & Bunz, H. (1985) A theoretical model of phase transitions in human hand movements. *Biol Cybern*, *51*, 347–356.

Hamill, J., Haddad, J.M. & Mcdermott, W.J. (2000) Issues in Quantifying Variability From a Dynamical Systems Perspective. *J Appl Biomech*, *16*, 407.

Kelso, J.A. (1984) Phase transitions and critical behavior in human bimanual coordination. *Am J Physiol*, *246*, R1000–R1004.

Kelso, J.A.S. (1995) *Dynamic patterns: The self-organization of brain and behavior*. MIT Press: Cambridge, MA.

Kelso, J.A.S. (2009a).Coordination dynamics. In M. R.A. Springer (Ed.), *Encyclopedia of Complexity and Systems Science* (pp. 1537–1564).

Kelso, J.A.S. (2009b) Synergies: atoms of brain and behavior. *Adv Exp Med Biol*, *629*, 83–91.

Kelso, J.A.S., Holt, K.G., Rubin, P. & Kugler, P.N. (1981) Patterns of human interlimb coordination emerge from the properties of non-linear, limit cycle oscillatory processes: theory and data. *J Mot Behav*, *13*, 226–261.

Kittler, J. & Illingworth, J. (1986) Minimum error thresholding. *Pattern Recogn*, *19*, 41–47. doi: http://dx.doi.org/10.1016/0031-3203(86)90030-0.

Lagarde, J., Kelso, J.A., Peham, C. & Licka, T. (2005) Coordination dynamics of the horse-rider system. *J Mot Behav*, *37*, 418–424.

Lee, T.D., Swinnen, S.P. & Verschueren, S. (1995) Relative Phase Alterations during Bimanual Skill Acquisition. *J Mot Behav*, *27*, 263–274.

Liesens, L. (2010) Endurance, Le livre d'un cavalier, Pour les cavaliers. www.endurance-belgium.com

Marsh, K.L., Richardson, M.J. & Schmidt, R.C. (2009) Social Connection Through Joint Action and Interpersonal Coordination. *Topics in Cognitive Science*, *1*, 320–339.

Minetti, A.E. (1998) A model equation for the prediction of mechanical internal work of terrestrial locomotion. *J Biomech*, *31*, 463–468.

Munsters, C.C., Visser, K.E., Van Den Broek, J. & Sloet Van Oldruitenborgh-Oosterbaan, M.M. (2012) The influence of challenging objects and horse-rider matching on heart rate, heart rate variability and behavioural score in riding horses. *Vet J*, *192*, 75–80.

Munz, A., Eckardt, F. & Witte, K. (2014) Horse-rider interaction in dressage riding. *Hum Mov Sci*, *33*, 227–237.

Oullier, O. & Kelso, J.A.S. (2009) Social coordination from the perspective of Coordination Dynamics. In M.R.A. Springer (Ed.), *The Encyclopedia of Complexity and Systems Science* (pp. 8198–8212). Berlin: Springer verlag.

Peham, C., Licka, T., Schobesberger, H. & Meschan, E. (2004) Influence of the rider on the variability of the equine gait. *Hum Mov Sci*, *23*, 663–671.

Peham, C. & Schobesberger, H. (2004) Influence of the load of a rider or of a region with increased stiffness on the equine back: a modelling study. *Equine Vet J*, *36*, 703–705.

Pfau, T., Spence, A., Starke, S., Ferrari, M. & Wilson, A. (2009) Modern riding style improves horse racing times. *Science*, *325*, 289.

Robinson, I.H. (1999) The human-horse relationship: how much do we know? *Equine Vet J Suppl*, 42–45.

Rosenblum, M.G., Pikovsky, A.S. & Kurths, J. (2001) Comment on Phase synchronization in discrete chaotic systems. *Phys Rev E*, *63*, 058201.

Salesse, R., Temprado, J.J. & Swinnen, S.P. (2005) Interaction of neuromuscular, spatial and visual constraints on hand-foot coordination dynamics. *Hum Mov Sci*, *24*, 66–80.

Schmidt, R.C., Carello, C. & Turvey, M.T. (1990) Phase transitions and critical fluctuations in the visual coordination of rhythmic movements between people. *J Exp Psychol Hum Percept Perform*, *16*, 227–247.

Schöllhorn, W.I., Peham, C., Licka, T. & Scheidl, M. (2006) A pattern recognition approach for the quantification of horse and rider interactions. *Equine Vet J Suppl*, 400–405.

Summers, J. (2002) Practice and Training in Bimanual Coordination Tasks: Strategies and Constraints. *Brain Cognition*, *48*, 166–178.

Swinnen, S.P., Lee, T.D., Verschueren, S., Serrien, D.J. & Bogaerds, H. (1997) Interlimb coordination: Learning and transfer under different feedback conditions. *Hum Mov Sci*, *16*, 749–785.

Temprado, J.J. & Laurent, M. (2004) Attentional load associated with performing and stabilizing a between-persons coordination of rhythmic limb movements. *Acta Psychol*, *115*, 1–16.

Trowbridge, E.A., Cotterill, J.V. & Crofts, C.E. (1995) The physical demands of riding in National Hunt races. *Eur J Appl Physiol O*, *70*, 66–69.

Viry, S., De Graaf, J.B., Frances, J.P., Berton, E., Laurent, M. & Nicol, C. (2014) Combined influence of expertise and fatigue on riding strategy and horse-rider coupling during the time course of endurance races. *Equine Veterinary Journal*.

Viry, S., Sleimen-Malkoun, R., Temprado, J.-J., Frances, J.-P., Berton, E., Laurent, M. and Nicol, C. (2013) Patterns of horse-rider coordination during endurance race: A dynamical system approach. *PLoS ONE*, *8*, e71804.

Visser, E.K., Van Reenen, C.G., Blokhuis, M.Z., Morgan, E.K., Hassmen, P., Rundgren, T.M. & Blokhuis, H.J. (2008) Does horse temperament influence horse-rider cooperation? *J Appl Anim Welf Sci, 11*, 267–284.

Visser, E.K., Van Reenen, C.G., Rundgren, M., Zetterqvist, M., Morgan, K. & Blokhuis, H.J. (2003) Responses of horses in behavioural tests correlate with temperament assessed by riders. *Equine Vet J, 35*, 176–183.

Westerling, D. (1983) A study of physical demands in riding. *Eur J Appl Physiol O, 50*, 373–382.

Wickler, S.J., Hoyt, D.F., Cogger, E.A. & Hall, K.M. (2001) Effect of load on preferred speed and cost of transport. *J Appl Physiol, 90*, 1548–1551.

Witte, K., Schobesberger, H. & Peham, C. (2009) Motion pattern analysis of gait in horseback riding by means of Principal Component Analysis. *Hum Mov Sci, 28*, 394–405.

Wolframm, I.A., Bosga, J. & Meulenbroek, R.G. (2013) Coordination dynamics in horse-rider dyads. *Hum Mov Sci, , 32*(1), 157–170.

Wolframm, I.A. & Meulenbroek, R.G.J. (2012) Co-variations between perceived personality traits and quality of the interaction between female riders and horses. *Applied Animal Behaviour Science, 139*, 96–104.

Zanone, P.G. & Kelso, J.A.S. (1997) Coordination dynamics of learning and transfer: collective and component levels. *J Exp Psychol Hum Percept Perform, 23*, 1454–1480.

Part III

Factors that influence interpersonal coordination

17 Affordances and interpersonal coordination

Kerry L. Marsh and Benjamin R. Meagher

Note

This work was completed while the first author was serving at the National Science Foundation. The views expressed in this paper are solely those of the authors and do not necessarily represent those of NSF. The authors thank an anonymous reviewer for comments on an earlier version of this paper.

Introduction

Coordinating our movements with others around us to achieve some goal, whether it is in the course of well-practiced actions involving highly honed skills on a football pitch or in the context of trying to unexpectedly evacuate a crashed railway carriage, makes incredible, time-constrained demands on us. We must rapidly see obstacles and opportunities for action in the environment surrounding us, seeing what is possible for ourselves and noting what is possible for others as well. Moreover, as we act and events unfold, our awareness of how new possibilities are opening up and others are closing down must be continually "updated" and adjusted to dynamically.

That we are able to do so rapidly and effectively is strong evidence for a central tenet of research on perception-action processes that has its origins in Gibson's ecological theory of perception (Gibson, 1979): for an animal in motion in their environment, information is available to it that specifies what Gibson termed *affordances* – possibilities for action, that hold under certain occasions for animals that have certain action capabilities (Michaels & Carello, 1981; Turvey, 1992). Possibilities are innumerable: that an aperture is "pass-through-able" rather than too small for us to fit our body through, that one can move under an obstacle if one ducks one's head, that an object is "grasp-able" and lightweight enough to be wielded. Once one takes up a detached object, such as a wooden rod, the affordance that a (previously unreachable) light switch is now "reach-able" is specified by information that is available for detection by the perceiver who is now operating as an individual+tool system. As these examples reveal, affordances are real (not constructed), *relational* features of the environment; they exist *with respect to* an individual's perception-action system, where that system may include other tools. Thus clothing can constrain us or assist us, as do knee pads, helmets, and high-heeled or cleated shoes; other tools that are wielded also extend our action capabilities, as would a hockey stick, rifle or broom.

The affordances that are possible for humans can be objectively quantified by finding the right dimensions of the environment taken with respect to the right features of the animal. For instance, the affordance of stairs being "climb-able" (or specifically, climbable using a normal walking stride) has been empirically demonstrated to be precisely determined by the height of the stair riser, with respect to a person's leg length (Warren, 1984). Research has also confirmed other central predictions of Gibson's approach (Michaels & Carello, 1981). Namely, active perceivers are sensitive to information that specifies affordances. They can readily verbalize their awareness of the affordances, and more importantly, their actions demonstrate this. For instance, people's natural tendencies to shift the manner in which they complete the action of, say, reaching for or moving an object are precisely determined by measures that assess an objective feature of the environment, scaled to a body (Cesari & Newell, 1999; Mark et al., 1997; van der Kamp, Savelsbergh & Davis, 1998).

Extensive research demonstrates that people are sensitive to affordances for themselves, one necessary condition for interpersonally coordinated action. Three other conditions are necessary as well. The first is that the individual must be sensitive to others in their environment, in terms of the action possibilities they present to the individual. Because people have inner lives and hidden motives, humans pose greater complexity to affordance perception than do inanimate tools in our environment. Are people accurate in determining crucial features of others that would be essential for being able to coordinate with them (e.g. their attributes and their intentions)?

A second condition is that individuals must be sensitive to how others' action capabilities fit with the environment, even if those other individuals differ from themselves. What evidence is there that we are able to see what actions are possible for another person whose body may differ substantially from our own? The third condition is that there should be new affordances that are available to individuals who are operating as an *interpersonally coordinated*, social perception-action system – that is, who are acting as a team. Do the principles about affordances that have been established for solo actors also hold for the flexible, coordinative assembly of multiple individuals that make up what has been termed a "social synergy" (Marsh, Richardson, Baron & Schmidt, 2006)?

In this chapter we begin by addressing current research on each of these three conditions. Following this review, we scale our discussion up beyond the time and space of a single dyad in order to consider the role of affordances within broader social contexts and settings, which inevitably involves situational, group-level and cultural factors. We address each of these issues in separate sections.

Evidence

Evidence regarding detecting what others afford me

Other people are particularly important and salient features of the world, as they attract a disproportionate percentage of human perceptual attention (Downing, Bray, Rogers & Childs, 2004; Fletcher-Watson, Findlay, Leekam & Benson, 2008). Importantly, other people may afford new and potentially unique opportunities for action. Gibson (1979) himself noted that other people are for humans the "richest and most elaborate affordances of the environment" (p. 135), and successful functioning within human groups, cities and societies requires attunement to the social information provided by others. Presumably, humans should have the capacity to detect information about another person that indicates what opportunities are available for social interaction (McArthur & Baron, 1983). For researchers seeking to evaluate this claim in the context of social perception, two questions can be posed: do people accurately detect the attributes of other people, and what is the information they are picking up to make these judgements?

Evidence suggests that humans can accurately detect a number of personal attributes after exposure to even minimal information, such as photographs or video recordings (for a review, see Zebrowitz & Montepare, 2006). Examples of highly accurate trait and identity judgements include intelligence (Zebrowitz & Rhodes, 2004), power (Berry, 1991a;), gender (Berry, 1991b) and sexual orientation (Rule, Ambady & Hallett, 2009). However, accurately detecting certain attributes requires having access, or being attuned, to the type of information that specifies these traits. It is for this reason that judgements of extraversion, a trait with clear and overt behavioural manifestations, tend to be far more accurate than ratings of the other Big Five personality traits (Kenny, 1994). Moreover, people differ in the extent to which they are "good targets", who actively communicate their personality to others (Funder, 1995; Human & Biesanz, 2013).

Other research has focused on the capacity to detect the current mental states of another person, such as their emotions and intentions. The challenge is determining whether particular patterns of motion can directly communicate what are traditionally thought of as internal mental processes. For example, Charles (2011) describes how tau (τ), measuring time to impact with an object (Lee, 1976), can be understood as expressing the intentions of a moving organism: "The way that different behaviours are coordinated with the time to impact with an object *is* the difference between a hawk, say, intending to land on a branch, or intending to crash through it" (p. 138). In the same way, perceivers can detect intentionality from certain physical behaviour. For example, Hodges and Lindhiem (2006) found that perceivers, when shown point-light-displays of people carrying a variety of objects, could detect carefulness in people's gait patterns. This finding is consistent with Runeson and Frykholm's (1983) earlier use of point-light films, which showed that perceivers could detect people's intention to disguise the weight of a box while lifting. Such findings have interesting implications for sport contexts. A skilled opponent may become highly attuned to the information that conveys an athlete's intentions to pass a ball to another player. Overt information such as change in gaze direction might be suppressed, but more subtle shifts in body weight and position might be detectable.

Evidence that people are sensitive to action-relevant person-environment fit for others

Understanding the social world and coordinating effectively with others also requires being capable of detecting what actions are possible for *them* in an environment, or adopting what is called an *allocentric* perspective (Stoffregen, Gorday, Sheng & Flynn, 1999) and scaling the environment with respect to another person's body. Research indicates that people are in fact able to adopt such a perspective. For example, Stoffregen et al. (1999) found that perceivers were able to judge accurately the maximum (i.e. the critical action boundary) and preferred sitting height of actors of varying heights by scaling to the leg length of the sitter. Mark (2007) found similar results for judgements of others' ability to reach, climb and step across gaps. In all cases, critical action boundaries were invariant when scaled to the body of the person being observed, a finding that suggests a process very similar to what occurs when judgements are made about one's own affordances. Beyond just scaling to body dimensions, perceivers can also utilize dynamic information about others' physical abilities to form these judgements. In one study, participants observed a person's walking patterns after they were discretely encumbered by ankle weights (Ramenzoni, Riley, Davis, Shockley & Armstrong, 2008). This information was sufficient to significantly lower estimates of that actor's maximum jumping height, relative to watching unencumbered walkers. Other researchers have examined the accuracy of judging what actions are possible for dyads. Chang, Wade and Stoffregen (2009) found that perceivers are able to accurately assess whether adult-child dyads are able to fit through apertures.

Critically, an ecological framework for how people can detect and understand the actions of others explains this capacity in terms of a perceiver's attunement to particular optical information, not with regard to internal, embodied simulation mechanisms. As a result, accuracy depends on perceptual learning, or becoming better at detecting relevant information and becoming more familiar with the actor being observed. Consistent with this expectation, several studies have revealed differences in adults' ability to judge the affordances of children, based on the perceivers' relationship and experience with that child. For example, research by Cordovil and colleagues finds that parents and trained caregivers are more accurate in their estimations than are those with less experience interacting with children, and mothers have been found to be more accurate than fathers (e.g. Cordovil et al., 2012; 2013).

All of these findings make it clear why watching a sporting event involving skilled perceiving-and-acting athletes is so exciting. Not only are athletes able to enact affordances at the very boundaries of capability, for example to deflect, catch, hit or throw a ball in situations that are at the edge of possibility, but simultaneously they can detect in real time the action possibilities of their opponents, particularly ones they know well. Consequently, they dynamically adjust their actions to attempt to remove such action possibilities for their opponent.

Emergence of new affordances through joint action

Although there is a tendency for psychologists to assume that all people act like positivist scientists, observing and forming judgements as a detached third party, in reality humans are actively and constantly engaged with the social world. Picking up information about others' traits, intentions and behavioural capacities is done in the context of interaction, and the ability to do so is what allows for continued social engagement and coordination (Creem-Regehr, Gagnon, Geuss & Stefanucci, 2013). Making physical contact, taking part in a conversation or even making eye contact is enough to pull individuals into a "social eddy" (Marsh, Johnston, Richardson & Schmidt, 2009) with the other person, as they switch from being autonomous agents to being a spontaneously coupled social unit. These social synergies create a whole new perception-action system, distinct from the mere individuals constituting it (Marsh et al., 2006), which provides new affordances in the environment scaled to the joint unit. For example, a heavy sofa may afford lifting and carrying for a pair of individuals, but not for any single individual. In this section, we first discuss evidence that the pull to be a coordinating unit with other individuals is fundamental. Next we discuss evidence from a theoretical approach that views others as providing resources for interacting with the world, before moving to evidence supporting a more wholly affordance-based account of the emergence of social units.

The pull toward social unity is fundamental

One of the most important social affordances in our world is the possibility for joint action – the opportunity to participate in becoming some acting entity that is at a higher (more macro) level than the self, even if it is just a temporary pull into orbit with another. Bids for joint action are difficult to ignore. Lab experiments in which participants are asked to engage in a simple solo action (placing an object on a table in repeated trials) find that some participants fail to complete the action correctly when an unexpected nonverbal gesture from a confederate (holding their hand out) occurs (Becchio, Sartori, & Castiello, 2010). Even when participants are overtly ignoring the apparent bid, the trajectory of their hand movement is measurably pulled somewhat in the direction of the bidding hand. Other neuroscience research provides evidence of our sensitivity to potential joint action, by examining neural responses during task contexts in which expectations for joint action have been induced (Kourtis, Sebanz & Knoblich, 2013). Intriguingly, it is clear that the body is responding not in a simple imitative manner, internally "mirroring" the actions of a potential interactant, but rather, when individuals see the behaviour of others who provide the opportunity for joint action, their bodies respond to the specific affordance implied (to complete the joint action). This point is illustrated by a study where participants watched a videotape in which an actor faces the perceiver and extends her arm toward the perceiver, holding out coffee as if to pour it (using a whole hand grip) or proffering a sugar shaker (using a precision grip). In the key experimental conditions, the video included a cup near the perceiver, implying a request for the participant to lift and move the cup so as to receive coffee or sugar. The cup that the participant was being invited to pick up required a handgrip that was different from that of the actor. In the key conditions where joint action was afforded, neurophysiological evidence indicated that the hand was readied for a complementary grip in the key conditions – not a mirroring one. This is striking evidence of the readiness to respond to joint action affordances given that the implied request was from a videotaped behaviour (Sartori, Cavallo, Bucchioni & Castiello, 2012).

Environmental demands are diminished by anticipated social help

Humans presumably have a rich understanding of why joint action might often be beneficial. Several researchers have argued that other people provide both physiological and psychological resources that are capable of influencing how one relates to the physical environment. This claim is supported by studies showing how perceptual judgements, typically involving distance estimates and judgements of hill slopes, are influenced by the presence of another person. For example, Schnall, Harber, Stefanucci and Proffit (2008) found that perceivers saw hills as less steep when they were with a supportive friend. In another study, researchers experimentally induced a feeling of understanding between participants, which led to less steep slope estimates and shorter judgements of distance to a target (Oishi, Schiller & Gross, 2013). The mere presence of a close other, or even just thinking of that other person, appears to be capable of easing the anticipated load or costs of action, making the environment appear more amenable. Similar results have been demonstrated for those actually expecting to work cooperatively. Doerrfeld, Sebanz and Shiffrar (2012) found that weights were viewed as lighter when people expected to lift them jointly relative to when they anticipated lifting individually. Perceptual judgements were viewed as tied to a perceiver's "energetic economy", the amount of effort required to act in the environment. Thus, having a partner to lift with reduces the anticipated physical costs of the action, thereby lowering judgements of that object's weight. Notably, research conducted from an energetic economy perspective focuses on individual judgements of the perceiver, not affordance judgements per se – that is, the judgements are affected by implied energy expenditure, not measures that are precisely scaled to both environment and body explicitly. On the other hand, perceptual judgements like metric and steepness estimates have been found to indirectly reflect people's attunement to affordance information (Lee, Lee, Carello & Turvey, 2012), and people's assessments of affordance action boundaries (e.g. the point when stairs become less comfortable to climb) are veridically tied to the increased energy expenditures at those points (Warren, 1984).

Joint action emerges at crucial affordance boundaries

However, an ecological account of the emergence of joint action requires explicitly using objective measures that should capture the boundaries at which solo action is more challenging and joint action becomes a stronger pull. Affordance theory makes specific predictions about when an individual shifts action modes, e.g. from picking up an object with one hand to shifting to two hands (Cesari & Newell, 1999; van der Kamp et al., 1998). According to affordance theory, cooperation is distributed across the person-person-environment system; it is a behaviour that emerges as a direct consequence of the environment making demands of a solo actor that cannot be met alone. Environmental constraints are crucial in the emergence of new action modes. Cooperative action should easily and rapidly emerge with little planning when objects become too long or too heavy to carry alone. Richardson, Marsh and Baron (2007) tested these hypotheses in a series of experiments in which participants were asked to move wooden planks of various sizes by touching the ends only (or were asked to make judgements about moving them). Across the experiments, the plank lengths included sizes that could be comfortably moved with only one hand, others that required two hands, and some that could only be comfortably moved by two people. Included as well were a continuous range of other sizes that could be stably carried in multiple modes (e.g. with two hands or with two people). Figure 17.1 illustrates the basic paradigm of experiments involving dyads.

The results of the experiments revealed clear similarities between how individual action-mode transitions are made and how transitions to cooperative action are made. Specifically, transitions were predicted by an action-scaled ratio, that is, a variable that captured the relationship between the length of the plank and the relevant physical attribute of the actor(s) (i.e. the hand span or arm span of the pair).

Figure 17.1 Illustration of the procedure from the two person plank-moving experiment (illustration: Daryl Lanzendorfer).

The design of the experiments also allowed for crucial tests of hypotheses that come from research on the dynamical emergence of affordances. In particular, the interpersonal perception-action system had the characteristic of "multi-stability", regions where multiple attractors were present (solo moves and joint action moves were both comfortable) and where the past trajectory of action (the history of the system) determined which state the system was in. So, in a condition in which planks were presented in descending size, people by necessity started off by moving planks together; when they reached the region where solo action was possible (a plank size at which people in the ascending condition, for instance, would have been still moving planks alone), people persisted somewhat longer in joint action. This characteristic of hysteresis, plus other dynamical features explored in these experiments, replicated phenomena found in other research on the dynamics of switching action modes (for solo actors).

In light of these results, one could easily interpret the empirical literature as indicating that others provide a purely additive contribution to one's own capabilities. A perspective such as this essentially conceptualizes other people as tools that can be used to ease one's load. However, joint action is clearly far more complex than this. The fit between those working cooperatively is critically important. In a follow-up study to Richardson et al. (2007), Isenhower, Richardson, Carello, Baron and Marsh (2010) found that transitions to joint action depended on the fit between the pair of actors. The behaviour of dyads with mismatching arm lengths was constrained more by the person with shorter arms, who cannot continue to move planks alone at an earlier point than their partner. In addition, one difference did occur between experiments examining shifts of action modes within solo behaviour (one and two hand moving) and those examining the emergence of joint action (one and two person moving). Namely, when planks were presented in descending size, the tendency to persist in joint action (before switching to solo) was stronger than the parallel tendency to stay in two hand mode (before switching to one hand) for solo actors. Whether this reflects an inherently stronger pull to joint action or social motives such as social awkwardness about going back to working alone is unclear.

More generally, although working jointly with another person may often provide new behavioural abilities and certain psychological benefits, it also introduces a number of additional challenges unique to social interaction. For example, in an experiment seemingly similar to Doerrfeld et al.'s (2012) on loads being "lightened" by anticipating joint lifting, Meagher and Marsh (2014) asked participants to estimate their distance to a target location when expecting to carry a heavy object either alone or jointly. They found that when coordinative challenges were made salient, participants expecting to work jointly actually gave larger estimates of distance than those planning to carry alone. This finding indicates that individuals are sensitive to more than just the reduction in physiological cost provided by lifting simultaneously. Instead, they are also sensitive to the fact that carrying an object with another person also involves new costs and obligations, such as needing to attend to the partner's movement, synchronize one's steps, and uphold the social motivations of being a valuable group member.

Such studies point to how relatively limited affordance research in the joint action realm has been to date, leaving numerous avenues for further study. For instance, in the Richardson et al. (2007) study, the new possibilities for action that were available to the team were not different in quality from the type of possibilities available to the solo actor. It would be interesting to extend such work to look at how becoming a "we" provides qualitatively new means of completing actions, not just affordances that extend our capabilities in magnitude.

Sporting contexts provide numerous such contexts and also illustrate an additional limitation: few studies motivated by an ecological perspective have used tasks that inherently require different roles for the different actors (e.g. defender versus offensive teammate). One exception is a task by Ramenzoni and colleagues (Ramenzoni, Davis, Riley, Shockley & Baker, 2011) in which individuals' actions are complementary. Each member of the dyad has a different tool and a different aspect of the task to accomplish. Successful achievement of the task requires one individual to pierce a ring that the other person holds. More broadly, it would be fruitful to examine the affordance boundaries and dynamics of actions that involve a wider range of joint action tasks. such as those that have been studied from mental simulation theoretical perspectives (e.g. Vesper, van der Wel, Knoblich & Sebanz, 2013).

Broader implications and applications

The imprint of socially coordinated activities on settings

In the discussions thus far, affordances in interpersonal coordination were examined exclusively in the interactions of a dyad or a team against a backdrop of environmental constraints quantified along a simple, single dimension. Such a limited look at the environment relegates it to a relatively circumscribed role. However, the influences that aid and constrain coordinated social behaviour clearly go beyond concrete, single dimensional features of the present. Behaviour is also determined by temporal and macro dimensions at a much larger scale. The imprint of the past and the values of the present, especially as carried in the cultural features of the physical world in which we navigate, are reflected in the environment and have a significant impact on behaviour.

The difficulty is that cultural meaning is carried not by the physical features of the environment alone, but rather is distributed across organism and environment. An individual who opens a door and happens upon a religious ceremony, or who stumbles into an unfamiliar sport competition, will be able to detect, even without prior experience, that he has entered a setting with a particular function and meaning. Barker developed the term "behaviour setting" to reflect such real situations, places that have detectable geographic, temporal and behavioural boundaries or constraints, and which involve a "milieu": artefacts (ball, chalklines, chalices, robes uniforms) that help support the behaviour of individuals in the setting (e.g. Barker & Wright, 1954; Heft, 2001). Though such behaviour settings as a game being played, or a religious service being held, have a reality independent of any single person's experience of them (a different leader may serve, substitutions are allowed), the behaviour settings are interdependent with – they cause and are in turn caused by – the behaviour of those in the setting (the fans, the audience, the religious leaders, the assistants, the referees). Moreover, behaviour settings are quasi-stable – if there is a temporary perturbation (a streaker interrupting play, a fainting participant, a misstep or rule violation), disruptions are handled and then there is a return to normality.

An important area for future research is understanding what the information is that specifies to an observer that she is in one behaviour setting rather than another. Boundaries should be apparent even from the behaviour of individuals alone, stripped of any environmental cues. The pattern of movement of individuals in an environment, the rate of turnover of individuals in a space, the relative patterning of their movement in relation to others, the extent to which some are in specific areas (the customers in a café) with others having more complete access (waiters) and other details will tell an observer whether some competitive game is in operation, whether a restaurant is in dinner service, or whether a bank is open for business. Heft and colleagues (Heft, Hoch, Edmunds, & Weeks, 2014) tested the hypothesis that the mere kinematics of people's motion (in terms of where in a space they went, how quickly, and how long they stayed in different areas before leaving, as well as overall density and complexity of motion relative to others) would inform observers as to whether they were viewing behaviour in a bank, a library, a basketball game, a café or an ice cream shop. After making observations of the trajectories of individuals in a sample of settings in a small Ohio town, the researchers created animations that simulated the motion of people, with no other visual information available about the setting, the objects and so forth. In three experiments, people were very good at determining what a setting was, assessing its similarity to other kinds of related settings, and understanding what behaviours were appropriate in a depicted setting (Heft et al., 2014).

For humans, the social world is intertwined with the physical world

Social systems are capable of constraining how one functionally relates to the environment. For example, a mug, as a physical object, is designed to be scaled to a typical adult's hand, thereby making it afford grasping and drinking from. However, when this mug is the property of another person, one's repertoire of behaviours become far more limited with respect to it (Schmidt, 2007). Intriguingly, experimental evidence suggests that this social information actually makes individuals less efficient at detecting the affordances of objects. Constable, Kritikos and Bayliss (2011) demonstrated that the Stimulus-Response Compatibility effect, the finding that response times are faster when stimuli objects are oriented to be compatible with how participants respond to it (e.g. participants respond faster with the right hand to a knife with its handle on the right side), could be abolished when the presented object belonged to someone else. Meagher (2015) recently found similar results at the intergroup level.

Participants were less efficient at detecting and responding to the affordances of objects when these objects were associated with the opposite gender. In addition to objects, social information relevant to behavioural opportunities also exist with respect to entire environments. For example, a visitor in another person's territory is embedded within an ambient array of not only functional information, but also social information. The visitor's behavioural repertoire is highly dependent on the actions of the host, as well as the pervasive cultural expectations that provide a historical framework for what a visitor can and should do in someone else's territory. In a pair of studies, Meagher (2014) found that visitors judged a room to be less spacious than did hosts, suggesting that social constraints can actually bleed into people's impressions of physical constraint.

Thus, perception and action for humans in many ways involve a type of social coordination even when functioning as a single individual. Detecting and realizing the opportunities for action available in the environment is always a situated act, which is both guided and constrained by the larger society to which one belongs. Considering such issues in future research would provide an interesting direction for future research in interpersonal coordination to pursue.

References

Barker, R.G. & Wright, H.F. (1954) *Midwest and its children: The psychological ecology of an American town*. Row, Peterson and Company: New York.

Becchio, C., Sartori, L. & Castiello, U. (2010) Toward you: The social side of actions. *Current Directions in Psychological Science, 19*, 183–188. doi:10.1177/ 0963721410370131

Berry, D.S. (1991a) Accuracy in social perception: contributions of facial and vocal information. *Journal of Personality and Social Psychology, 61*, 298–307. doi: 10.1037/0022-3514.61.2.298

Berry, D.S. (1991b) Child and adult sensitivity to gender information in patterns of facial motion. *Ecological Psychology, 3*, 349–366. doi:10.1207/s15326969eco0304_3

Blanch-Hartigan, D., Andrzejewski, S.A. & Hill, K.M. (2012) The effectiveness of training to improve person perception accuracy: A meta-analysis. *Basic and Applied Social Psychology, 34*, 483–498. doi:10.1080/01973533.2012.728122

Cesari, P. & Newell, K.M. (1999) The scaling of human grip configurations. *Journal of Experimental Psychology: Human Perception and Performance, 25*, 927–935. doi:10.1037/0096-1523.25.4.927

Chang, C., Wade, M.G. & Stoffregen, T.A. (2009) Perceiving affordances for aperture passage in an environment-person-person system. *Journal of Motor Behavior, 41*, 495–500.

Charles, E. P.(2011) Ecological psychology and social psychology: It is Holt, or nothing! *Integrative Psychological & Behavioral Science, 45*, 132–153. doi:10.1007/ s12124-010-9125-8

Constable, M.D., Kritikos, A. & Bayliss, A.P. (2011) Grasping the concept of personal property. *Cognition, 119*, 430–437. doi:10.1016/j.cognition.2011.02.007

Cordovil, R., Andrade, C. & Barreiros, J. (2013) Perceiving children's affordances: Recalibrating estimation following single-trial observation of three different tasks. *Human Movement Science, 32*, 270–278. doi:10.1016/j.humov.2013.01.001

Cordovil, R., Santos, C. & Barreiros, J. (2012) Perceiving children's behavior and reaching limits in a risk environment. *Journal of Experimental Child Psychology, 111*, 319–330. doi:10.1016/j.jecp.2011.09.005

Creem-Regehr, S.H., Gagnon, K.T., Geuss, M.N. & Stefanucci, J.K. (2013) Relating spatial perspective taking to the perception of others' affordances: Providing a foundation for predicting the future behavior of others. *Frontiers in Human Neuroscience, 7*, 596. doi:10.3389/fnhum.2013.00596

Doerrfeld, A., Sebanz, N. & Shiffrar, M. (2012) Expecting to lift a box together makes the load look lighter. *Psychological Research, 76*, 467–475. doi:10.1007/s00426-011-0398-4

Downing, P.E., Bray, D., Rogers, J. & Childs, C. (2004) Bodies capture attention when nothing is expected. *Cognition, 93*, B27–B38. doi:10.1016/j.cognition.2003.10.010

Fletcher-Watson, S., Findlay, J.M., Leekam, S.R. & Benson, V. (2008) Rapid detection of person information in a naturalistic scene. *Perception, 37*, 571–583. doi:10.1068/p5705

Funder, D.C. (1995) On the accuracy of personality judgment: A realistic approach. *Psychological Review, 102*, 652–670. doi:10.1037/0033-295x.102.4.652

Gibson, J.J. (1979) *The ecological approach to visual perception.* Houghton-Mifflin: Boston, MA.

Heft, H. (2001) *Ecological psychology in context: James Gibson, Roger Barker, and the legacy of William James's radical empiricism.* Lawrence Erlbaum Associates Publishers: Mahwah, NJ.

Heft, H. (2007) The social constitution of perceiver-environment reciprocity. *Ecological Psychology, 19*, 85–105. doi:10.1080/10407410701331934

Heft, H., Hoch, J., Edmunds, T. & Weeks, J. (2014) Can the identity of a behavior setting be perceived through patterns of joint action? An investigation of place perception. *Behavioral Sciences, 4*, 371–393. doi:10.3390/bs4040371

Hodges, B.H. & Lindhiem, O. (2006) Carrying babies and groceries: The effect of moral and social weight on caring. *Ecological Psychology, 18*, 93–111. doi:10.1207/s15326969eco1802_2

Human, L.J. & Biesanz, J.C. (2013) Targeting the good target: An integrative review of the characteristics and consequences of being accurately perceived. *Personality and Social Psychology Review, 17*, 248–272. doi:10.1177/1088868313495593

Isenhower, R.W., Richardson, M.J., Carello, C., Baron, R.M. & Marsh, K.L. (2010) Affording cooperation: Embodied constraints, dynamics, and action-scaled invariance in joint lifting. *Psychonomic Bulletin & Review, 17*, 342–347. doi:10.3758/pbr.17.3.342

Kenny, D.A. (1994) *Interpersonal perception: A social relations analysis.* Guilford Press: New York.

Kourtis, D., Sebanz, N. & Knoblich, G. (2013) Predictive representation of other people's actions in joint action planning: An EEG study. *Social Neuroscience, 8*, 31–42.

Lee, D.N. (1976) A theory of visual control of braking based on information about time-to-collision. *Perception, 5*, 437–459. doi:10.1068/p050437

Lee, Y., Lee, S., Carello, C. & Turvey, M.T. (2012) An archer's perceived form scales the "hitableness" of archery targets. *Journal of Experimental Psychology: Human Perception and Performance, 38*, 1125–1131. doi:10.1037/a0029036

Mark, L.S. (2007) Perceiving the actions of other people. *Ecological Psychology, 19*, 107–136. doi:10.1080/10407410701331967

Mark, L.S., Nemeth, K., Gardner, D., Dainoff, M.J., Paasche, J., Duffy, M. & Grandt, K. (1997) Postural dynamics and the preferred critical boundary for visually guided reaching. *Journal of Experimental Psychology: Human Perception and Performance, 23*, 1365–1379. doi:10.1037/0096-1523.23.5.1365

Marsh, K.L., Johnston, L., Richardson, M.J. & Schmidt, R.C. (2009) Toward a radically embodied, embedded social psychology. *European Journal of Social Psychology, 39*, 1217–1225. doi:10.1002/ejsp.666

Marsh, K.L., Richardson, M.J., Baron, R.M. & Schmidt, R.C. (2006) Contrasting approaches to perceiving and acting with others. *Ecological Psychology, 18*, 1–38. doi:10.1207/s15326969eco1801_1

McArthur, L.Z. & Baron, R.M. (1983) Toward an ecological theory of social perception. *Psychological Review, 90*, 215–238. doi:10.1037/0033-295x.90.3.215

Meagher, B.R. (2014) *The emergence of home advantage from differential perceptual activity* (Unpublished doctoral dissertation). University of Connecticut, Storrs, CT.

Meagher, B.R. (2015) *Interacting with out-group objects: Social associations inhibit the detection of affordances.* Unpublished manuscript.

Meagher, B.R. & Marsh, K.L. (2014) The costs of cooperation: Action-specific perception in the context of joint action. *Journal of Experimental Psychology: Human Perception and Performance, 40*, 429–444. doi:10.1037/a0033850

Michaels, C.F. & Carello, C. (1981) *Direct perception*. Prentice-Hall: Englewood Cliffs, NJ.

Oishi, S., Schiller, J. & Gross, E.B. (2013) Felt understanding and misunderstanding affect the perception of pain, slant, and distance. *Social Psychological and Personality Science, 4*, 259–266. doi: 10.1177/1948550612453469

Ramenzoni, V.C., Davis, T.J., Riley, M.A. & Shockley, K. (2010) Perceiving action boundaries: Learning effects in perceiving maximum jumping-reach affordances. *Attention, Perception, & Psychophysics, 72*, 1110–1119. doi:10.3758/APP.72.4.1110

Ramenzoni, V.C., Davis, T.J., Riley, M.A., Shockley, K. & Baker, A.A. (2011) Joint action in a cooperative precision task: Nested processes of intrapersonal and interpersonal coordination. *Experimental Brain Research, 211*, 447–457. doi:10.1007/s00221-011-2653-8

Ramenzoni, V.C., Riley, M.A., Davis, T., Shockley, K. & Armstrong, R. (2008) Tuning in to another person's action capabilities: Perceiving maximal jumping-reach height from walking kinematics. *Journal of Experimental Psychology: Human Perception and Performance, 34*, 919–928. doi:10.1037/0096-1523.34.4.919

Richardson, M.J., Marsh, K.L. & Baron, R.M. (2007) Judging and actualizing intrapersonal and interpersonal affordances. *Journal of Experimental Psychology: Human Perception and Performance, 33*, 845–859. doi:10.1037/0096-1523.33.4.845

Rochat, P. (1995) Perceived reachability for self and for others by 3- to 5-year-old children and adults. *Journal of Experimental Child Psychology, 59*, 317–333. doi:10.1006/jecp.1995.1014

Rule, N.O., Ambady, N. & Hallett, K.C. (2009) Female sexual orientation is perceived accurately, rapidly, and automatically from the face and its features. *Journal of Experimental Social Psychology, 45*, 1245–1251. doi:10.1016/j.jesp.2009.07.010

Runeson, S. & Frykholm, G. (1983) Kinematic specification of dynamics as an informational basis for person-and-action perception: Expectation, gender recognition, and deceptive intention. *Journal of Experimental Psychology: General, 112*, 585–615. doi:10.1037/0096-3445.112.4.585

Sartori, L., Cavallo, A., Bucchioni, G. & Castiello, U. (2012) From simulation to reciprocity: The case of complementary actions. *Social Neuroscience, 7*, 146–158. doi:10.1080/17470919.2011.586579

Schmidt, R.C. (2007) Scaffolds for social meaning. *Ecological Psychology, 19*, 137–151. doi:10.1080/10407410701332064

Schnall, S., Harber, K.D., Stefanucci, J.K. & Proffitt, D.R. (2008) Social support and the perception of geographical slant. *Journal of Experimental Social Psychology, 44*, 1246–1255. doi:10.1016/j.jesp.2008.04.011

Stoffregen, T.A., Gorday, K.M., Sheng, Y.-Y. & Flynn, S.B. (1999) Perceiving affordances for another person's actions. *Journal of Experimental Psychology: Human Perception and Performance, 25*, 120–136. doi:10.1037/0096-1523.25.1.120

Tucker, M. & Ellis, R. (1998) On the relations between seen objects and components of potential actions. *Journal of Experimental Psychology: Human Perception and Performance, 24*, 830–846. doi:10.1037/0096-1523.24.3.830

Turvey, M.T. (1992) Affordances and prospective control: An outline of the ontology. *Ecological Psychology, 4*, 173–187. doi:10.1207/s15326969eco0403_3

van der Kamp, J., Savelsbergh, G.J.P. & Davis, W.E. (1998) Body-scaled ratio as a control parameter for prehension in 5- to 9-year-old children. *Developmental Psychobiology, 33*, 351–361. doi:10.1002/(SICI)1098-2302(199812)33:4<351::AID-DEV6>3.0.CO;2-P

Vesper, C., van der Wel, R.P.R.D., Knoblich, G. & Sebanz, N. (2013) Are you ready to jump? Predictive mechanisms in interpersonal coordination. *Journal of Experimental Psychology: Human Perception and Performance, 39*, 48–61. doi:10.1037/a0028066

Warren, W.H. (1984) Perceiving affordances: Visual guidance of stair climbing. *Journal of Experimental Psychology: Human Perception and Performance, 10*, 683–703. doi:10.1037/0096-1523.10.5.683

Zebrowitz, L.A. & Montepare, J. (2006). The ecological approach to person perception: Evolutionary roots and contemporary offshoots. In M. Schaller, J.A. Simpson & D.T. Kenrick (Eds.), *Evolution and social psychology* (pp. 81–113). Psychosocial Press: Madison, CT.

Zebrowitz, L.A. & Rhodes, G. (2004) Sensitivity to "bad genes" and the anomalous face overgeneralization effect: Cue validity, cue utilization, and accuracy in judging intelligence and health. *Journal of Nonverbal Behavior, 28*, 167–185. doi:10.1023/B:JONB.0000039648.30935.1b

18 Social coordination of verbal and nonverbal behaviours

Alexandra Paxton, Rick Dale and Daniel C. Richardson

Social coordination of verbal and non-verbal behaviours

Interpersonal coordination is key to our everyday experiences. This is apparent in iconic acts of coordination like dancing, but even everyday communication is an intensely coordinative act. From what we say to how we sit, conversation weaves together numerous verbal and nonverbal systems. The diverse behaviours that contribute to communication are distributed across multiple timescales and across physical, cognitive and social systems. The interconnectedness of these systems slips into the background during effortless conversation, but in this chapter we will bring them to the fore. We argue that the interdependence of these systems during communication should be reflected in our study of interpersonal coordination.

We can think of coordination in two related but distinct ways: *coordination as joint action* (e.g. Clark, 1996; Harris, 1996) and *coordination as convergence* (e.g. Giles et al., 1991; Pickering & Garrod, 2004), though further distinctions can be made. Coordination-as-joint-action assumes that two people are intentionally engaged in a common goal, such as cooking a dinner or moving a table, and their actions become aligned and intertwined to reach that goal. In coordination-as-convergence, behaviours become more similar as a consequence of co-presence – like a yawn spreading through a room or romantic partners becoming more similar over time.

In this chapter we will use the term *coordination* to refer to coordination-as-convergence, our primary focus here.[1] Other terms for this phenomenon include adaptation, alignment, mimicry and synchrony, to name but a few (for a review, see Paxton & Dale, 2013c). However, because we will occasionally discuss coordination-as-joint-action, we will point this out to the reader where appropriate.

Coordination is a growing research area that explores the ways that people affect one another over time as a result of their contact. The phenomena and methods are diverse, investigating a range of related questions about emotion (e.g. Neumann & Strack, 2000), posture (e.g. Shockley, Baker, Richardson, & Fowler, 2007) and more. Our chapter reviews empirical work on and extends theoretical explorations of the emergence of interpersonal coordination between verbal and nonverbal systems during interaction. Specifically, we focus on how coordination might occur across various *levels* of communication.

Throughout the chapter we will conceptualize communication systems as constituted by two levels of description or analysis. We will refer to these two levels as *systems*, but we will emphasize that their ontological status is open to continued empirical investigation. *Top-level systems* are relatively *slower* processes that occur on a lower frequency and have fewer degrees of freedom. Examples of top-level systems might include interaction goals, interpersonal relationships and conversational context. *Bottom-level systems*, on the other hand, are relatively *faster* processes, operate on a higher frequency, and have more degrees of freedom available to them. Phonetics, gaze and body movement are examples of these kinds of bottom-level systems.

During conversation, interlocutors balance the needs and pressures of each of these systems. Basic bottom-level demands shape top-level systems, while the top-level constraints feed back into the interaction, moulding the interaction landscape available to the bottom levels (cf. Van Orden, Hollis & Wallot, 2012). To this view, nothing is considered in isolation: all cognitive, physical and social systems are highly interconnected and interdependent during communication. We here investigate interaction as a series of interconnected and interdependent systems, arguing that the bidirectional influence across different levels of communication will provide this domain with a deeper understanding of the integrative aspect of human interaction.

Prominent theories of verbal coordination

Below we highlight four theoretical perspectives on coordination: communication accommodation theory, interactive alignment theory, partner-specific adaptation, and synergies. Though there are many others, these have been perhaps the most influential.

Communication accommodation theory

One of the first accounts of linguistic coordination was *communication accommodation theory* (CAT), also known as *speech accommodation theory*, *accommodation theory*, or *accommodation*. CAT explores the effects of social forces on speech at multiple scales, from speech production to social perception (e.g. Giles, Taylor & Bourhis, 1977; Giles, Coupland & Coupland, 1991; Babel, 2010). One of the defining features of CAT is its focus on the strategic *convergence* and *divergence* of speech behaviours according to social pressures. In this view, individuals are more likely to converge (or engage in similar speech behaviours) when trying to strengthen social ties and are more likely diverge (or engage in dissimilar speech behaviours) when trying to increase social distance (Giles, 1973).

Interactive alignment theory

Pickering and Garrod (2004) proposed the *interactive alignment theory* (IAT) – often known simply as *alignment* – to explain linguistic coordination as largely unintentional. Under IAT, linguistic coordination can be explained primarily through priming rather than conscious choice (e.g. Branigan, Pickering & Cleland, 2000; Ferreira & Bock, 2006). From the seminal work (Pickering & Garrod, 2004), alignment has been targeted as a multi-timescale (e.g. phonetics, diction, syntax) and multi-modality (e.g. speech, cognition) phenomenon in which alignment along one timescale or modality can increase alignment along other dimensions (e.g. Reitter, Moore & Keller, 2006). Perhaps due to this explicit multiscale and multimodal focus, alignment has become a highly influential theory not only for linguistic coordination (e.g. Richardson, Taylor, Snook, Conchie & Bennell, 2014) but also for coordination along a number of other modalities (e.g. Hasson, Ghazanfar, Galantucci, Garrod & Keysers, 2012).

Partner-specific adaptation

Partner-specific adaptation – also known as *talker-specific adaptation* or *adaptation* – advocates for coordination (primarily coordination-as-joint-action) as an intentional process (Brennan & Hanna, 2009). Interlocutors begin communication with a set of shared goals and information called their *common ground*, which grows over time through interaction in a process called *grounding* (e.g. Brennan, Galati, & Kuhlen, 2010; Clark, 1996; Clark & Krych, 2004). As interlocutors increase their common ground, they begin to *adapt* their behaviours to their partner's specific needs (e.g. Brennan, 1991; Rogers, Fay, & Maybery, 2013).

Adaptation is generally most concerned with linguistic communication. In this viewpoint, interlocutors' speech production and comprehension are highly sensitive to one another's needs and understanding. Speakers adapt utterances to facilitate their listeners' understanding through *audience design*, choosing to include or exclude information based on the listeners' needs and their common ground (e.g. Galati & Brennan, 2010; Clark & Krych, 2004). Meanwhile, listeners actively engage in *partner-specific processing* to adapt to speakers' idiosyncrasies and the information available through common ground (e.g. Trude & Brown-Schmidt, 2012).

Synergies

The view of *interpersonal synergies* applies ideas from the motor coordination literature (e.g. Bernstein, 1967; Haken, 1983; Turvey, 1990) to linguistic coordination (e.g. Dale, Fusaroli, Duran, & Richardson, 2013; Riley, Richardson, Shockley, & Ramenzoni, 2011; Schmidt & Richardson, 2008; Shockley, Richardson, & Dale, 2009). This relatively new approach posits that just as muscles in a single body come together to achieve different physical goals, different communicative systems can come together across people to create *synergies* or *coordinative structures.* These structures manage variability in task-relevant domains by reducing the functional degrees of freedom in the system. This facilitates communication by creating on-the-fly groups of systems that are coupled together to achieve interaction goals more easily.

One influential idea emerging from this account is the suggestion that strict synchrony or strong coupling – that is, simply becoming more similar in behaviour and cognition – may not be the most optimal configuration of interpersonal dynamics. Optimality should instead be determined by functional pressures. Therefore, synchronous behaviour across communication systems may be optimal for some types of interactions, while weak coupling or even complementarity across systems may be more optimal for other types of interactions. Recent empirical work supporting this view has centred mostly on synergies within movement (e.g. Black, Riley & McCord, 2007; Schmidt & Richardson, 2008), with an increasing emphasis on interaction (e.g. Abney, Paxton, Dale & Kello, 2015; Fusaroli et al., 2012).

Quantifying verbal coordination

The methodological landscape of verbal coordination research is as rich as its theoretical landscape. Again, we will only introduce a few prominent examples, pointing to in-depth resources for interested readers (cf. Bakeman & Quera, 2011; Kenny, Kashy, & Cook, 2006; Richardson, Dale & Marsh, 2014; Riley & Van Orden, 2005).

Linguistic Inquiry and Word Count (LIWC) and Language Style Matching (LSM)

Linguistic Inquiry and Word Count (LIWC; Pennebaker, Booth & Francis, 2007) is a bag-of-words style text analysis tool that looks beyond word usage to investigate the underlying meaning or style of the text. LIWC scans corpora and categorizes each unit of text (e.g. sentence, paragraph, document) into a number of classes based on default or user-built dictionaries. LIWC provides a context-agnostic evaluation of text composition by percentage that can then be subjected to statistical analysis. While other quantifications of verbal communication may be applied to various types of data, LIWC focuses exclusively on linguistic analyses.

LIWC has been applied to questions of linguistic coordination using large-scale text analysis, from transcripts of face-to-face interactions to large-scale analyses of online data (for review, see Tausczik & Pennebaker, 2010). *Language* (or *linguistic*) *style matching* (LSM) builds on data derived from LIWC to quantify linguistic coordination between individuals, as measured by similarities in usage across LIWC categories. Researchers can use LIWC categories to target more syntactic or structural coordination (e.g. with function words) or broader discourse-level coordination (e.g. with specific content categories), even within the same dataset. LSM measures similar word usage along each LIWC category, allowing researchers to target various levels of linguistic coordination with a frequency-based text-analysis approach.

Distributional analyses

Distributional analyses complement dynamic analyses (described below) by investigating the degree to which statistical properties of behaviours match across individuals over a period of time. Distributional analyses are grounded in the idea that although individuals may differ in behaviour at the local level, interacting individuals should come to display similar frequencies of behaviours during interaction. For example, many bag-of-words analyses do not take into account the fine-grained dynamics of language use but do measure how much interlocutors tend to use similar language across larger chunks of time (see Tausczik & Pennebaker, 2010).

These analyses can be useful for targeting behaviours while abstracting somewhat from time, quantifying longer-scale trends beyond turn-adjacent coordination. For example, various work has investigated verbal coordination through mean speech rate (Webb, 1969) and choices of syntactic construction (Bock, 1986). A subtype of distributional analysis called *complexity matching* (West, Geneston, & Grigolini, 2008) has been used to compare interlocutors' distributions of speech behaviours, showing that individuals tend to produce clustering patterns of speech during interaction (Abney, Paxton, Dale, & Kello, 2014). These kinds of analyses provide a global-level companion to local-level, dynamic analyses of coordination.

Cross-recurrence quantification analysis

Cross-recurrence quantification analysis (CRQA) is an extension of methods originally developed for the natural sciences (Marwan et al., 2007; Marwan, 2008) and is now used to study patterns of coordination over time (for review, see Coco & Dale, 2014). In addition to quantifying temporal patterns of behavioural influence, the method can be used to visualize the interpersonal system in cross-recurrence plots and provide unique insights into recurring patterns of behaviour in the dyad. CRQA can be applied to both continuous and categorical data, providing quantification of coordination in various aspects of verbal (and nonverbal) communication.

Essentially, CRQA quantifies coordination by identifying all possible intersections of identical behaviours between two participants over the course of their interaction. Time series of behaviours (e.g. linguistic contributions) for each participant are recorded. When participants make the same action at the same point in time it is plotted along the $y=x$ diagonal of a recurrence plot, the *line of coincidence*. The two time series are then aligned with a lag of t time points (e.g. milliseconds) between them. Occurrences of the same behaviour are now plotted along $y=x+t$ diagonal. A full recurrence plot consists of all values of t, at whatever granularity is required. Because the cross-recurrence plots include comparisons of *all* possible time points, CRQA allows for the investigation of patterns of influence across long delays instead of simply comparing behaviours as they occurred in time. CRQA can also highlight recurring dyadic states and identify periodic behaviour at the dyadic level.

Multiple resources across various platforms facilitate CRQA. Researchers can turn to the crqa package in R (Coco & Dale, 2014) or the crptoolbox toolbox for MATLAB (Marwan, 2013). The B(eo)W(u)LF data structure (Paxton & Dale, 2013b) can help format linguistic data for CRQA analyses using Python and MATLAB. The computer software Discursis (Angus, Smith & Wiles, 2012) provides a programming-free approach to analysing and visualizing recurrence based on underlying content (i.e. *conceptual recurrence*) instead of lexical choice.

Perhaps the feature of CRQA that makes it so suitable for discussion in this chapter is that it can easily handle both discrete and continuous signals. This means it can accommodate a wide variety of behaviours, creating a common analysis environment in which to explore the dynamics of interdependent behaviours. In the following section we review a series of experiments that feature CRQA as a measure of coordination across verbal and nonverbal systems, highlighting the ways in which different levels and systems constrain and influence one another during communication.

Bridging top- and bottom-level systems: gaze, communication and coordination

Both top- and bottom-level systems have often been studied as distinct entities, perhaps viewed as affecting one another only incidentally. Visual attention provides an excellent example of this. The classic understanding of visual attention holds that it is a veridical information-gathering perceptual system. Its dynamics are determined primarily by features of the world (such as motion and visual contrast) and processes of cognition, such as memory and expectation (e.g. Henderson, 2003). However, this section presents support for an alternative view of visual attention and, by extension, other bottom-level systems: When visual attention is embedded in the *social* world, what emerges is a more complex interplay between interpersonal communication, visual context and the relationship between the people who share it.

For example, one study (Richardson, Dale & Tomlinson, 2009) asked two participants in neighbouring booths to have a political discussion over an intercom while each looked at a blank grid on a computer screen. Despite their lack of shared physical location and their inability to see one another, both partners systematically coordinated their vision during their discussion, looking moment by moment at the same empty regions of the screen. Again, there was *nothing* to see on-screen. In contrast to the traditional views of visual attention, these individuals were not using their eyes to gather information, as there was no information to gather. The only thing moving their eyes was the social context – their interaction and shared common ground.

Additional work supports this same notion: that bottom-level perceptual mechanisms – in this case, visual attention – interact with top-level systems like social context during communication. The studies described next highlight the bidirectional influences that top- and bottom-level systems exert on one another. They exemplify how top-level systems – including beliefs, memory and social context – interact with visual attention, serving as an exemplar for other bottom-level systems (cf. posture in Shockley et al., 2007; overall body movement in Paxton & Dale, 2013a). Taken together, these findings provide compelling evidence for the interconnectedness and interdependence of verbal and nonverbal systems during interaction.

Expectations of context constrain bottom-level systems

We will first look at a case of how even "minimal social context" (von Zimmermann & Richardson, 2014) influences bottom-level systems. That is, when participants simply believe that they are looking at a stimuli at the same time as another individual, it changes how they perceive it – without any interaction taking place between them.

A series of studies (Richardson et al., 2012) asked pairs of participants to look at sets of pictures, some with positive valence and some with negative valence. Half of the time they participants believed that they were looking at the same images as their partner, and half of the time they believed that they were looking at different images. This social context changed randomly on a trial-by-trial basis, and participants reported that they mostly ignored the information about their partner's condition. Despite this reported behaviour, however, simply knowing that another person was attending to the same stimulius – even though they could not see each other or have any verbal interaction – shifted participants' attention. When participants believed that they were looking at the images together with another person, they tended to look towards the more negative images.

In another experiment (also reported in Richardson et al., 2012), participants were told to either (a) search a set of pictures for an "X", or (b) memorize a set of pictures. Each participant was given one of these tasks and was told which of these tasks their partner would be doing as well. In this study we again see the powerful effects that social context and belief can have on lower-level behaviour: Believing their partner was experiencing the same *stimulus* but not sharing the same *task* did not result in joint perception. Joint perception only occurred when participants believed that their partner was engaged in exactly the same task (Richardson et al., 2012). One explanation is that when the stimuli were believed to be shared, participants looked towards the images that they thought their partner would also be looking at. In other words, even with this minimal social context of no interaction, participants were seeking to coordinate their visual attention.

Bottom-level coordination improves cognitive performance

The previous studies on simple expectations of context build on other work that quantifies gaze coordination between people under various conditions. In the following studies the social context becomes richer, as participants are allowed to communicate with each other.

In the first of these quantifications of gaze coordination (Richardson & Dale, 2005), communication is only one-way. The speech and eye movements of one set of participants were recorded as each looked at pictures of TV sitcom cast members and spoke spontaneously about their favourite episode and characters. From these monologues, one-minute segments were cut and played back to a separate set of participants. The listeners looked at the same visual display of the cast members, and their eye movements were also recorded as they listened to the segments of speech. CRQA was used to quantify the degree to which speaker and listener eye positions overlapped at successive time lags. From the moment a speaker looked at a picture, and for the following six seconds, a listener was more likely than chance to be looking at that same picture. The listener was *most* likely to be looking at the same cast member two seconds after the speaker fixated it. The amount of recurrence between the speaker-listener pairs correlated with the listeners' accuracy on comprehension questions that the listeners answered.

A second experiment then showed that gaze coordination and comprehension were causally connected. Pictures flashing in time with the speakers' fixations caused the listeners' eye movements to look more like the speakers', compared to a randomly flashing control condition. This experimental manipulation improved the speed of listeners' performance when answering comprehension questions. This highlights the bidirectional interconnectedness of multiple levels of communicative systems: low-level perceptuo-motor coordination – that is, simply following the gaze patterns of a conversational partner – significantly affects high-level cognitive systems, improving memory and understanding between individuals.

Shared knowledge shapes bottom-level coordination

Conversations are typically interactive, of course. In a study by Richardson, Dale and Kirkham (2007), both participants were able to communicate. They first listened separately to a ninety-second passage describing either the meaning of a specific painting or facts from the painter's biography. Participants then saw the painting together and discussed it while their gaze was tracked. Conversational partners who heard the same information had higher gaze coordination than those who heard different information. These results reinforce the ideas presented earlier from the studies of participants in isolation: Even in completely interactive contexts, higher-level systems – in this case, shared factual knowledge – shape how coordination unfolds in lower-level systems like visual attention.

Concurrent top- and bottom-level coordination

In many of the examples we have reviewed thus far, one system – either the top- or bottom-level system under consideration – has led the other. In Dale et al.'s (2011) gaze coordination experiment, however, coordination emerged from both top- and bottom-level systems, as pairs of participants completed a computerized version of the tangram task (Krauss & Weinheimer, 1964) while being eye-tracked. This task asks each pair to work with a set of six unfamiliar, abstract shapes. Each participant sees the same shapes but arranged in a different order, and each participant is unable to see her partner or her partner's shapes. By talking to each other, the "matcher" must arrange her shapes to match the order of the "director". Once all six shapes are correctly re-ordered, the pair repeats the task.

In the tangram task, a robust pattern of change occurs as the same set of shapes are used repeatedly. Solutions take less time, require fewer words, and are facilitated by a jointly constructed scheme of descriptions for the shapes (Clark & Wilkes-Gibbs, 1986). After multiple rounds the pair is capable of effectively identifying tangrams and completing the task quite rapidly. In this sense, the two people become a coherent, functional unit (Hutchins, 1995).

Another experiment (Dale et al., 2011) showed that during the tangram task the gradual construction of a shared vocabulary – a form of linguistic coordination – filters down to affect the fine-grained dynamics of the partners' eyes and hands, as quantified by CRQA. At the start of the experiment the director's eye movements led the matcher's, demonstrating a lagged but coordinated relation. Intriguingly, this coupling changed over rounds of the tangram task. By the final round, systematic cross-modal coordination emerged: The director and matcher now *synchronized* their gaze and hand movements, with no clear leader or follower. The director and matcher did not simply achieve the task faster; they strongly synchronized their perceptuo-motor activity. With their emerging coordination across multiple top- and bottom-level systems (e.g. linguistic, visual, conceptual), the two participants came to act as a single, coordinated "tangram recognition system" with richly interconnected verbal and nonverbal behaviours.

Putting it together: solving the coordination problem

At the outset of this chapter we argued that language is likely based on a rich process of feedback, and we have presented a series of findings that support this interactivity. Low-level nonverbal systems like motor control and perception weave into longer time scales such as conversations and their topics, which in turn constrain what combinations of actions and perceptions are viable.[2] Guy Van Orden and his colleagues conceived of this process of feedback across time scales as fundamental to the way the cognitive system operates:

> Slower dynamics thus constrain faster dynamics, which allows the flow of visible or audible, or otherwise available, context to constrain the dynamics

the brain. The flow of invariants across perception occurs on the slower time scales of change in brain activity[…], supplying constraints that reduce the degrees of freedom for what may happen next.

(Van Orden et al., 2012, p. 6)

Van Orden and colleagues intend to describe a relationship between the body or environment and the brain. However, the same kind of relationships can be articulated between fast-changing behaviours in interaction and slower systems like the goals and intentions of our interaction partners. The general idea, from Van Orden and others, is that the system must be integrated in its dynamics for it to function successfully (cf. interaction-dominant dynamics; Van Orden et al., 2003). The implications of this multiscale, multicomponent perspective have not yet been borne out by interaction researchers (cf. Dale et al., 2014).

Simply observing this multiscale and multimodal organization alone cannot *"explain"* interaction. Any theory of interaction must be highly specific if it is to render more compelling mechanistic descriptions and make compelling targeted predictions. Our explanation cannot rest purely on our generic reflections in this chapter. Indeed, most proposals about the computational basis of language – from phonetic cues to syntactic structures – are arguably simpler than the neural makeup of seemingly simple creatures without linguistic communication systems (like, say, squirrels, with brains and bodies that reflect elaborate and impressive engines of evolution).

A concept that may be useful to develop computationally – and dynamically – is that of multimodal synergy. When two people interact they generate a wide array of behaviours. These shape the behaviour of both interacting individuals together. Interaction is thus a coordination problem that is specific to our ecology and that is solved, in our evolution, by placing a wide array of subtle signals in interdependent relationships. In some ways it is astonishing to think that at the surface of one of our most common behaviours is a fundamental scientific mystery: how do we solve this coordination problem? We are solving synergies not only among muscle groups (Bernstein, 1967; Turvey, 1990) but also at interrelated time scales that, importantly, must fluidly interact to get the system right.

We believe an important next step will be using network formalisms to understand these patterns of interdependence. They can be devised concretely and without strong representational commitments, while allowing researchers to be explicit about relationships among levels. These networks can have the sophistication to capture synergies (Sporns & Edelman, 1993) and perhaps even tensegrities that could be vital to understanding perception and action (Turvey & Carello, 2011). Such an approach, while risking certain theoretical assumptions or ontological simplifications, would permit explorations of multilevel relationships. Recent work on deep learning neural networks, for example, would allow exploration of behavioural modes at different spatial or temporal levels (Hinton et al., 2012). We have already conducted some initial discussion and visualization in terms of networks (see Figure 18.1; see also Bergmann & Kopp,

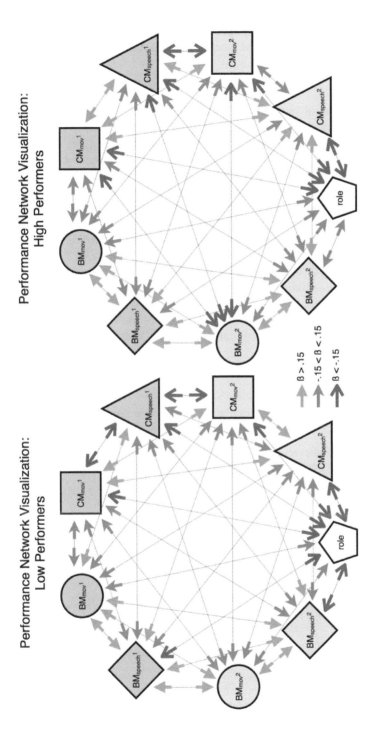

Figure 18.1 Network visualization of top- and bottom-level communication systems during a cooperative dyadic task (Paxton et al., 2014).

Notes: Network analyses, visualizations and models can spur new insights into the relations across top- and bottom-level systems – for instance, by seeing how the connections between various systems change according to the performance on a cooperative task. (In this figure: BM = behaviour matching of speech [subscript *speech*] and movement [subscript *mov*]; CM = complexity matching of speech [subscript *speech*] and movement [subscript *mov*]; role = beliefs about distribution of power during task; superscript numbers = first or second half of the task.)

2010; Dale et al., 2014; Dale & Louwerse, 2012; Paxton, Abney, Dale & Kello, 2014). These initial forays offer a glimpse at what may be possible in the future: a more *integrative* understanding of the function and structure of human interaction.

Conclusion

The recent explosion of interest in coordination has led to a host of new and interesting questions about verbal and nonverbal communication. The goal of this chapter has been to act as a guide to theories and analyses of verbal coordination to ground this new interest in the history of investigations into how and why individuals affect one another's communicative behaviours. These theories and methods serve as an important foundation for exciting new ways of seeing verbal coordination, viewing the verbal systems as simply *part* of a larger network of communication systems. Seeing each of these systems as interconnected leads to a richer – and, arguably, more situated – picture of communication and coordination.

As we have noted throughout this chapter, however, this line of research is far from complete. Additional research must continue to map out this interconnectivity across levels of communication. These new questions extend old lines of enquiry, bridging traditionally distinct research areas to more fully understand the complex interactions across multimodal, multi-timescale levels of communication.

Notes

1   Of course, convincing arguments can be made to include many of the phenomena discussed here as coordination-as-joint-action, but these are outside the scope of the current chapter.
2   We thank an anonymous reviewer for pointing out the similarity between this view and the *enslaving principle* from synergetics (e.g. Haken, 1983).

References

Abney, D., Paxton, A., Dale, R. & Kello, C. (2014) Complexity matching in dyadic interaction. *Journal of Experimental Psychology: General, 143*(6), 2304–2315.

Abney, D., Paxton, A., Dale, R. & Kello, C. (2015) Movement dynamics reflect a functional role for weak coupling and role structure in dyadic problem solving. *Cognitive Processing, 16*(4), 325–332.

Angus, D. Smith, A. & Wiles, J. (2012) Conceptual recurrence plots: Revealing patterns in human discourse. *IEEE Transactions on Visualization and Computer Graphics, 18*(6), 988–997.

Babel, M. (2010) Dialect divergence and convergence in New Zealand English. *Language in Society, 39*(4), 437–456.

Bakeman, R. & Quera, V. (2011). *Sequential analysis and observational methods for the behavioral sciences.* Cambridge University Press: Cambridge.

Bernstein, N.A. (1967) *Coordination and regulation of movement.* Pergamon Press: New York.

Bergmann, K. & Kopp, S. (2010) Modeling the production of coverbal iconic gestures by learning Bayesian decision networks. *Applied Artificial Intelligence, 24*(6), 530–551.

Black, D.P., Riley, M.A. & McCord, C.K. (2007) Synergies in intra-and interpersonal interlimb rhythmic coordination. *Motor Control,* 11(4), 348–373.

Bock, J.K. (1986) Syntactic persistence in language production. *Cognitive Psychology, 18,* 355–387.

Bourhis, R.Y., Giles, H. & Lambert, W.E. (1975) Social consequences of accommodating one's style of speech: A cross-national investigation. *International Journal of the Sociology of Language, 13*(166), 55–72.

Branigan, H.P., Pickering, M.J. & Cleland, A.A. (2000) Syntactic co-ordination in dialogue. *Cognition, 75*(2), B13–B25.

Brennan, S.E. (1991) Conversation with and through computers. *User Modeling and User-Adapted Interaction, 1*(1), 67–86.

Brennan, S.E., Galati, A. & Kuhlen, A.K. (2010) Two minds, one dialog: Coordinating speaking and understanding. In B. H. Ross (Ed.), *The psychology of learning and motivation* (Vol. 53, pp. 301–344). Academic Press: Burlington.

Brennan, S.E. & Hanna, J.E. (2009) Partner-specific adaptation in dialog. *Topics in Cognitive Science, 1*(2), 274–291.

Clark, H.H. (1996) *Using language.* Cambridge University Press: Cambridge.

Clark, H.H. & Krych, M.A. (2004) Speaking while monitoring addressees for understanding. *Journal of Memory and Language, 50*(1), 62–81.

Clark, H.H. & Wilkes-Gibbs, D. (1986) Referring as a collaborative process. *Cognition, 22*(1), 1–39.

Coco, M.I. & Dale, R. (2014) Cross-recurrence quantification analysis of categorical and continuous time series: an R package. *Frontiers in Quantitative Psychology and Measurement,* 5.

Dale, R., Fusaroli, R., Duran, N.D. & Richardson, D.C. (2014) The self-organization of human interaction. In B.H. Ross (Ed.), *The psychology of learning and motivation* (Vol. 59, pp. 43–95). Academic Press: Amsterdam.

Dale, R., Kirkham, N. & Richardson, D. (2011) The dynamics of reference and shared visual attention. *Frontiers in Cognition, 2,* 355.

Dale, R. & Louwerse, M.M. (2012, May) *Multimodal communication as a dynamic network.* Paper presented at the 11th Conceptual Structure, Discourse, and Language Conference. Vancouver, BC, Canada.

Ferreira, V.S. & Bock, K. (2006) The functions of structural priming. *Language and Cognitive Processes, 21*(7–8), 1011–1029.

Fusaroli, R., Bahrami, B., Olsen, K., Roepstorff, A., Rees, G., Frith, C. & Tylen, K. (2012) Coming to terms: Quantifying the benefits of linguistic coordination. *Psychological Science, 23*(8), 931–939.

Galati, A. & Brennan, S.E. (2010) Attenuating information in spoken communication: For the speaker, or for the addressee? *Journal of Memory and Language, 62*(1), 35–51.

Giles, H. (1973) Accent mobility: A model and some data. *Anthropological Linguistics, 15*(2), 87–105.

Giles, H., Coupland, N. & Coupland, J. (1991) Accommodation theory: Communication, context, and consequence. In H. Giles, N. Coupland & J. Coupland (Eds.), *Contexts of accommodation: Developments in applied sociolinguistics* (pp. 1–68). Cambridge University Press: New York.

Giles, H., Taylor, D.M. & Bourhis, R.Y. (1977) Dimensions of Welsh identity. *European Journal of Social Psychology, 7*(2), 165–174.

Haken, H. (1983) *Advanced synergetics*. Springer: Berlin.

Harris, R. (1996) *Signs, language, and communication: integrational and segregational approaches*. Psychology Press: New York.

Hasson, U., Ghazanfar, A.A., Galantucci, B., Garrod, S. & Keysers, C. (2012) Brain-to-brain coupling: A mechanism for creating and sharing a social world. *Trends in Cognitive Sciences*, *16*(2), 113–120.

Henderson, J.M. (2003) Human gaze control during real-world scene perception. *Trends in Cognitive Sciences*, *7*(11), 498–504.

Hinton, G., Deng, L., Yu, D., Dahl, G.E., Mohamed, A.R., Jaitly, N., ... & Kingsbury, B. (2012) Deep neural networks for acoustic modeling in speech recognition: The shared views of four research groups. *IEEE Signal Processing Magazine*, *29*(6), 82–97.

Hutchins, E. (1995) *Cognition in the wild*. MIT Press: Cambridge, MA.

Kenny, D.A., Kashy, D.A. & Cook, W.L. (2008) *Dyadic data analysis*. Guilford Press: New York.

Krauss, R.W. & Weinheimer, S. (1964) Changes in reference phrases as a function of frequency of usage in social interaction: A preliminary study. *Psychonomic Science*, *1*(1–12), 113–114.

Marwan, N. (2013) *Cross recurrence plot toolbox* [computer software]. Retrieved from http://tocsy.pik-potsdam.de/CRPtoolbox.

Marwan, N. (2008) A historical review of recurrence plots. *The European Physical Journal Special Topics*, *164*(1), 3–12.

Marwan, N., Carmen Romano, M., Thiel, M. & Kurths, J. (2007) Recurrence plots for the analysis of complex systems. *Physics Reports*, *438*(5), 237–329.

Neumann, R. & Strack, F. (2000) "Mood contagion": The automatic transfer of mood between persons. *Journal of Personality and Social Psychology*, *79*(2), 211–223.

Paxton, A. & Dale, R. (2013a) Argument disrupts interpersonal synchrony. *Quarterly Journal of Experimental Psychology*, *66*(11), 2092–2102.

Paxton, A. & Dale, R. (2013b) *B(eo)W(u)LF: Facilitating recurrence analysis on multi-level language*. arXiv:1308.2696 [cs.CL].

Paxton, A. & Dale, R. (2013c) Frame-differencing methods for measuring bodily synchrony in conversation. *Behavior Research Methods*, *45*(2), 329–343.

Paxton, A., Abney, D., Kello, C.K. & Dale, R. (2014) Network analysis of multimodal, multiscale coordination in dyadic problem solving. In P.M. Bello, M. Guarini, M. McShane & B. Scassellati (Eds.), *Proceedings of the 36th Annual Meeting of the Cognitive Science Society*. Austin, TX: Cognitive Science Society.

Pennebaker, J.W., Booth, R.J. & Francis, M.E. (2007) *Linguistic Inquiry and Word Count (LIWC): A computerized text analysis program*. LIWC.net: Austin, TX.

Pickering, M.J. & Garrod, S. (2004) Toward a mechanistic psychology of dialogue. *Behavioral and Brain Sciences*, *27*(2), 169–190.

Reitter, D., Moore, J.D. & Keller, F. (2006) Priming of syntactic rules in task-oriented dialogue and spontaneous conversation. In R. Sun (Ed.), *Proceedings of the 28th Annual Meeting of the Cognitive Science Society* (pp. 685–690). Cognitive Science Society: Austin, TX.

Richardson, B.H., Taylor, P.J., Snook, B., Conchie, S.M. & Bennell, C. (2014) Language style matching and police interrogation outcomes. *Law and Human Behavior*, *38*(4), 357–366.

Richardson, D.C. & Dale, R. (2005) Looking to understand: The coupling between speakers' and listeners' eye movements and its relationship to discourse comprehension. *Cognitive Science*, *29*(6), 1045–1060.

Richardson, D.C., Dale, R. & Kirkham, N.Z. (2007) The art of conversation is coordination common ground and the coupling of eye movements during dialogue. *Psychological Science, 18*(5), 407–413.

Richardson, D.C., Dale, R. & Tomlinson, J.M. (2009) Conversation, gaze coordination, and beliefs about visual context. *Cognitive Science, 33*(8), 1468–1482.

Richardson, D.C., Street, C.N.H., Tan, J.Y.M., Kirkham, N.Z., Hoover, M.A. & Cavanaugh, A.G. (2012) Joint perception: Gaze and social context. *Frontiers in Human Neruoscience, 6*, 194.

Richardson, M.J., Dale, R. & Marsh, K.L. (2014) Complex dynamical systems in social and personality psychology: Theory, modeling, and analysis. In H.T. Reis & C.M. Judd (Eds.), *Handbook of Research Methods in Social and Personality Psychology* (pp. 251–280). Cambridge University Press: New York.

Riley, M.A., Richardson, M.J., Shockley, K. & Ramenzoni, V.C. (2011) Interpersonal synergies. *Frontiers in Psychology, 2*, 38.

Riley, M.A. & Van Orden, G.C. (2005) *Tutorials in contemporary nonlinear methods for the behavioral sciences.* Retrieved from http://www.nsf.gov/sbe/bcs/pac/nmbs/nmbs.jsp.

Rogers, S.L., Fay, N. & Maybery, M. (2013) Audience design through social interaction during group discussion. *PloS ONE, 8*(2), e57211.

Schmidt, R. & Richardson, M.J. (2008) Dynamics of interpersonal coordination. In A. Fuchs & V. Jirsa (Eds.), *Coordination: Neural, Behavioral, and Social Dynamics* (pp. 281–308). Springer: Berlin.

Shockley, K., Baker, A.A., Richardson, M.J. & Fowler, C.A. (2007) Articulatory constraints on interpersonal postural coordination. *Journal of Experimental Psychology: Human Perception and Performance, 33*(1), 201–208.

Shockley, K., Richardson, D.C. & Dale, R. (2009) Conversation and coordinative structures. *Topics in Cognitive Science, 1*(2), 305–319.

Sporns, O. & Edelman, G.M. (1993) Solving Bernstein's problem: A proposal for the development of coordinated movement by selection. *Child Development, 64*(4), 960–981.

Tausczik, Y.R. & Pennebaker, J.W. (2010) The psychological meaning of words: LIWC and computerized text analysis methods. *Journal of Language and Social Psychology, 29*(1), 24–54.

Trude, A.M. & Brown-Schmidt, S (2012). Talker-specific perceptual adaptation during online speech perception. *Language and Cognitive Processes, 27*(7-8), 979–1001.

Turvey, M.T. (1990) Coordination. *American Psychologist, 45*(8), 938–953.

Turvey, M.T. & Carello, C. (2011) Obtaining information by dynamic (effortful) touching. *Philosophical Transactions of the Royal Society B: Biological Sciences, 366*(1581), 3123–3132.

Van Orden, G., Hollis, G. & Wallot, S. (2012) The blue-collar brain. *Frontiers in Physiology, 3*, 207.

Van Orden, G.C., Holden, J.G. & Turvey, M.T. (2003) Self-organization of cognitive performance. *Journal of Experimental Psychology: General, 132*(3), 331.

Von Zimmermann, J. & Richardson, D.C. (2014) Joint perception. In S.S. Obhi & E.S. Cross (Eds.), *Shared representations: Sensorimotor foundations of social life.* Cambridge: Cambridge University Press.

Webb, J.T. (1969) Subject speech rates as a function of interviewer behaviour. *Language and Speech, 12*(1), 54–67.

West, B.J., Geneston, E.L. & Grigolini, P. (2008) Maximizing information exchange between complex networks. *Physics Reports, 468*, 1–99.

Part IV

Methods, tools and devices

19 Measuring interpersonal coordination

A selection of modern analysis techniques

Robert Rein

Introduction

Numerous studies show that interacting people start to coordinate their movements (Neda, Ravasz, Brechet, Vicsek & Barabasi, 2000; Oullier, de Guzman, Jantzen, Lagarde, & Kelso, 2008; Schmidt, Fitzpatrick, Caron, & Mergeche, 2011). This phenomenon occurs both explicitly as well as implicitly or spontaneously (Lopresti-Goodman, Richardson, Silva & Schmidt, 2008). In the present chapter, several analysis approaches will be presented to analyse data from empirical investigation studying interpersonal coupling effects. The level of the presentation is intended to be as accessible as possible. Naturally, the presented methods represent only a selection of all available or currently used approaches in the literature.

The chapter roughly follows a historical route, starting with more traditional measures investigating the relative phasing in oscillator-like behaviours, including the discrete and relative phase and an introduction to the analytical signal and the related Hilbert transform. Subsequently, two relatively novel approaches to study interpersonal coordination, principal component analysis and uncontrolled manifold analysis, will be presented, which offer the opportunity to gain some interesting new insights into interpersonal coordination. The chapter concludes with a presentation of the cluster-phase method, which is also a relatively novel approach that allows the study of interpersonal coordination in groups involving more than two actors – an emerging topic in the area of interpersonal coordination. As already mentioned, and as will become evident during the chapter, most of the techniques applied to interpersonal coupling follow what Schmidt et al. (2011) call a "principle of similitude". Techniques first developed to study intra-personal coordination are being applied to study interpersonal coordination. One of the more prominent theoretical paradigms for the study of interpersonal coordination is dynamical systems theory (DST), and the principle of similitude is in itself a nice continuation of the DST approach which started with small scale finger-wagging studies and continuously advanced to more complex movements, eventually generalizing to interpersonal dynamics (Haken, Kelso & Bunz, 1985; Kelso, de Guzman, Reveley & Tognoli, 2009; Riley, Richardson, Shockley & Ramenzoni, 2011).

Methods

Relative phase

One of the most widely applied measurement tools in studies of interpersonal coupling is the discrete relative phase (DRP). The DRP describes the phasing relationship between two oscillators (Hamill, Haddad & McDermott, 2000) by measuring a succession of three time points: two instances of signal peaks in a reference signal $s_1(t)$ (t_{1i-1} and t_{1i}) and the time of peak t_{2i} in a second signal $s_2(t)$ which falls in between the two events in $s_1(t)$. Thereby, T_i is the period of the ith oscillation of $s_1(t)$. In the context of interpersonal coordination the two signals could represent joint angles during elbow flexion-extension movements performed by two actors (for a similar example see Schmidt, Carello & Turvey, 1990). In order for the discrete relative phase to be applicable the basic shape of the signals should follow a sinusoidal pattern:

$$s_i(t) = A\sin(\omega t + \psi)$$
(Eq. 1)

where A = amplitude, ω = frequency (in radians) and ψ = phase shift. Accordingly, the ith DRP measurement can be calculated using:

$$T_i = t_{1i} - t_{1i-1}$$
$$\phi_i = \frac{t_{1i} - t_{2i}}{T_i} 2\pi$$
(Eq. 2)

(compare also Figure 19.1). By exchanging the factor 2π in (Eq. 1) with 360°, one obtains the same result in degrees rather than radians. A DRP value close to zero (or 2π) indicates an in-phase pattern, whereas π indicates an anti-phase pattern.

As DRP measurement results in one measurement point for each oscillation, the average discrete relative phase $\bar\phi$ and the standard deviation of the DRP SD are commonly used for further analysis. Thus, the SD ϕ is often seen as a measurement of coupling stability (Schmidt et al., 1990). One potential problem with DRP occurs when the frequencies between the two oscillators are widely

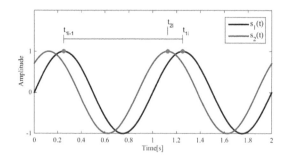

Figure 19.1 Example calculating of the discrete relative phase between two sinusoidal signals.

different and multiple DRP measures of the faster signal are obtained for each cycle of the slower oscillator. DRP also does not provide any information about the amplitude of the signal, and its use is restricted to periodic movement models because otherwise the determination of the required peak points becomes difficult, if not arbitrary.

A common alternative to the discrete relative phase is the continuous relative phase (CRP). In contrast to the discrete relative phase, which measures the phasing relationship during specific events, the continuous relative phase measures the continuous phase space deviation between two oscillators. The phase space is a concept derived from oscillator theory where typically two variables, the phase and its velocity, are used to describe the dynamics of the system, since these parameters are sufficient to describe the state of an oscillator (Pikovsky, Rosenblum, & Kurths, 2003) (compare also Figure 19.2a).

Using these two variables the continuous phase angle of the system can be calculated using:

$$\varphi(t) = \frac{d\omega(t)}{dt}$$
$$\Phi(t) = \arctan\left(\frac{\varphi(t)}{\omega(t)}\right)$$

(Eq. 3)

$$CRP(t) = \Phi_1(t) - \Phi_2(t)$$

(Eq. 4)

If one does not want to use the phase angle it is also possible to use displacement and velocity directly.

$$\varphi(t) = \arctan\left(\frac{v(t)}{f_0 \triangle x(t)}\right)$$

(Eq. 5)

Similar to DRP, the continuous relative phase has the constraint of being mainly useful when applied to sinusoidal-like signals. Using simulated data, Peters, Haddad, Heiderscheid, van Emmerik and Hammill (2003) showed that the CRP can be highly misleading when the signals differ strongly from a sinusoidal pattern. To remedy this situation, several normalization approaches have been

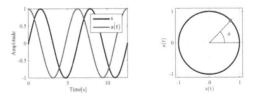

Figure 19.2 a) Position and velocity of a oscillator. b) Phase plane representation of the oscillator and the calculation of the continuous phase.

suggested to account for deviations from pure sinusoidal patterns (compare also Lamb & Stockl, 2014; Varlet & Richardson, 2011). However, to the best knowledge of the author, no silver bullet is available; care has to be taken when analysing non-sinusoidal signals and normalization routines have to be applied on a case-by-case basis. Because one obtains a continuous measurement of the relative phasing of the same lengths as the original data, as with to DRP, the average CRP and the standard deviation of the CRP are often used for further analysis. Examples of the use of CRP in studies of interpersonal coordination can be found in Schmidt, Christianson, Carello, and Baron (1994) and Richardson, Marsh, Isenhower, Goodman, and Schmidt (2007).

An implicit assumption underlying CRP is that the studied signals are nearly stationary, which means in casual terms that the process generating the signal does not change during the observation period. Unfortunately, this assumption is seldom satisfied for human movements (Varlet & Richardson, 2011). To circumvent this problem the Hilbert transform has been suggested as an alternative approach. The basic approach underlying the Hilbert transform was formulated by Gabor (1946) using the notion of a complex analytical signal a(t). A complex analytical signal does not have any negative frequency components, unlike the Fourier transform of a real signal for example, where frequency components are typically conjugate complex pairs. The analytical signal is therefore based on the construction of a unique complex signal from a real signal s(t). The analytical signal can be constructed by setting all negative frequencies of the signal to zero and doubling the amplitudes of all positive frequencies (Boashash, 1992; Gabor, 1946). Formally, the analytical signal a(t) of a real signal s(t) is obtained through:

$$a(t) = s(t) + is_H(t) \qquad \text{(Eq. 6)}$$

Here, s(t) is the original real signal and s_H is the Hilbert transform of the signal (Gabor, 1946). The Hilbert transform of a signal is defined as:

$$S_H(t) = \frac{1}{\pi} P.V. \int_{-\infty}^{\infty} \frac{s(\tau)}{t-\tau} d\tau \qquad \text{(Eq. 7)}$$

The term P.V. refers to the Cauchy Principle value, and is a method to deal with the singularity at t = 0 which makes (Eq. 7) an improper integral. As (Eq. 7) describes the convolution between the signal s(t) and the kernel $k(t) = \frac{1}{\pi t}$, a somewhat more intuitive insight about the Hilbert transform is obtained when calculating the Fourier transforms of k(t), yielding (King, 2009, p. 252):

$$F\{k\}(\omega) = -i \, sng(\omega)$$
$$sng(x) = \begin{cases} -1 & \text{if } x < 0 \\ 1 & \text{if } x < 0 \end{cases} \qquad \text{(Eq. 8)}$$

where i := imaginary unit $\sqrt{-1}$ and sng := signum function. Using (Eq. 8) the Hilbert transform can thus be performed using a simple multiplication in the frequency domain. This results in a constant phase lag of $\pi/2$ of the original signal, yielding the suppression of all negative frequency components and the doubling of the amplitudes of the positive frequencies (compare Figure 19.3 for an example with a sinusoidal signal) in the analytical system.

As only a single positive frequency component is obtained, the somewhat involved derivation of the analytical system is therefore actually closer to the intuitive understanding of the frequency representation of a sinusoidal signal with a single peak at the base frequency of the oscillator.

The utility of the complex analytical signal stems from the polar expression of complex numbers, from which it follows that the signal can be described as:

$$a(t) = s(t) + is_H(t) = A(t)e^{i\phi(t)}$$
$$A(t) = \sqrt{s(t)^2 + s_H(t)^2}$$
$$\phi(t) = \arctan\left(\frac{s_H(t)}{s(t)}\right)$$

(Eq. 9)

Using (Eq. 9) leads to the notion of an instantaneous amplitude A(t) and an instantaneous phase, and in addition an instantaneous frequency $f(t) = \frac{1}{2\pi}\frac{d\phi(t)}{dt}$ (Boashash, 1992). Using the instantaneous phase allows to calculate the relative phase between two signals by subtracting their instantaneous phases according to (Eq. 4). When using numerical packages (MATLAB or Python) one has to be aware of that these routines typically wrap the phase to the interval $[-\pi, \pi]$ (compare with Figure 19.4c). Thus, when calculating the relative phase the data can exhibit sudden jumps in the relative phase, and one therefore needs to unwrap the phase before calculating the relative phase. In addition, the numerical approximation of the Hilbert transform at the signal start and end is typically unreliable. Thus, one should anticipate these boundary effects (compare with Figure 19.4b).

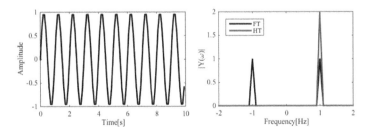

Figure 19.3 a) Sinusoidal signal. b) Frequency domain representation of the Fourier transform of the original real signal (FT) and the Fourier transform of the analytical system (HT).

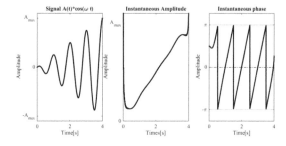

Figure 19.4 Example of the analytical signal of a) a simple sinusoidal signal with a linear increasing amplitude; b) the resulting instantaneous amplitude; c) the instantaneous phase.

Although the derivation is somewhat involved, the analytical signal allows us not only to study the differences with respect to phase relationships, but also to investigate instantaneous amplitude and frequency effects. A caveat when using the analytical signal is the fact that if the frequency bands of the amplitude variation and the phase variation are overlapping, the instantaneous amplitude and phase do not directly represent physical quantities, although the analytical signal will be unique (Boashash, 1992). Further, in order for the phase to be always defined the analytical signal should pass through the origin, which often can be achieved by subtracting the mean from the signal (Pikovsky et al., 2003). Examples of the use of the Hilbert transform in studies of interpersonal coordination can be found in Kelso et al. (2009), van Ulzen, Lamoth, Daffertshofer, Semin, and Beek (2008) and Palut and Zanone (2005).

In summary, the discrete relative phase, the continuous relative phase and the analytical signal are all appropriate when applied to oscillatory behaviour, and successively give more information about the underlying signal. In the following sections, novel approaches will be presented which target somewhat different experimental questions and are also more appropriate when studying non-sinusoidal patterns.

Principal component analysis

One analysis method which has become quite prominent in motor control research has been principal component analysis (PCA) (see for example Chen, Liu, Mayer-Kress & Newell, 2005; Deluzio, Wyss, Zee, Costigan, & Sorbie, 1997; Post, Daffertshofer & Beek, 2000; Santello, Flanders & Soechting, 1998). The basic idea behind PCA is to reduce the dimensionality of a high dimensional (multi-variate) dataset by extracting the information between interrelated variables. PCA looks for correlations between the variables and calculates new variables, the so-called principal components which capture most of the variance in the original dataset using fewer dimensions (Jolliffe, 2002). In effect, this approach is heuristically similar to the techniques described above, because for example the relative phase also can be interpreted as an information compression mechanism mapping the behaviour of two actors onto a shared single parameters (Riley et al., 2011).

PCA starts with a random dataset – for example, $X \in \Re^{nxp}$ consisting of p variables measured n times. The obtained principal components are then the eigenvectors λ, i = 1, 2, ..., p of the sample correlation matrix Σ of X, where the λs also indicates the magnitude of variance of the principal components they explain. The principal components are ordered according to their magnitude (Stuart, 1982). Thus, the first principal component explains the greatest variances, the second principal component the second largest, and so on. By using only the first few components it is therefore possible to obtain a compressed representation of the information contained in the original data X.

An example of the necessary steps to perform a PCA using synthetic data is given in Figure 19.5. Three different signals are used: $s_1(t) = 3t+\sin(t)+\xi$, $s_2(t) = 4 + 0.3s_1(t) + \xi$, $s_3(t) = 10 + 0.5(t-6)^2 + \xi$ (ξ = Gaussian random noise). Based on these formulas it can be seen that signal $s_2(t)$ is just a scaled version of $s_1(t)$, which is a linear function plus an added sinusoidal component. Thus, the correlation between $s_1(t)$ and $s_2(t)$ will be very high – or, stated differently, most of the information contained in $s_2(t)$ will be already available in $s_1(t)$. In contrast, signal $s_3(t)$ follows a right-shifted parabola function (see Figure 19.5a). One would therefore expect that two components would be sufficient to describe the main information contained in the data. Inspection of the obtained principal components (see Figure 19.5b) shows that only the first two components contain some information, whereas the third principal component contains mainly noise fluctuations around zero. Accordingly, the first component captures the information of $s_1(t)$ and $s_2(t)$, namely the linear increase with time plus the sinusoidal trend, whereas the second component reflects the squared term in $s_3(t)$. Given a set of principal components, it is necessary to decide how many to include for further analysis. A typical rule of thumb used in the literature regarding the number of

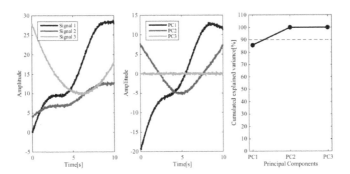

Figure 19.5 a) Simulated signals: $s1(t) = 3t+\sin(t)+\xi$, $s2(t) = 4 + 0.3s1(t) + \xi$, $s3(t) = 10 + 0.5(t-6)2 + \xi$, b) Principal components. c) Explained variance by cumulated principal components.

principal components is to use enough components to account for 90% of the explained variance. To calculate the explained variance one uses the ordered eigenvalues λ_i, which reflect the variance of the respective principal components and calculates the cumulated sum divided by their total sum:

$$\text{var}(i)_{cumulated} = \frac{\sum_{n=1}^{i} \lambda_n}{\sum_{n=1}^{p} \lambda_n}, i = 1, ..., p \qquad (\text{Eq. 10})$$

Plotting the cumulated variance against the number of principal components gives an indication of how many components should be included for further analysis (see Figure 19.5c). One additional set of parameters obtained from the PCA are the factor loadings. They describe the contribution of each original variable to the according principal component.

In Table 19.1 the loadings for the present example are given. The values show that the first principal component obtains its main input from $s_1(t)$, whereas the third principal component loads high on $s_2(t)$, but only after all the redundant information contained in $s_1(t)$ is extracted leaving mainly the Gaussian random noise. Finally, the second component shows a high loading for signal $s_2(t)$, extracting the parabola shape of the function. A good start to further explore the application of PCA is a tutorial paper by Daffertshofer, Lamoth, Meijer, and Beek (2004) (compare also Chau, 2001; Forner-Cordero, Levin, Li & Swinnen, 2005; Rein, 2012).

Ramenzoni, Riley, Shockley and Baker (2012) used PCA to study interpersonal coordination in a joint supra-postural precision task between two actors. One actor held their index finger within a circle, which in turn was held by another actor. This condition was compared with a control condition where the circle was fixed to a stand. The positions of the arm, forearm and hand, together with the postural sway of both actors, were recorded. The hypothesis stated that the actors would couple their actions to each other and therefore reduce the number of degrees of freedom used by the combined system. Indeed, the results showed that only four components were necessary during the interaction condition to explain more than 90% of the variance in the data, whereas six components were needed during the control condition (Ramenzoni et al., 2012; Riley et al., 2011).

Table 19.1 Factor loadings from PCA as calculated from $S_1(t)$, $S_2(t)$ and $S_3(t)$.

	PC1	*PC2*	*PC3*
s1(t)	0.91	0.30	−0.29
s2(t)	0.27	0.09	0.96
s3(t)	−0.31	0.95	−0.01

Recently a variation of the traditional principal component approach has been introduced, known as functional principal component analysis (Ramsay & Dalzell, 1991; Ramsay & Silverman, 2005). The key difference between normal PCA and functional principal component analysis (fPCA) lies in the fact that fPCA does not treat the input dataset as vectors, but rather as realizations of functions. Thus, for example, all data points stemming from a single joint belong to the same function. It follows that this approach does not calculate eigenvectors but eigenfunctions, which describe the information shared across the data through functions. The obtained eigenfunctions have the advantage that they reside in the same domain as the initial data and therefore are easier to interpret in comparison with the principal components obtained through PCA (Donoghue, Harrison, Coffey & Hayes, 2008; Ryan, Harrison & Hayes, 2006). However, fPCA involves several processing steps which require informed choices by the researcher. First, the data must be smoothed and interpolated. Typically a spline-interpolation is used for this task (Ramsay & Silverman, 2005), but other choices are available. Subsequently, the choice of an appropriate basis set of functions is necessary. Although some guidelines are available in the literature it is not clear to what extent this choice impacts on the results. Currently there are no applications of fPCA to interpersonal data; most applications are in biomechanics (Dona, Preatoni, Cobelli, Rodano, & Harrison, 2009; Donoghue et al., 2008; Ryan et al., 2006). However, treating the data from a functional standpoint seems appropriate in many cases, and it will be interesting to see what information can be obtained from this kind of approach.

Uncontrolled manifold analysis

Another recent analysis technique from the domain of intra-individual analysis which has found its way into studies of interpersonal coordination is uncontrolled manifold analysis (UCM). This approach was introduced by Scholz and Schöner (1999) and is based on concepts from robotic motion-planning. The key idea behind uncontrolled manifold analysis lies in mapping the variability of so-called elementary variables describing the behaviour of the system under investigation to the variability of a system performance variable. Through this mapping it is possible to uncover hidden structural features of behavioural variability (Scholz & Schöner, 1999; Schöner & Scholz, 2007). To this end, UCM separates the observed variability of the elementary variables into two different components. One component describes those parts of the behavioural variability which leave the behaviour of the systems' performance unchanged, whereas the other, orthogonal component describes the variability which varies the performance variable. For example, the elementary variables might be joint segment angles, joint torques or some other quantity the researchers believes the system is influencing to control its behaviour. The sum of all elementary performance variables spans the degrees of freedom available to the system (Schöner & Scholz, 2007). The performance or task variables, on the other hand, are the variables which are necessary to accomplish a task (Schöner & Scholz, 2007). Both elementary and performance variables are derived from a priori theoretical considerations and represent the starting point for every UCM analysis. For example, when investigating an upright standing task the elementary variables could be the joint angles, and the task variable could be the horizontal position of the centre of mass which needs to be controlled by the system to prevent falling over (Hsu, Scholz, Schoner, Jeka & Kiemel, 2007). Based on the hypothesized elementary and performance variables the researcher has to derive a model to map the domain of the elementary variables to the domain of the performance variable. In case of the upright standing example, a geometrical model of the body segments could be established which maps changes in joint configuration to changes of the position of the centre of mass (Hsu et al., 2007; Reisman, Scholz, & Schöner, 2002). The application of uncontrolled manifold analysis allows us to identify so-called synergies (Latash, 2000, 2008). Multiple constituting elements of a system form a synergy by contributing behaviour to achieve a specific task. The components thereby exhibit a reciprocal sharing pattern, where one element may reciprocally compensate the actions of another element (Latash, 2008, p. 13). Thus, applied to interpersonal coordination, synergies are formed by coupled actors – for example to perform a postural task (Riley et al., 2011).

As the application of UCM is somewhat involved, in the following a more in-depth presentation of the necessary steps will be given. An example of the application of UCM to interpersonal coordination was given by Black, Riley and McCord (2007). In their experiment the authors investigated the coupling between two actors swinging one pendulum each in either an in-phase or an anti-phase condition. Thus, the actors were instructed to form an interpersonal synergy. Based on previous results (for example, Schmidt et al., 1990) the position of the pendulum and its velocity were used as the elementary variables and the relative phase as the performance variable. The authors predicted that the two actors would form a synergy through compensatory behaviour such that fluctuations made by one actor would be compensated for by the other.

Thus, the mapping between the elementary variables and the performance variable was achieved using:

$$\varphi_i = \arctan\left(\frac{v_i}{f_{0i}\Delta x_i}\right)$$
$$\phi = \varphi_1 - \varphi_2 \qquad\qquad \text{(Eq. 11)}$$

which is variation (Eq. 3), where v = velocity of the hand swinging the pendulum, f_0 = frequency = $1/T$, and Δx = displacement from the mean position of the hand, $i = 1, 2$. To be able to calculate the UCM the Jacobian J of the mapping from the elementary variables to the performance variable is needed. Described somewhat imprecisely, the Jacobian is the multi-dimensional equivalent of the (partial) derivate of a function:

$$J = \begin{bmatrix} \dfrac{d\theta}{dv_1} & \dfrac{d\theta}{dx_1} & \dfrac{d\theta}{dv_2} & \dfrac{d\theta}{dx_2} \end{bmatrix}$$

$$= \begin{bmatrix} \dfrac{1}{f_{01}\left(1+\dfrac{v_1^2}{f_{01}^2 x_1^2}\right)x_1} & \dfrac{-v_1}{f_{01}\left(1+\dfrac{v_1^2}{f_{01}^2 x_1^2}\right)x_1^2} & \dfrac{1}{f_{02}\left(1+\dfrac{v_2^2}{f_{02}^2 x_2^2}\right)x_2} & \dfrac{-v_2}{f_{02}\left(1+\dfrac{v_2^2}{f_{02}^2 x_2^2}\right)x_2^2} \end{bmatrix} \quad \text{(Eq. 12)}$$

Subsequently, it is necessary to calculate the null space of the Jacobian. Remembering that one variability component collects all the variability which does not change the performance variable, it becomes intuitively clear what the Jacobian is needed for. As the Jacobian is similar to the derivate f' of the function f from elementary variables to performance variable, all those values of x which do not change the performance variable represent the null space.

In the present example the space of the elementary variables has four dimensions ($n = 4$; v_1, x_1, v_2, x_2) whereas the space of the performance variable has only a single dimension ($d = 1$). Thus, the null space has three ($n - d$) dimensions. To obtain numerical values for the Jacobian, Scholz and Schöner (1999) suggested using the grand mean $\overline{\Theta}=(\overline{v}_1,\overline{x}_2,\overline{v}_2,\overline{x}_2)'$ of the data over all trials. In the present experiment when calculating the UCM at the beginning of the movement, all trials are used to calculate the grand mean $\overline{\Theta}$. The obtained values are than used and plugged into the formula for the Jacobian (Eq. 11). The Jacobian will have a different set of values for different mappings between elementary and performance variables, and a formal derivation may be too complex or impossible. To address this issue, de Freitas and Scholz (2010) presented a method to calculate the Jacobian using a simple regression approach. After numerical values for the Jacobian have been calculated, the null space and its according basis vectors can be obtained. To actually calculate the null space in praxis, numerical packages like MATLAB, NumPy or R are used. To separate the variability of the elementary variables one subsequently projects the deviations of the individual trials from the grand mean $\overline{\Theta} - \Theta_i$, $i = 1, ..., N$ (where N = number of trials) into the null space $\theta_{\parallel}$ and the space orthogonal to the null space $\theta_{\perp}$ using:

$$\theta_{\parallel} = \sum_{i=1}^{n-d} e_i^T (\overline{\Theta} - \Theta_i) e_i \qquad \text{(Eq. 13)}$$

$$\theta_{\perp} = (\overline{\Theta} - \Theta_i) - \theta_{\parallel}$$

where e_i = basis vector of the null space, $\theta_{\parallel}$ describes the component of the variability which leaves the value of the performance variable unchanged or compensated variability (Var_{comp}), whereas $\theta_{\perp}$ represents the variability which changes the performance variable or the uncompensated variability (Var_{uncomp}). As the two components span different dimensions they should be normalized according to their sizes (the cardinality of the vector space).

$$Var_{comp} = \frac{1}{(n-d)N} \sum_N \theta_{\parallel}^2$$

$$Var_{uncomp} = \frac{1}{dN} \sum_N \theta_{\perp}^2$$

(Eq. 14)

By dividing these two measurements one obtains a ratio value R describing which component of the variability is greater:

$$Ratio = \frac{Var_{comp}}{Var_{uncomp}}$$

(Eq. 15)

Accordingly, if the ratio > 1, the amount of compensated variability is greater and the components interact so as to stabilize the performance variable, whereas when the ratio < 1 the performance variable is not stabilized. The results in Black et al. (2007) indicated that the two actors did indeed reciprocally adapt their actions in order to stabilize the instructed phase relationship between the pendulums (Black et al., 2007; Riley et al., 2011).

In summary, although the calculation of the uncontrolled manifold appears to be somewhat involved, the information gain one gets through its application is quite significant, and therefore hopefully it will find more utility in the field of interpersonal coupling studies. UCM analysis per se is not confined to the application to dyadic interactions, but easily generalizes to studies of group behaviour.

Cluster phase method

The last method in this overview is the cluster phase method developed by Frank and Richardson (2010). Currently, in the literature only few experiments study more than two actors (Richardson, Garcia, Frank, Gergor & Marsh, 2012). In parts this may be due to the complexities in analysing such data, which may be overcome through the cluster phase method. The principal idea behind the cluster phase is to calculate the deviations of individual actors from the group average pattern and use these deviations to estimate group synchrony.

The cluster phase method starts with N time series measured from N actors at T time points. The data is therefore initially of the form $x_n(t_t) = y_{nt}$, where n = 1, ..., N and t = 1, ..., T. These N time series have to be transformed into phase time series $\theta_n(t_i)$ for each actor, for example by using the analytical signal described above. Subsequently, the cluster phase $q(t_i)$ for the whole group can to be calculated using:

$$q'(t_i) = \frac{1}{n}\sum_{k=1}^{n} e^{i\theta_k(t_i)} \in$$

(Eq. 16)

$$q(t_i) = \text{atan} 2\left(\frac{\Re\{q'(t_i)\}}{\Im\{q'(t_i)\}}\right) \in (-\pi,\pi)$$

(' denotes complex numbers.) For , $\Re(z)$ returns the real part of the complex z, and $\Im(z)$ returns the imaginary part of z. Accordingly, the cluster phase $q(t_i)$ describes the average phase of all actors at time t_i. Next, for each actor $k = 1, ..., n$ the deviations of the individual phases from the group phase are calculated according to:

$$\phi_k(t_i) = \theta_k(t_i) - q(t_i)$$

(Eq. 17)

From this deviation time series the average deviation for actor k is calculated according to:

$$\bar{\phi}_k' = \frac{1}{N}\sum_{i=1}^{N} e^{i\phi_k(t_i)} \in$$

(Eq. 18)

$$\bar{\phi}_k = \arctan(\bar{\phi}_k') \in (-\pi,\pi]$$

where N = number of time steps. Another useful measure is the individual dispersion of the individual angles:

$$\rho_k = |\bar{\phi}_k'|$$

(Eq. 19)

$\rho_k \in [0.1]$, where 0 denotes when the actor is completely unsynchronized with the group because her phases are distributed across the whole phase, whereas when $\rho_k = 1$ the actor is perfectly synchronized with the group (Mardia & Jupp, 2000; Richardson et al., 2012). To calculate the whole group synchronization at time t_i:

$$\rho_{group,i} = \left|\frac{1}{n}\sum_{k=1}^{n} e^{i(\phi_k(t_i)-\bar{\phi}_k)}\right| \in [0,1]$$

(Eq. 20)

$$\rho_{group} = \frac{1}{N}\sum_{i=1}^{n} \rho_{group,i} \in [0,1]$$

Again, values close to 1 indicate close synchronization.

Figure 19.6 shows the results of calculating the group cluster phase with synthetic data simulating five oscillators. In Figure 19.6a the five oscillators all had the same frequency, individual phases were randomly sampled from the

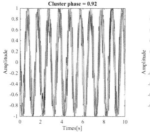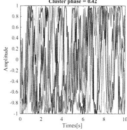

Figure 19.6 Group cluster-phase of synthetic data.

Notes: a) Five oscillators with equal frequency and random phases. b) Five oscillators with random frequencies and random phases.

interval $\left[-\dfrac{\pi}{4}, \dfrac{\pi}{4}\right]$ and random noise sampled from a uniform distribution was added. Thus, the five oscillators were quite similar. Performing a cluster phase analysis resulted in a value of $\rho_{\text{group}} = 0.92$. In contrast, in Figure 19.6b both the frequency and the phase of the five oscillators were randomly chosen, which resulted in a group cluster-phase value of $\rho_{\text{group}} = 0.42$.

Speaking in stricter terms, this example actually shows that the result that some actors display strong synchronization does not necessarily provide evidence for a relevant interaction between actors. In the example shown in Figure 19.6a the five oscillators did not really interact with each other, but were deliberately set up so that they appear synchronized (Pikovsky et al., 2003).

Using cluster phase analysis in a rocking chair experiment with six participants the cluster-phase showed significantly greater values when actors were rocking with eyes open and thus were able to see each other, whereas group synchronization was reduced when individuals were instructed to close their eyes (Frank & Richardson, 2010; Richardson et al., 2012). Another recent example applied cluster phase analysis to soccer data, showing highly synchronized movement behaviour between the teams (Duarte et al., 2013).

Summary

As stated in the introduction, the presented methods are only a selection. One prominent analysis technique which has not been included due to space constraints is cross-recurrence quantification (Shockley, Butwill, Zbilut, & Webber Jr., 2002; Shockley, Santana, & Fowler, 2003; Zbilut, Giuliani, & Webber Jr, 1998). Fortunately, for the interested reader there is an excellent tutorial available on the web (Riley & van Orden, 2005). Research analysis methods are a bit of a moving target as experimental paradigms mature or new paradigms become available; the same is also true for the analysis methods supporting these paradigms. It is the hope of the author that the present selection will lower the entry barriers for researchers interested in interpersonal coordination, an area which is gaining more and more support in the research community as a whole.

Note

The source code to produce all figures used in this chapter can be found on http://github.com/robertreingit/interpersonal-analysis-methods/

References

Black, D.P., Riley, M.A. & McCord, C.K. (2007) Synergies in intra- and interpersonal interlimb rhythmic coordination. *Motor Control, 11*(4), 348–373.

Boashash, B. (1992) Estimating and interpreting the instantaneous frequency of a signal. I. Fundamentals. *Proceedings of the IEEE, 80*(4), 520–538.

Chau, T. (2001) A review of analytical techniques for gait data. Part 1: fuzzy, statistical and fractal methods. *Gait and Posture, 39*(1), 49–66.

Chen, H.-H., Liu, Y.-T., Mayer-Kress, G. & Newell, K.M. (2005) Learning the pedalo locomotion task. *Journal of Motor Behavior, 37*(3), 247–256.

Daffertshofer, A., Lamoth, C.J.C., Meijer, O.G. & Beek, P.J. (2004) PCA in studying coordination and variability: a tutorial. *Clinical Biomechanics, 19*(4), 415–428.

de Freitas, S.M. & Scholz, J.P. (2010) A comparison of methods for identifying the Jacobian for uncontrolled manifold variance analysis. *Journal of Biomechanics, 43*(4), 775–777.

Deluzio, K.J., Wyss, U.P., Zee, B., Costigan, P.A. & Sorbie, C. (1997) Principal component models of knee kinematics and kinetics: Normal vs. pathological gait patterns. *Human Movement Science, 16*(2-3), 201–217.

Dona, G., Preatoni, E., Cobelli, C., Rodano, R. & Harrison, A.J. (2009) Application of functional principal component analysis in race walking: an emerging methodology. *Sports Biomech, 8*(4), 284–301.

Donoghue, O.A., Harrison, A.J., Coffey, N. & Hayes, K. (2008) Functional data analysis of running kinematics in chronic Achilles tendon injury. *Med Sci Sports Exerc, 40*(7), 1323–1335.

Duarte, R., Araújo, D., Correia, V., Davids, K., Marques, P. & Richardson, M.J. (2013) Competing together: Assessing the dynamics of team-team and player-team synchrony in professional association football. *Hum Mov Sci, 32*(4), 555–566.

Forner-Cordero, A., Levin, O., Li, Y. & Swinnen, S.P. (2005). Principal component analysis of complex multijoint coordinative movements. *Biological Cybernetics, 93*(1), 63–78.

Frank, T.D. & Richardson, M.J. (2010) On a test statistic for the Kuramoto order parameter of synchronization: An illustration for group synchronization during rocking chairs. *Physica D: Nonlinear Phenomena, 239*(23–24), 2084–2092.

Gabor, D. (1946) Theory of communication. Part 1: The analysis of information. *Electrical Engineers - Part III: Radio and Communication Engineering, Journal of the Institution of, 93*(26), 429–441.

Haken, H., Kelso, J.A.S. & Bunz, H. (1985) A theoretical model of phase transitions in human hand movements. *Biological Cybernetics, 51*, 347–356.

Hamill, J., Haddad, J.M. & McDermott, W.J. (2000) Issue in quantifying variability from a dynamical systems perspective. *Journal of Applied Biomechanics, 16*, 407–418.

Hsu, W.L., Scholz, J.P., Schoner, G., Jeka, J.J. & Kiemel, T. (2007) Control and estimation of posture during quiet stance depends on multijoint coordination. *Journal of Neurophysiology, 97*(4), 3024–3035.

Jolliffe, I.T. (2002) *Principal Component Analysis* (2nd ed.). Springer: New York.

Kelso, J.A., de Guzman, G.C., Reveley, C. & Tognoli, E. (2009) Virtual Partner Interaction (VPI): exploring novel behaviors via coordination dynamics. *PLoS One, 4*(6), e5749.

King, F.W. (2009) *Hilbert Transforms* (Vol. 1). Cambridge University Press: Cambridge.

Lamb, P.F. & Stockl, M. (2014) On the use of continuous relative phase: Review of current approaches and outline for a new standard. *Clin Biomech (Bristol, Avon), 29*(5), 484–493.

Latash, M.L. (2000) There is no motor redundancy in human movements. There is motor abundance. *Motor Control, 4*(3), 259–260.

Latash, M.L. (2008) *Synergy*. Oxford University Press: New York.

Lopresti-Goodman, S.M., Richardson, M.J., Silva, P.L. & Schmidt, R.C. (2008) Period basin of entrainment for unintentional visual coordination. *Journal of Motor Behavior, 40*(1), 3–10.

Mardia, K.V. & Jupp, P.E. (2000) *Directional Statistics*. John Wiley & Sons: Chichester.

Neda, Z., Ravasz, E., Brechet, Y., Vicsek, T. & Barabasi, A.L. (2000) The sound of many hands clapping. *Nature, 403*(6772), 849–850.

Oullier, O., de Guzman, G.C., Jantzen, K.J., Lagarde, J. & Kelso, J.A. (2008) Social coordination dynamics: measuring human bonding. *Social Neuroscience, 3*(2), 178–192.

Palut, Y. & Zanone, P.G. (2005) A dynamical analysis of tennis: concepts and data. *J Sports Sci, 23*(10), 1021–1032.

Peters, B.T., Haddad, J.M., Heiderscheit, B.C., van Emmerik, R.E.A. & Hammill, J. (2003) Limitations in the use and interpretation of continuous relative phase. *Journal of Biomechanics, 36*, 271–274.

Pikovsky, A., Rosenblum, M. & Kurths, J. (2003) *Synchronization: A universal concept in nonlinear sciences*. Cambridge University Press: Cambridge.

Post, A.A., Daffertshofer, A. & Beek, P.J. (2000) Principal components in three-ball cascade juggling. *Biological Cybernetics, 82*(2), 143–152.

Ramenzoni, V.C., Riley, M.A., Shockley, K. & Baker, A.A. (2012) Interpersonal and intrapersonal coordinative modes for joint and single task performance. *Human Movement Science, 31*(5), 1253–1267.

Ramsay, J.O. & Silverman, B.W. (2005) *Functional Data Analysis*. Springer: New York.

Rein, R. (2012). Measurement methods to analyze changes in coordination during motor learning from a non-linear perspective. *The Open Sports Sciences Journal, 5*, 36–48.

Reisman, D.S., Scholz, J.P. & Schöner, G. (2002) Coordination underlying the control of whole body momentum during sit-to-stand. *Gait and Posture, 15*, 45–55.

Richardson, M.J., Garcia, R.L., Frank, T.D., Gergor, M. & Marsh, K.L. (2012) Measuring group synchrony: a cluster-phase method for analyzing multivariate movement time-series. *Front Physiol, 3*, 405.

Richardson, M.J., Marsh, K.L., Isenhower, R.W., Goodman, J.R. & Schmidt, R.C. (2007) Rocking together: dynamics of intentional and unintentional interpersonal coordination. *Hum Mov Sci, 26*(6), 867–891.

Riley, M.A., Richardson, M.J., Shockley, K. & Ramenzoni, V.C. (2011) Interpersonal synergies. *Frontiers in Psychology, 2*, 38.

Riley, M.A. & van Orden, G.C. (2005) Tutorials in contemporary nonlinear methods for the behavioral sciences. WWW.NSF.GOV/SBE/BCS/PAC/NMBS/NMBS.JSP

Ryan, W., Harrison, A. & Hayes, K. (2006) Functional data analysis of knee joint kinematics in the vertical jump. *Sports Biomech, 5*(1), 121–138.

Santello, M., Flanders, M. & Soechting, J.F. (1998) Postural hand synergies for tool use. *Journal of Neuroscience, 18*(23), 10105–10115.

Schmidt, R.C., Carello, C. & Turvey, M.T. (1990) Phase transitions and critical fluctuations in the visual coordination of rhythmic movements between people. *Journal of Experimental Psychology: Human Perception and Performance, 16*(2), 227–247.

Schmidt, R.C., Christianson, N., Carello, C. & Baron, R. (1994) Effects of social and physical variables on between-person visual coupling. *Ecological Psychology, 6*(3), 159–183.

Schmidt, R.C., Fitzpatrick, P., Caron, R. & Mergeche, J. (2011) Understanding social motor coordination. *Hum Mov Sci, 30*(5), 834–845.

Scholz, J.P. & Schöner, G. (1999) The uncontrolled manifold concept: identifying control variables for a functional task. *Experimental Brain Research, 126*(3), 289–306.

Schöner, G. & Scholz, J.P. (2007) Analyzing multi-degree-of-freedom movement system based on variance: Uncovering structure vs. extracting correlations. *Motor Control, 11*(3), 259–275.

Shockley, K., Butwill, M., Zbilut, J.P. & Webber Jr., C.L. (2002) Cross recurrence quantification of coupled oscillators. *Physics Letters A, 305*(1-2), 59–69.

Shockley, K., Santana, M.V. & Fowler, C.A. (2003) Mutual interpersonal postural constraints are involved in cooperative conversation. *Journal of Experimental Psychology: Human Perception and Performance, 29*(2), 326–332.

Stuart, M. (1982) A geometric approach to principal components analysis. *The American Statistician, 36*(4), 365–367.

van Ulzen, N.R., Lamoth, C.J., Daffertshofer, A., Semin, G.R. & Beek, P.J. (2008) Characteristics of instructed and uninstructed interpersonal coordination while walking side-by-side. *Neurosci Lett, 432*(2), 88–93.

Varlet, M. & Richardson, M.J. (2011) Computation of continuous relative phase and modulation of frequency of human movement. *J Biomech, 44*(6), 1200–1204.

Zbilut, J.P., Giuliani, A. & Webber Jr, C.L. (1998) Detecting deterministic signals in exceptionally noisy environments using cross-recurrence quantification. *Physics Letters A, 246*(1–2), 122–128.

20 Modelling interpersonal coordination

Ana Diniz and Pedro Passos

Introduction

Many human actions occur in social contexts in a variety of situations. Simple examples of interpersonal interaction involve establishing a conversation, manipulating objects, dancing or walking together. Previous research in cognitive and psychological sciences has shown that interpersonal interaction tends to be *coordinated* (Schmidt & O' Brien, 1997). This phenomenon implies several co-adaptive behaviours from one person relative to another person, including for instance eye or limb movements. Interestingly, people seem to achieve interpersonal coordination in a very natural way without much effort, even unintentionally (Richardson et al., 2007). So how does interpersonal coordination work, and how can we model it?

Numerous studies have reported that the mechanisms that govern human coordination can be understood through a dynamical theory, known as *coordination dynamics* (Kelso & Engström, 2006; Kelso, 2009). An interdisciplinary field, strongly related to this theory, that deals with systems composed of different parts and the self-organized emergence of new properties is named *synergetics* (Haken, 1969, 1977). An important idea within synergetics is that a system is characterized by order parameters (i.e. collective variables) that determine the behaviours and the corresponding fluctuations of its parts at each time. By definition, collective variables are "relational quantities that are created by the cooperation among the individual parts of a system. Yet they, in turn, govern the behavior of the individual parts" (Kelso, 2009). On the other hand, a system is under the influence of control parameters (external or internal) and when these control parameters reach certain critical values the system may become unstable and take on a new state. By definition, control parameters are "naturally occurring environmental conditions or intrinsic endogenous factors that move the system through its repertoire of patterns and cause them to change" (Kelso, 2009). Moreover, the dynamics of the system can be described by differential equations (i.e. motion equations) for the variables of interest, in which the temporal evolution of the collective variables is related to the present state and the control parameters. Briefly, a differential equation is a mathematical equation involving a function and some of its derivatives which represent rates of change. Finally, for the solution of the differential equations, it is assumed that the system evolves towards attractors for specific values of the control parameters. In simple terms, an attractor is a set of values (a state) towards which a system tends to go and where it remains in time for a large number of starting positions (e.g. an attracting fixed point).

Another field with connections to the theory of coordination dynamics (and synergetics) that handles the behaviour of dynamical systems over time, usually through differential equations, is so-called *dynamical systems theory* (e.g. Abraham, Abraham & Shaw, 1990). Concisely, a dynamical system is a mathematical notion where a fixed rule describes how a given point in a space evolves with time. This point represents the relation between system components (e.g. two players in a rugby union game). When the system can be solved, given an initial position it is possible to obtain all the future positions (i.e. its trajectory). In general terms, the stability of a dynamical system means that there are close initial conditions for which the trajectories remain unchanged. If $x = x(t)$ is a trajectory and $f = f(x)$ is a smooth function, then a simple example of a differential equation of motion is the equation:

$$dx/dt = f(x).$$

In this equation, the time derivative of the trajectory x equals the function f, and so the equation expresses exactly how the trajectory x changes over time t. In order to illustrate how the former equation works, consider the simple situation in which the function f is given by $f(x) = k\,x$, where k is a constant, implying that:

$$dx/dt = k\,x.$$

More specifically, this means that the rate of change of x is proportional to x at each time t. Using some properties of integrals and logarithms, it can be shown that the solution of the equation is:

$$x(t) = x(0)\,e^{kt}.$$

Therefore, in this particular situation it is possible to solve the equation and to obtain an exact, analytic solution which expresses the position $x(t)$ of the system at each time t. However, in more complex situations this is not feasible and the use of graphical and numerical methods only leads to approximated solutions.

Previous dynamical models in human movement

Pioneering work on the dynamics of intrapersonal (bimanual) coordination showed that when individuals start to cycle their two index fingers in an anti-phase (anti-symmetrical) mode, an increase in the cycling frequency generates a sudden switch to an in-phase (symmetrical) pattern (Kelso, 1984). In simpler terms, when subjects start to move both index fingers at a given tempo in the same direction (both to the left, then both to the right, and so on), if they enlarge the movement velocity, then unexpectedly the fingers start to move in opposite directions (engaging homologous muscles). The experimental observations revealed properties of self-organization which led to an emblematic model (the HKB model) with two attractors (i.e. the anti-phase and the in-phase modes) and transitions from one attractor to the other at specific values of movement frequency, derived from nonlinear interactions among the moving elements (Haken, Kelso, & Bunz, 1985). Joining concepts of synergetics (e.g. order parameters, control parameters etc.) and tools of nonlinear dynamical systems (e.g. instability, multistability etc.), the emergent behaviour of the system was formally modelled using a potential function and a differential equation (a motion equation). More precisely, the order parameter was the relative phase of the two fingers describing the position of both fingers on a cycle, whereas the control parameter was the cycling frequency responsible for the changes in the fingers' relative position. The selected potential function defined an attractor "landscape" in which the valleys functioned as attractors (i.e. the anti-phase and the in-phase modes) reflecting relatively stable behavioural states and the correspondent differential equation offered a mathematical description of the system's behaviour as time passes and parameters change. Denoting the order parameter by x and the potential function by V, the evolution of x over time was expressed by the differential equation:

$$dx/dt = - dV/dx.$$

From this equation, a value of x for which the derivative dx/dt (and thus -dV/dx) is equal to zero corresponds to a steady state, with a minimum of V representing an attractor (a stable state) and a maximum of V representing a repellor (an unstable state). Also, the slope of dx/dt at the zero crossing is negative for an attractor and positive for a repellor (Figure 20.1). In the index fingers coordination experiment, this means that when the two finger system is attracted to a minimum of V, the system remains stable and theoretically resists perturbations (such as changes in the cycling frequency), whereas if the system is close to a maximum of V, the system becomes unstable and a slight change in the cycling frequency pushes it to one of the two attractors.

As previously mentioned, in the HKB model the order parameter was the relative phase ϕ of the two oscillating fingers and the potential function V, which was supposed to be periodic and symmetric, was defined as $V(\phi) = - a \cos(\phi) - b \cos(2\phi)$, where a and b were two control parameters. By rescaling, the parameters a and b were expressed as a single parameter k = b/a representing the inverse of the frequency of oscillation (Figure 20.2).

Thus, the motion equation was written as:

$$d(\phi)/dt = - a \sin(\phi) - 2b \sin(2\phi).$$

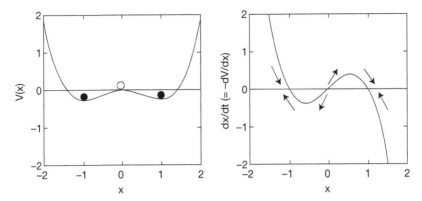

Figure 20.1 Graphical representation of an exemplar potential function (on the left) and the corresponding motion equation (on the right) showing two attractors (black circles) and one repellor (white circle).

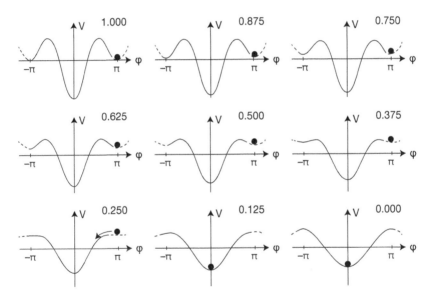

Figure 20.2 Graphical representation of the potential function V/a for varying values of b/a (Haken, Kelso, & Bunz, 1985 – adapted with kind permission of Springer Science + Business Media).

Beyond deterministic and symmetric interlimb coordination, the HKB model was later extended to different motor coordination tasks and mathematical formalizations. For example, a subsequent study suggested the addition of a stochastic part to account for inherent random fluctuations of the system that cannot be measured, such as attention or fatigue (Schöner, Haken & Kelso, 1986). The new equation included a random white noise process ε_t with mean 0 and variance Q, giving:

$$d(\phi)/dt = -a\sin(\phi) - 2b\sin(2\phi) + \varepsilon_t.$$

Another study on sensorimotor coordination proposed a broken symmetry version of the model to reflect the fact that the two oscillators may have different preferred frequencies of oscillation (Kelso, DelColle & Schöner, 1990). This new model incorporated a detuning term $\Delta\omega$ defined as the arithmetic difference between the uncoupled frequencies of the rhythmic units providing:

$$d(\phi)/dt = -a\sin(\phi) - 2b\sin(2\phi) + \Delta\omega + \varepsilon_t.$$

Thereafter, the HKB model was used as a reference to formalize modes of coordination in human movement systems. For instance, another study also on intrapersonal coordination (e.g. interlimb coordination) examined subjects oscillating two hand-held pendulums (left hand and right hand) parallel to the sagittal plane of motion (Schmidt, Shaw & Turvey, 1993). Here there were manipulations of the frequency of oscillation $((b/a)^{-1})$ as well as the frequency competition $(\Delta\omega)$ between the two pendulums to observe the shifts in the attractor points (again, the anti-phase and the in-phase modes of coordination defined on the HKB model).

Though initially developed for intrapersonal coordination, this type of dynamical model also seems to successfully explain the features of interpersonal coordination. A previous study on interpersonal coordination investigated pairs of participants performing a rhythmic task of visual coordination of their outer legs (Schmidt, Carello & Turvey, 1990). Again, the results presented a switching between two modes of coordination, anti-phase to in-phase, due to changes in the values of the frequency of oscillation of the legs. A further study on this theme examined pairs of subjects establishing visual coordination using hand-held pendulums swung with their outer hands (Schmidt & Turvey, 1994). The outcomes were in line with previous research on intrapersonal (interlimb) coordination with hand-held pendulums within a person. Thus, the same laws of movement can be applied to describe and explain intrapersonal (interlimb) and interpersonal (between subjects) coordination.

The theoretical structure of *coordination dynamics* has also been used in other fields of research (besides human movement coordination), such as speech perception and its relation to sound. Significant work on the dynamics of speech perception suggested that listeners perceive the word "say" with short silent gaps between the "s" and the "ay" and perceive the word "stay" with long silent gaps (Tuller, Case, Ding & Kelso, 1994). The empirical results, with properties of self-organization, motivated a formal model with two attractors and switches from one attractor (i.e. the word 'say') to the other (i.e. the word 'stay') at particular values of gap duration (Tuller et al., 1994; Case, Tuller, Ding & Kelso, 1995). In the proposed speech perception model, the order parameter x was a variable illustrating the perceptual form, and the potential function V was written as $V(x) = kx - x^2/2 + x^4/4$, where k was a control parameter specifying the silent gap duration (Figure 20.3).

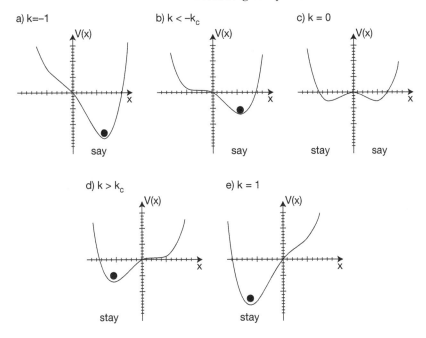

Figure 20.3 Graphical representation of the potential landscape V for five values of k (Tuller, Case, Ding & Kelso, 1994 – adapted with permission of American Psychological Association).

So, the motion equation was written as:

$$dx/dt = -k + x - x^3.$$

Based on the observed patterns, the control parameter k was then written as a function of the gap duration and other pertinent parameters.

In summary, dynamical models are obviously simplified representations of real situations, but they are extremely useful from theoretical and practical points of view. On the one hand, they give a description of the behaviour of systems over time and their dependence on significant (control) parameters. On the other hand, they allow for predictions of future states of the systems from known values of the relevant (control) parameters.

Modeling the dynamics of competitive social systems – the example of rugby union

As previously stated, during the last decades some important dynamical models have been presented to describe and explain human motor coordination (and other human behaviours). These models have been focused on intrapersonal coordination and also on interpersonal coordination, mainly involving limb rhythmic tasks (index finger wiggling or hand-held pendulum oscillation).

More recently, there has been a growing interest in creating and developing formal models which describe and explain subjects' interactions in social systems such as competitive team sports environments. For this purpose, it has been assumed that two (or more) players involved in a competitive team sports task may be considered as a dynamical system (McGarry et al., 2002). However, due to the large number of variables that may influence the individual parts (i.e. subjects), social systems such as team sports are characterized as complex systems. A consequence of this complexity is the nonlinear behaviour displayed by the players, which means that it is not easy to accurately describe where the system will be in the next moment. This implies that there are increased difficulties in identifying order and control parameters that capture the dynamics of the system and can lead to mathematically defined attractors (i.e. preferred states of the system).

A challenging case of dynamical modelling in team sports is the case of rugby union. In this particular situation, some novel studies have suggested formal models with two (or more) attractors to describe the competitive behaviour between ball carrier and defender (Araújo, Diniz, Passos & Davids, 2014; Diniz, Barreiros & Passos, 2014).

A study reflecting a first effort to model competitive relations in rugby union inspected simple 1v1 systems – that is, attacker-defender dyads (Araújo, Diniz, Passos & Davids, 2014). Based on empirical results and their specific properties, three coordination patterns were proposed as attractor states: clean try, tackle where the attacker passes the defender, and effective tackle. Thus, to formally describe the behavioural dynamics of the dyadic system, a three-attractor model was suggested (Araújo, Diniz, Passos & Davids, 2014). Moreover, the order parameter was the angle between the attacker-defender vector and an imaginary line parallel to the try line, the control parameters were the interpersonal distance and the relative velocity, and the potential function was a sixth-order polynomial. Through mathematical simulations, it was shown that the model successfully reproduced the dynamical performance of two players in the dyadic system.

A more recent study on competitive interactions in rugby union investigated 2v1 systems composed of a ball carrier, a support player and a defender (Diniz, Barreiros & Passos, 2014). The following text is dedicated to presenting a step-by-step tutorial on how the formal model for this specific situation was built (and, more generally, on how to build a formal model for a given social system). Thus, the first step was to define the order parameter (i.e. a collective variable), meaning that it was necessary to choose a single variable (relational quantity) that accurately describes the interactive behaviour between players as a system. Grounded in previous research, the selected order parameter was the ball carrier-defender interpersonal distance (Passos et al., 2008). The empirical observations revealed that to create any advantage over the defender, the ball carrier decision must occur within a critical region defined by 4 metres of interpersonal distance between players; outside this region the defender rebalances the system.

Then, the second step was to define the control parameter (i.e. an endogenous factor that causes the system to change). Based on previous research, the control parameter was the ball carrier-defender relative velocity, given by the ball carrier velocity minus the defender velocity (Passos et al., 2008). However, it is noteworthy that changes in the system due to changes in the relative velocity values can only occur within the critical region described by 4 metres of interpersonal distance between ball carrier and defender.

Furthermore, within the critical regions two different outcomes may occur: i) when the ball carrier-defender interpersonal distance is small and the relative velocity is negative (the defender has an advantage), the ball carrier frequently passes the ball to the support player – *attractor 1*; ii) when the ball carrier-defender interpersonal distance is large and the relative velocity is positive (the ball carrier has an advantage), the ball carrier usually goes forward – *attractor 2*. So, to describe the dynamics of the interactions between two attackers and one defender, a dynamical model with two attractors seemed plausible (Diniz, Barreiros & Passos, 2014). It is worth mentioning that there are other possible attractive states in this kind of system (e.g. the dummy pass), which were not incorporated in this model due to a lack of empirical information. As previously stated, a critical region of 4 metres of interpersonal distance was set. Therefore, for symmetry reasons related to the potential function, the order parameter x was re-expressed as two minus the interpersonal distance from the ball carrier to the defender, the control parameter k was related to the ball carrier-defender relative velocity ($k(v) = (v / (v_m/r))^3$), and the potential function V was the same as the one used in Tuller et al.'s model ($V(x) = x^4/4 - x^2/2 + k\,x$) (Figure 20.4).

Thus, the motion equation was written as:

$$dx/dt = -x^3 + x - k.$$

Moreover, and interestingly, this work also presented the exact values of the critical points at which a transition to one of the two attractors occurs. More precisely, by studying the extremes of V, i.e. the roots of dV/dx, it was proven that the shifting points at which the system theoretically evolves to one of the two alternative states correspond to relative velocities of ± 2 ms^{-1}.

So, if the value of the order parameter x at which the minimum of the potential function V occurs is positive and larger than the positive *critical point*, then the value of the control parameter k is negative and the value of the relative velocity v is also negative. This means that the ball carrier-defender interpersonal distance is small and the defender velocity is larger than the ball-carrier velocity, leading to attractor 1 – the ball carrier passes the ball (Figure 20.4 top). In contrast, if the value of the order parameter x at which the minimum of the potential function V occurs is negative and smaller than the negative *critical point*, then the value of the control parameter k is positive and the value of the relative velocity v is also positive. This means that the ball carrier-defender interpersonal distance is large and the ball-carrier velocity is larger than the defender velocity, leading to attractor 2 – the ball carrier goes forward (Figure 20.4 bottom).

This dynamical model, which can be improved by including other control parameters (e.g. the angle between ball carrier and defender) or other relevant

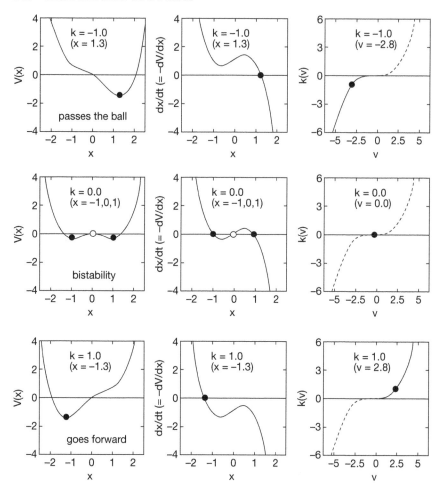

Figure 20.4 Graphical representation of the potential function V (on the left), the motion equation -dV/dx (on the centre), and the control parameter function k (on the right) for three values of k (Diniz, Barreiros & Passos, 2014 – adapted with permission of Taylor & Francis).

restrictions, is a fine example of the dynamical switching of behavioural states in a complex real life environment.

Brief conclusion

Within this chapter we have discussed numerous models that have been used across the years to formally describe intrapersonal and interpersonal coordination. Our major goal was to present the several stages that are needed to model interpersonal coordination in team sports (or coordination in a social system more generally). More precisely, the first step is to define an order parameter (i.e.

a collective variable) which accurately describes the interactive behaviour between system components (e.g. the interpersonal distance between ball carrier and defender in rugby union). Then, the second step is to identify possible steady states towards which the system is attracted (e.g. passing the ball or going forward in rugby union). Moreover, it is necessary to define the control parameter which leads the system to each one of the steady states (e.g. relative velocity in rugby union). Finally, it is *only* required to select the suitable potential function and differential equation to formally describe, explain and predict the dynamical system behaviour.

However, of course, there are still many future challenges and big open questions that need to be solved. An interesting issue for further research related to human interaction modelling is the use of nested control parameters. This point was already mentioned in a previous study on rugby union, in which the authors suggested that dyadic attacker-defender interpersonal distance and both players' relative velocity may work as nested control parameters (Passos et al., 2008). The two main ideas are: i) to formalize the existence of nested control parameters; and ii) to formalize the weight that each control parameter has in relation to the system outcome. Another suggestion for further work is to specify when and where it is true that the coordination dynamics of intrapersonal situations hold almost exactly for interpersonal situations. It is important to remember that subjects (e.g. players) are independent entities that may, under some conditions (i.e. constraints), couple to react as a single entity, and as such to stabilize performance (Kugler & Turvey, 1987). It is the behavioural adjustment at the subject (micro) level that stabilizes performance at a collective (macro) level (Scholz & Schoner, 1999; Riley et al., 2011) and this micro-macro level relation still needs considerable investigation.

References

Abraham, F.D., Abraham, R. & Shaw, C.D. (1990) *A Visual Introduction to Dynamical Systems Theory for Psychology*. Aerial Press: Santa Cruz, CA.

Araújo, D., Diniz, A., Passos, P. & Davids, K. (2014) Decision making in social neurobiological systems modeled as transitions in dynamic pattern formation. *Adaptive Behavior*, *22*(1), 21–30.

Case, P., Tuller, B., Ding, M. & Kelso, J.A.S. (1995) Evaluation of a dynamical model of speech perception. *Perception and Psychophysics*, *57*, 977–988.

Diniz, A., Barreiros, J. & Passos, P. (2014) To pass or not to pass: A mathematical model for competitive interactions in Rugby Union. *Journal of Motor Behavior*, *46*(5), 293–302.

Haken, H. (1969). Lectures at Stuttgart University. In H. Haken & R. Graham (1971) (Eds.), *Umschau*, *6*, 191.

Haken, H. (1977) *Synergetics*. Springer: Berlin.

Haken, H., Kelso, J.A.S. & Bunz, H. (1985) A theoretical model of phase transitions in human hand movements. *Biological Cybernetics*, *51*, 347–356.

Kelso, J.A.S. (1984) Phase-transitions and critical-behavior in human bimanual coordination. *American Journal of Physiology*, *246*, 1000–1004.

Kelso, J.A.S. (2009) Coordination dynamics. In R.A. Meyers (Ed.), *Encyclopedia of Complexity and System Science* (pp. 1537–1564). Springer: Heidelberg.

Kelso, J.A.S., Delcolle, J.D. & Schöner, G. (1990) Action-perception as a pattern formation process. In M. Jeannerod (Ed.), *Attention and performance XIII* (pp. 139–169). Erlbaum: Hillsdale, NJ.

Kelso, J.A.S. & Engstrom, D.A. (2006) *The Complementary Nature*. Bradford Books: Cambridge, MA.

Kugler, P. & Turvey, M. (1987) *Information, natural law and the self-assembly of rhythmic movement*. Erlbaum: Hillsdale: NJ.

McGarry, T., Anderson, D.I., Wallace, S.A., Hughes, M.D. & Franks, I.M. (2002) Sport competition as a dynamical self-organizing system. *Journal of Sports Sciences, 20,* 771–781.

Passos, P., Araújo, D., Davids, K., Gouveia, L., Milho, J. & Serpa, S. (2008) Information-governing dynamics of attacker-defender interactions in youth rugby union. *Journal of Sports Sciences, 26*(13), 1421–1429.

Richardson, M.J., Marsh, K.L., Isenhower, R.W., Goodman, J.R. & Schmidt, R.C. (2007) Rocking together: dynamics of intentional and unintentional interpersonal coordination. *Human Movement Science, 26*(6), 867–891.

Riley, M.A., Richardson, M.J., Shockley, K. & Ramenzoni, V.C. (2011) Interpersonal synergies. *Front Psychol, 2,* 38.

Schmidt, R.C., Carello, C. & Turvey, M.T. (1990) Phase transitions and critical fluctuations in the visual coordination of rhythmic movements between people. *Journal of Experimental Psychology: Human Perception and Performance, 16,* 227–247.

Schmidt, R.C. & O'Brien, B. (1997) Evaluating the dynamics of unintended interpersonal coordination. *Ecological Psychology, 9,* 189–206.

Schmidt, R.C., Shaw, B.K. & Turvey, M.T. (1993) Coupling dynamics in interlimb coordination. *Journal of Experimental Psychology: Human Perception and Performance, 19*(2), 397–415.

Schmidt, R.C. & Turvey, M.T. (1994) Phase-entrainment dynamics of visually coupled rhythmic movements. *Biological Cybernetics, 70,* 369–376.

Scholz, J.P. & Schoner, G. (1999) The uncontrolled manifold concept: identifying control variables for a functional task. *Exp Brain Res, 126*(3), 289–306.

Schöner, G., Haken, H. & Kelso, J.A.S. (1986) A stochastic theory of phase transitions in human hand movement. *Biological Cybernetics, 53*(4), 247–257.

Tuller, B., Case, P., Ding, M. & Kelso, J.A.S. (1994) The nonlinear dynamics of speech categorization. *Journal of Experimental Psychology: Human Perception and Performance, 20*(1), 3–16.

21 Technology for studying interpersonal coordination in social collectives

John Kelley, David Higham and Jon Wheat

Introduction

As other chapters in this book demonstrate, interpersonal coordination has been studied in a variety of contexts, using many analysis techniques. Whether investigating gross movements of players in a sports team, or the smaller movements of a group of musicians, sufficiently accurate estimates of raw data (often position) are required. The effectiveness of the various analyses of interpersonal coordination is limited by the accuracy of input data. Obtaining input data for these analyses often requires the use of technology. Although interpersonal coordination has been studied in a variety of sports, physical activity and experimental contexts, this chapter will mainly focus on technologies for monitoring interpersonal coordination in team sports as this is an area that is receiving growing interest. Fundamentally, this most often requires the measurement of player position.

Simple technologies for the measurement of player position have existed for many years. For example, basic measures with tape and trundle wheels have been used frequently and tape measures are still used to measure performance in the majority of long jump competitions. Such simple technologies are often regularly used as gold standard devices against which to compare new technologies. However, they quickly become impractical when a large number of measurements are required in a short space of time. Similarly, these technologies are not well-suited to measure changes in position during movement.

When considering the dynamics of interpersonal coordination, we often need the position of players over time. Technological advances over recent years have enabled the collection of these data in an increasingly automated way. Several technologies exist based on different measurement principles, with associated benefits and limitations. They can be categorized into those based on: 1) signal propagation sensing, 2) inertial sensors, 3) vision/image-based systems, and 4) electro-magnetic tracking. The availability of these technologies is increasing rapidly. Indeed, some devices are now ubiquitous (e.g. mobile phones), offering the possibility of large scale and frequent analysis of interpersonal coordination.

Importantly, though, an appreciation of the principles by which these technologies function, together with the associated benefits, drawbacks and likely accuracy for different scenarios, is important to facilitate the appropriate choice of measurement technique. The aim of this chapter is to provide an overview of the technologies available for studying interpersonal coordination, highlighting the key measurement principles. Important benefits and limitations are identified, along with examples of their use in the literature and indicative accuracy information.

Signal propagation sensing

Signal propagation sensing is the use of the behaviour of waves as they propagate. The following sections focus on radio wave-based techniques.

Global Navigation Satellite System

A Global Navigation Satellite System (GNSS) is a system of satellites that provides positioning over the entire globe. The United States' Global Positioning System (GPS) is the best known GNSS. GPS is a thirty-two-satellite network (US Naval Observatory, 2014) which provides location information anywhere on earth that has line of sight to four or more satellites. Typically there is a line of sight to nine satellites from any point on the earth's surface, assuming no interference from terrain and structures. The line of sight requirement restricts performance in certain situations: 1) indoors, 2) in large sports stadiums, and 3) in built-up areas.

GPS estimates the position of the receiver using the *time of flight* (ToF) – the time between when the signal is transmitted and when it is received – of signals transmitted from four or more satellites, using a method known as *trilateration*. Synchronization between all the satellites is ensured using atomic clocks.

Wisbey et al. (2010) used GPS to quantify movement demands in Australian League Football (AFL). GPS was chosen since coverage was widely available on the various fields of play and GPS devices are permitted in the competitive environment by the AFL rules, making GPS the most practical method to collect the data.

MacLeod et al. (2009) assessed the validity of the use of GPS positions for field hockey player movement and found that the mean distances and speeds were in agreement with trundle wheel and timing gate measurements. Coutts and Duffield (2010) tested six different GPS devices for tracking movement at a range of speeds over a 128.5-metre circuit. All devices measured within 5 metres of the lap distance and all the devices were in good agreement for the total distance covered over six laps. There was good inter-device agreement when measuring slow speeds, but for higher running speeds there was large variation between the measurements of the different devices. This indicates that GPS devices cannot be reliably used to measure the speed of activities at a running pace or higher. Duffield et al. (2010) proposed that GPS is not suitable for measuring the position of a person travelling at high speed. However, for slower activities such as swimming, GPS can provide useful position data. For example, Beanland et al.

(2014) reported no significant differences between position data from GPS and position data measured using video cameras.

GPS devices typically have update rates between 1 and 15 Hz. Higher update rates are likely to provide more accurate distance and velocity measurements due to increased sensitivity to small deviations from a linear path that may be missed at lower update rates. Coutts and Duffield (2010) found 1 Hz GPS was less accurate than 5 Hz for high intensity movements.

The widespread availability and cheap cost of GPS units has made it an economical method for tracking people. GPS chipsets are now commonly integrated into objects such as cars and mobile phones, as well as specific tracking units. Benson, Bruce and Gordon (2012) compared the distance measured by a sport-specific GPS unit and by an iPhone to the known distance of six laps of an athletics track (2,400 metres). They found that both devices significantly underestimated the distance covered and the accuracy of the iPhone's GPS decreased as the distance increased. They suggest that while the error is acceptable for everyday use, the iPhone is not suitable for high accuracy applications. A popular use of GPS is to track activities such as running and cycling. Such activities can be compared across multiple users using smartphone apps such as Strava (Strava, 2015).

Ultra-wideband

Ultra-wideband (UWB) positioning exploits UWB radio technology, which is very low power but limited to a short range. Multiple receivers are used to provide the relative position of 'tags'. Each tag sends a short UWB pulse, which is detected by receivers around the periphery of the tracking area. The relative location of the tag is calculated using either *multilateration* or *triangulation*. The tags are not synchronized, but the short nature of the pulse means collisions are unlikely and hundreds of tags can be processed almost simultaneously.

UWB systems are designed to overcome the problems associated with GPS indoors or in areas where satellite coverage is restricted, such as large stadiums. Obstacles in the tracking area can impede performance, so court and field-based team sports are most suited to this type of measurement. Leser et al. (2014) investigated the accuracy of the UWB positioning system for use in temporal position and movement analysis in sport. Basketball players were asked to perform movements that simulated match play while wearing a head mounted sensor and while operating a trundle wheel. The mean ± standard deviation of the system's difference to the trundle wheel was 3.45 ± 1.99% for the total distance of the movements, which was typically about 400 metres.

Non-sporting and health-related indoor tracking systems have been designed and proposed. Kearns et al. (2014) presented the Ubiwatch, a device which prompts activities to be completed dependent upon location. It is envisioned this may be useful for people with cognitive impairment, who may forget the task to be completed in the time taken to reach the required location. The authors also suggest that the vertical position of Ubiwatch could be used by carers to detect falls.

Radio-frequency identification

Radio-frequency identification (RFID) uses electromagnetic fields to identify 'tags' using a 'reader'. Often the tags do not have a power source and are powered wirelessly as they approach the reader. This restricts the use of the tags to an area in the close vicinity of the reader. A clear line of site is not always needed, although it can increase range. As such, RFID tags can be partially embedded in equipment. Mass participation races commonly provide each athlete with a RFID tag that starts and stops the clock as they cross the start and finish lines.

Alternatively, a tag may have a local power source that allows it to operate at hundreds of metres from the reader. These powered transponder tags are incorporated into Local Position Measurement (LPM) devices (Stelzer, Pourvoyeur and Fischer, 2004) created specifically for tracking athletes in sport. Multiple transponders are tracked by base stations placed around the edge of the tracking area, which can be up to 500 x 500 metres. The system operates at 1000 Hz, which is shared between all of the transponders. Therefore, as the number of transponders increases the update rate will reduce. As the measurements received from each transponder are sequential, the measurements are not synchronized. Further, a line of sight is required between transponder and base station.

Using LPM, Resch et al. (2012) found that the standard deviation of the difference to the actual position was x = 31.86 mm, y = 16.56 mm and z = 99.99 mm from ten thousand static measurements. Frencken, Lemmink and Delleman (2010) tested the accuracy and validity of the system specifically for soccer. They found an accuracy of 20–30 millimetres for a transponder worn by a participant during initial static tests, similar to Resch et al. (2012). In dynamic tests, Frencken, Lemmink and Delleman (2010) reported that LPM underestimates the distance travelled by up to 0.4 metres over a 25 metre course and the coefficient of variation increases with speed and turn angle. Both Ogris et al. (2012) and Frencken et al. (2010) suggested that LPM is suitable for estimating position and velocity, but Ogris et al (2012) suggested that care should be taken when considering accelerations and instantaneous velocities.

WiFi and Bluetooth

While the primary use of WiFi and Bluetooth is data transfer, they can also be used for localization. In the case of WiFi an existing network can be used, which can keep implementation costs low. For each base station in range, a device will obtain an associated Received Signal Strength (RSS). The strength of a propagated signal is proportional to the inverse-square of the distance from the source. RSS uses this to estimate the distance of the receiver from a transmitter. These RSSs can be combined to calculate the position either by *trilateration* or *multilateration*. However, due to reflection (see Figure 21.1), the path from the transmitter to the receiver may be increased or multipath may occur, which will decrease the accuracy of the distance measurement.

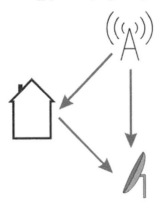

Figure 21.1 An illustration of 'multipath', showing two routes a signal may take from a transmitter to a receiver.

An alternative, fingerprinting, overcomes this by calculating the position directly using the RSS from a number of transmitters. In a calibration stage the RSS from each base station is collated for a set of calibration points distributed throughout the environment. A model for the distribution of RSS is created, which, in the online stage, can be consulted to provide an estimated position. However, the calibration phase means fingerprinting has limited use in an unknown or inaccessible environment.

As fingerprinting does not require line of sight, these methods are ideal for environments where there are many small, congested rooms. The update rate is dependent upon the refresh rate of the received signal strength, which in turn is dependent upon the device.

WiFi base stations typically have a range of 100 metres, but coverage can be achieved over a larger area by the overlapping of base stations as required. The errors are comparatively large for WiFi and Bluetooth-based tracking systems. Using fingerprinting in a multi-room, single storey 35 metre x 40 metre test area, Xiang et al. (2004) found that with a 90% probability, the error distance is less than 5 metres for walking. This is similar to the accuracy of 2 metres reported by Khoury and Kamat (2009) when tracking day-to-day activities. Xiang et al. (2004) suggested that accuracy increases with the number of access points.

Woo et al. (2011) investigated the feasibility of a fingerprinting system to track workers on a construction site. Each worker had a WiFi receiver mounted onto their hard hat. The authors tolerated the observed 5 metre error, suggesting the system is suitable for tracking workers across a large, harsh environment such as a construction site.

General techniques for signal propagation sensing

Several position calculation techniques were mentioned above in the Signal Propagation Sensing section. Here these techniques are described in detail.

Trilateration

Trilateration is the use of distances to an object from known distinct points in order to calculate the object's relative position. In two dimensions, at least three known points are required. If the distance R_a from a known location P_a is known, then the object must lie on the edge of Circle A, centred on P_a with a radius R_a (see Figure 21.2). If a second distance R_b from another location P_b is also known, then the object must lie on one of the two intersections of Circles A and B. An additional construction, Circle C, is required to determine which of the two intersections the object lies on.

The technique can equally be applied to three dimensions where spheres replace circles and four known distinct points are required. In the real world distance measurements often contain error; the object is therefore estimated to be somewhere in the region of intersection. By increasing the number of distance measurements this region can be reduced and the location accuracy increased.

Multilateration

Multilateration or Hyperbolic Location Technique is the use of the difference in distance to known locations to calculate a relative location. Given a difference in distance ΔD between an emitter and two points, a hyperboloid exists that satisfies:

$$\Delta D + D_1 = D_2$$

where *D1* is the distance from the emitter to receiver 1 and *D2* is the distance from the emitter to receiver 2.

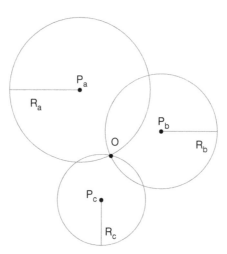

Figure 21.2 Using two-dimensional trilateration to determine the position of an object, O, using measured distances R_a, R_b and R_c, to known locations P_a, P_b and P_c.

Considering a set of N receivers, hyperboloids exist for the $\sum_{n=1}^{N-1}$ possible differences. The location of the transmitter is the intersection of these hyperboloids. Due to measurement error or resolution, the hyperboloids are more likely to identify regions rather than a specific point. This region will reduce in size as N (and n) increases, and in practice the small region is reported as a point in space.

A method for calculating ΔD is *Time Difference of Arrival* (TDOA). Using synchronized receivers, the TDoA of a signal at two points can easily be calculated. Given a known signal speed the TDoA can be converted into a distance. Using asynchronous receivers the TDoA can be calculated using the cross-correlation of the two signals; however, this is limited to one phase difference. The greatest accuracy is achieved when a line of sight exists.

Triangulation

Triangulation is the process of determining an object's relative positon on a plane using angles from known locations. Given a baseline of length L and the two angles α and β measured between the ends of this baseline and the object, the perpendicular distance of the object from the baseline is:

$$d = \frac{L \sin \alpha \sin \beta}{\sin(\alpha + \beta)}$$

The distance along the baseline can then be calculated using simple trigonometry to give a relative object position. Triangulation cannot be used if the object is in line with the baseline, as no triangle can be formed. In this case the direction of the object can be determined but not its distance. Location systems may overcome this by combining triangulation with a measurement of distance such as ToF.

Angles α and β can be calculated using *Angle of Arrival* (AoA). AoA calculates each angle by measuring the TDoA using an array of receivers at each end of the baseline. Commonly two receivers are combined into a single unit which is used with an emitter in the triangulation process. As a result of using TDoA, AoA has a line of sight requirement.

Inertial sensors

Inertial sensors use an object's resistance to change of its state of motion to measure that change in motion. The two most relevant inertial sensors are accelerometers, which measure linear acceleration, and gyroscopes which measure angular velocity.

Accelerometers

An accelerometer deduces the acceleration an object experiences by measuring the force exerted on a known mass. Conceptually, an accelerometer uses a mass on a spring in combination with Hooke's law to give acceleration according to , where k is the spring's stiffness, x is displacement and m is mass (see Figure 21.3). In practice the spring is often replaced by piezoelectric or capacitive components to convert the mechanical motion into an electrical signal.

An accelerometer is sensitive to all acceleration except for acceleration due to gravity (Foxlin, 2001) since the same acceleration is exerted upon the mass and casing. Also, when the casing is at rest – e.g. lying on a surface – the reaction force exerted by the surface case will result in a reading of 1 g, assuming the accelerometer is aligned along the gravitational axis (an individual accelerometer measures the acceleration in a single plane). Generally three accelerometers are combined into a tri-axial accelerometer to determine the acceleration in the three axes. Modern accelerometers are often implemented using Micro-Electro-Mechanical systems (MEMS). MEMS tri-axial accelerometers can offer output data rates in the order of 1 kHz on chips of a size in the order of 1 mm.

Theoretically, velocity and positional data can be inferred from the measured acceleration using integration. However, the measured acceleration is subjected to error, such as bias. Bias can be considered a constant error in short-term data collection and can often be removed by calibration. However, when integrating acceleration to obtain velocity and position, acceleration bias becomes a linear error for velocity and a quadratic error for position. Figure 21.4 illustrates the impact of a constant 0.02 ms⁻² bias on a stationary accelerometer over 10 seconds. This bias is relatively small; technical specifications of typical accelerometers used in smartphones state potential bias in the range of $\pm$ 0.2 ms⁻². The effect of bias is commonly known as drift; over a 10-second period, a drift of approximately 0.9 metres is observed in Figure 21.4.

Filtering techniques can be employed to reduce the rate of drift. Eventually, however, without calibration the velocity and positional data will become unusable. To overcome this accelerometers are often used in hybrid systems, where the accelerometer provides a high frequency positional estimate and is corrected using a more accurate system running at a lower frequency. For example the SPI HPU (GPSPORTS, 2012) updates the positional estimate from a 100 Hz accelerometer with 15 Hz GPS.

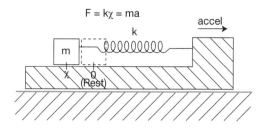

Figure 21.3 An illustration of an accelerometer.

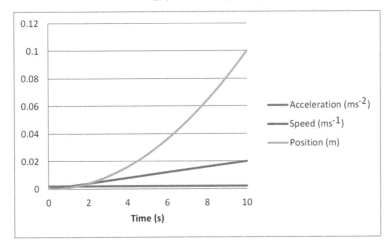

Figure 21.4 Measured Acceleration, calculated Speed and calculated Position over 10 seconds, assuming a stationary accelerometer with a 0.02 ms² accelerometer bias.

Gyroscopes

A gyroscope measures the orientation of a device. Classically, a gyroscope is a spinning disc with an axle that is free to assume any orientation. To achieve this, the spinning disc is mounted in an inner and an outer gimbal. The response of such a system to an external torque is much less than for a non-spinning object (because of the conservation of angular momentum), such that the orientation of the spin axis remains almost fixed. This provides a reference against which orientation can be measured.

Similar to accelerometers, modern gyroscopes are often implemented using MEMS. This cheap miniaturization means that gyroscopes are now commonly integrated into everyday devices such as mobile phones and laptop computers. As with accelerometers, gyroscopes can suffer from drift due to noise in the signal. Gyroscopes are also often combined with magnetometers, which measure alignment to the Earth's magnetic field, to correct for drift in the orientation measurements over time.

Accelerometers and gyroscopes are often combined if the orientation of the accelerometer cannot be guaranteed. This allows the direction of acceleration to be calculated.

Use of inertial sensors

Inertial sensors can be used to assess performance in individual training such as weightlifting or strength and conditioning. Sensorize (2015) produce such a sensor, which communicates with bespoke PC software to monitor strength and conditioning training, with specific feedback for a range of different lifts. Murgia et al. (2012) used the Sensorize FreePower Training system to measure bench-press performance with and without audio stimuli.

Combined GPS and inertial sensor systems

Systems designed for team sports analysis often incorporate GPS and inertial sensors. An example is the Catapult Outdoor System (Catapult, 2015), which incorporates GPS, accelerometers and magnetometer-corrected gyroscopes. The system provides real-time feedback on athlete performance. Gabbett, Jenkins and Abernethy (2012) used a Catapult system that incorporated a 5 Hz GPS unit with onboard 100 Hz tri-axial accelerometers and gyroscopes to assess the physical demands of professional rugby league in training and competition. They reported that the distance and the number of moderate and heavy collisions are playing-position dependent. They also reported that the physical demands of training were lower than those experienced in competition.

Combined systems such as this use the GPS to give accurate position and distance data and the inertial sensors to provide accurate acceleration data. Wu et al. (2007) introduced a method for combining GPS and inertial sensors to produce accurate velocity data, as well as improving position and acceleration measurements.

Vision systems

In vision systems, objects are tracked using one or more cameras. The tracking procedure may be in real time or happen after data collection. If multiple cameras are used, a position in three dimensions can be calculated. If only a single camera is used then it is often assumed that the object lies on the ground plane and a two-dimensional position is calculated. Using multiple cameras also allows greater coverage of the environment while reducing the likelihood of occlusion. Vision systems can be categorized as marker based or non-marker based. Most vision systems are automated to some extent, but manual intervention is required for some.

Marker based systems

In a marker based system, markers are attached to the object to be tracked. Commonly these markers are retroreflective spheres. High-end, off-the-shelf systems typically include multiple red or infra-red sensitive cameras that emit matching frequencies of light. The retroreflective markers will have a very high contrast to the background and ambient light when viewed by the cameras. Computer vision techniques are used to locate the centroid of any detected high contrast objects and also to form a temporal association for those objects detected. If any objects within the analysis volume other than the markers are reflective, artefacts can be introduced which can affect the temporal association. A manual intervention is often needed to correct mistakes in the temporal association caused either by artefacts or by the close proximity of multiple markers.

Marker based systems are highly accurate; a typical observed root mean square error is 0.1–1.0 millimetres (Richards, 1999). This means the systems are frequently used as a gold standard when assessing the accuracy and validity of alternative tracking technologies (Ogris, 2012). They can also provide very high frequency sampling rates of up to ~1000 Hz. As the position of each object is calculated from the same frame the positions are truly synchronous.

Since the cameras must be able to resolve the markers there is a limit on the size of the tracking area; with a ten-camera system this is approximately 10 m². This area can be increased by using additional cameras. In the past there have been issues using these systems outside due to natural light flooding the cameras, but this has been resolved in more contemporary systems.

Cost is a prohibitive factor in the adoption of the systems; typically a high-end system costs $150,000. Companies have recently introduced systems one order of magnitude cheaper than their top of the range offering. Thewlis et al. (2013) reported that the accuracy of these systems was comparable to their more expensive counterparts, and suggested that they provide "accuracy acceptable in research for laboratories with a limited budget".

Markers are also often used in non-automated methods. For example, Choppin, Goodwill and Haake (2011) studied the tennis ball and racket impact in warm up for the 2006 Wimbledon Qualifying Tournament using high-contrast tape markers attached to the racket. Two cameras were used and algorithms allowed the ball and racket movements to be reconstructed in three dimensions. Within each camera view, the ball and racket position were tracked manually before the reconstruction step. The reconstruction error was up to 2.5 mm in a 2 m³ volume.

Non-marker based systems

The attachment of markers on a participant is sometimes not possible or practical and can affect the movement to be tracked. Marker based systems are also not suitable or practical in sports where it is important to know the opponents' position as well. These issues can be overcome by using a non-marker system.

In a manual non-marker system, the video is 'digitized' by a user. This could be tracking an object by clicking on points of interest, or it could be more descriptive 'tagging' or 'coding' of the key events within the video. Many commercial systems add an element of computer vision to minimize the manual digitizing process. Computer vision techniques can be used to locate the object in the frame, associate objects temporally and re-identify objects after they have re-entered the field of view. The computer vision techniques greatly reduce the time taken to digitize a video and allow a larger set of events to be 'coded'.

Companies, such as ProZone® (ProZone, 2014), provide a service for teams to 'code' sporting matches. After a match, video footage from eight cameras distributed around the stadium is sent for analysis. Position, speed and key event data are provided to the client. Many different trained observers may be involved in the coding of a match, leading to concerns about the inter-observer reliability. Bradley et al. (2007) reported that when two trained observers coded the same events on ProZone®, they placed the events within 8.5 metres of one another 95% of the time.

The update rate of all vision based systems is dictated by the frame rate of the cameras, typically 25–50 Hz. Rather than a usable range, such systems are usable if the object of interest is within the view of the required number of cameras, and appears sufficiently large within the image to be analysed.

Edgecomb and Norton (2006) compared the use of a trundle wheel, GPS and manual video digitization tracking methods. While they found statistically significant differences between all methods, the measured distance was within 5% and both the GPS and manual video digitization tracking methods were deemed accurate.

Pallavi et al. (2008) used computer vision techniques to track players in broadcast video of soccer matches. The technique was also applied to other field sports such as hockey. Player detection rates were around 90–95% in all cases, with a false positive rate of around 5%.

Rampinini et al. (2007) used ProZone® to assess the variation in top level soccer match performance. They investigated the influence of seasonal variation, opposition quality and first half activity on the total movement distance, high-intensity running and very high-intensity running. Higher activity intensity was observed against better opposition and also towards the end of the season.

Magnetic tracking

Magnetic tracking systems function by measuring the strength of magnetic fields (Kindratenko, 2000). Six degrees of freedom can be calculated, meaning not only the three-dimensional position of the sensor but also the three-dimensional orientation can be reported. The system consists of a magnetic source and several receivers. The source contains three orthogonal coils, each of which creates an electromagnetic dipole field in sequence. A sensor comprises three orthogonal coils that measure the strength of the field generated by the source. The position of the sensor is determined by comparing the strength of the received field with that of the emitted field. The orientation of the sensor is determined by comparing the strength of the different fields received by the three orthogonal coils.

The accuracy and resolution of the system is proportional to the distance from the source to the sensor; therefore, the range suitability of the system is dependent upon the accuracy and resolution required for the task. Typically companies quote a range from source of 1.5 to 9 metres, but this can be extended by adding further sources. Systems normally have an update rate between 50 Hz and 240 Hz. Magnetic sensing does not have a line of sight requirement, making it ideal for densely populated tracking. However, conductive materials distort electromagnetic fields, introducing errors into the measurements. If the conductive material is static the effect of the distortion can be corrected using a mapping/calibration procedure, but this can be complex and time consuming.

The accuracy of the Polhemus Fastrak 3-SPACE (Polhemus, 2014) is quoted as being 0.08 cm and 0.15° at a range of 76 cm, and the system has a usable range up to 305 cm with reduced accuracy. Pelz et al. (1999) used the same system to track head movement in a head-mounted display based virtual environment system.

This allows the image displayed to change to match the direction in which the participant is looking.

Summary

The aim of the chapter was to provide an overview of technologies available for studying interpersonal coordination, highlighting the key measurement principles associated with four categories of technology: 1) signal propagation sensing, 2) inertial sensors, 3) vision/image-based systems, and 4) electro-magnetic tracking. Important benefits and limitations of each of these measurement approaches were identified, together with examples of their use in the literature and indicative accuracy information. The technologies are summarised in Table 21.1.

Signal propagation sensing methods can be used to track the position of large numbers of people without requiring infrastructure beyond a small receiver. They cannot be used if that infrastructure is not available, such as GPS indoors, and measures of velocity and acceleration often have large errors.

Inertial sensors are able to make accurate acceleration measures but are not accurate for position measurement. Combining them with GPS produces a powerful system, able to measure position, velocity and acceleration accurately.

Vision techniques can be used for either a high level of accuracy (marker based) or to provide data without having to interfere with the competitive environment (non-marker based). As such, marker based systems are often used in a laboratory environment or a practice environment, whereas non-marker based systems are often the only option in the competition environment if sensors and markers cannot be attached to the players.

The major advantage of magnetic tracking is that it does not require line of sight between the attached sensor and many parts of the rest of the system while providing a high level of accuracy. This allows complex movements and marker arrangements to be used. However, range is limited.

Finally, the capabilities and availability of technologies that can be used to assess interpersonal coordination are developing rapidly. Technologies such as mobile phones – containing GPS and inertial sensors – offer considerable potential. Other useful technologies are also emerging. For example, Choppin and Wheat (2013) recently explored the feasibility of using a consumer depth camera for tracking the position of a player on a badminton court. A depth camera is a single device that, in addition to the data available from a traditional camera, provides information about the distance from the camera to objects in the scene (depth information). This '2.5D' (Belhedi et al., 2012) information makes segmenting players from an image of the playing area considerably easier than in traditional images and enables three-dimensional reconstruction from a single camera. Depth cameras show great potential for tracking the two- (floor plane) and possibly three-dimensional player centre-of-mass position in badminton and similar sports (Choppin and Wheat, 2013). These and other developing technologies offer the possibility of extending the scale and frequency of interpersonal coordination analyses in both research and real-world contexts.

Table 21.1 A summary of the technologies described in this chapter.

Technology	Indicative sampling frequency	Indicative accuracy	Advantages	Disadvantages
Global Navigation Satellite System (GNSS/GPS)	1–15 Hz	4% error of true distance[a]	Global range Cheap and small chipsets integrated into many common devices, e.g. cars, smartphones	Requires line of sight to at least 4 satellites. Therefore cannot be used inside buildings and built up areas Significant inter-unit variability Requires the use of a GPS unit
Ultra-wideband (UWB)	Shared 133 Hz	3.5% ± 2.0% error of true distance[b]	Low power Can support hundreds of tags	Requires line of sight Requires the use of a tag Unsynchronized measurements
Radio Frequency Identification (RFID)	–	–	Often powered wirelessly as tag approaches reader No line of sight requirement	Short range Requires the use of a tag
Local Position Measurement (LPM)	Shared 1000 Hz	20–30 mm error for static test[c] 1.6% error of true distance[c]	Long range	Requires line of sight. Requires the use of a tag Unsynchronised measurements
WiFi and Bluetooth	Device dependent	2–5 m error[d e]	Fingerprinting – no line of sight requirement Can use existing WiFi infrastructure	Requires the use of a receiver Fingerprinting – calibration stage required
Inertial Motion Unit (IMU)	0.1–2 kHz	0.8°–1.3°[f]	High sample rate Cheap and small chipsets integrated into many common devices, e.g. laptops, smartphones	Requires the use of a MEMS unit Accelerometer bias and noise causes drift in the calculation of position and velocity

GPS and Inertial Sensors	100 Hz accelerometer with 5–15 Hz GNSS	4% error of true distance[a] 0.2°–1.3°[f]	Global range Position, velocity and acceleration data	Requires the use of a unit
Vision – Marker Systems	Dictated by camera frame rate	0.1–1 mm Root Mean Square Error[g]	Highly accurate Synchronous	Requires line of sight Requires markers Small range as markers must be resolved Expensive for top end systems
Vision – Non-marker Systems	Dictated by camera frame rate	5% error of true distance[h]	No marker required, so can capture opposing team	Requires line of sight Range dependent upon field of view
Magnetic	50–240 Hz	Proportional to the distance from source to receiver	Position and orientation can be calculated No line of sight requirement	Requires use of a receiver Short range Ferrous materials introduce errors in results

Notes:

[a] Coutts and Duffield (2010)
[b] Leser et al. (2014)
[c] Frencken, Lemmink and Delleman (2010)
[d] Xiang et al. (2004)
[e] Khoury and Kamat (2009)
[f] Brodie, Walmsley and Page (2008)
[g] Richards (1999)
[h] Edgecomb and Norton (2006)

References

Beanland, E., Main, L.C., Aisbett, B., Gastin, P. & Netto, K. (2014) Validation of GPS and accelerometer technology in swimming. *Journal of Science and Medicine in Sport, 17*(2), 234–238.

Belhedi, A., Bartoli, A., Gay-Bellile, V., Bourgeois, S. & Sayd, P. (2012) Depth Correction for Depth Cameras From Planarity. In *Proceedings British Machine Vision Conference* (pp. 1–10). Surrey, UK.

Benson, A., Bruce, L. & Gordon, B. (2012) Comparing the measurement of physical activity using sport specific GPS and an iPhone™'app'. *Journal of Science and Medicine in Sport, 15*, S293–S294.

Bradley, P., O'Donoghue, P., Wooster, B. & Tordoff, P. (2007) The reliability of ProZone MatchViewer: a video-based technical performance analysis system. *International Journal of Performance Analysis in Sport, 7*(3), 117–129.

Brodie, M.A., Walmsley, A. & Page, W. (2008) Dynamic accuracy of inertial measurement units during simple pendulum motion: technical note. *Computer Methods in Biomechanics and Biomedical Engineering, 11*(3), 235–242.

Catapult (2015) http://www.catapultsports.com/

Choppin, S., Goodwill, S. & Haake, S. (2011) Impact characteristics of the ball and racket during play at the Wimbledon qualifying tournament. *Sports Engineering, 13*(4), 163–170.

Choppin, S. & Wheat, J. (2013) The potential of the Microsoft Kinect in sports analysis and biomechanics. *Sports Technology, 6*(2), 78–85.

Coutts, A.J. & Duffield, R. (2010) Validity and reliability of GPS devices for measuring movement demands of team sports. *Journal of Science and Medicine in Sport, 13*(1), 133–135.

Duffield, R., Reid, M., Baker, J. & Spratford, W. (2010) Accuracy and reliability of GPS devices for measurement of movement patterns in confined spaces for court-based sports. *Journal of Science and Medicine in Sport, 13*(5), 523–525.

Edgecomb, S.J. & Norton, K.I. (2006) Comparison of global positioning and computer-based tracking systems for measuring player movement distance during Australian football. *Journal of Science and Medicine in Sport, 9*(1), 25–32.

Foxlin, E. (2002) Motion tracking requirements and technologies. *Handbook of Virtual Environment Technology, 8*, 163–210.

Frencken, W.G., Lemmink, K.A. & Delleman, N.J. (2010) Soccer-specific accuracy and validity of the local position measurement (LPM) system. *Journal of Science and Medicine in Sport, 13*(6), 641–645.

Gabbett, T.J., Jenkins, D.G. & Abernethy, B. (2012) Physical demands of professional rugby league training and competition using microtechnology. *Journal of Science and Medicine in Sport, 15*(1), 80–86.

GPS Sports (2014) http://gpsports.com/

Inmotio (2015) http://www.inmotio.eu/

Jennings, D., Cormack, S., Coutts, A., Boyd, L. & Aughey, R. (2010) The validity and reliability of GPS units for measuring distance in team sport specific running patterns. *International Journal of Sports Physiology and Performance, 5*, 328–341.

Kearns, W.D., Fozard, J.L., Webster, P. & Jasiewicz, J.M. (2014) Location Aware Smart Watch to Support Aging in Place. *Gerontechnology, 13*(2).

Kindratenko, V.V. (2000) A survey of electromagnetic position tracker calibration techniques. *Virtual Reality, 5*(3), 169–182.

Khoury, H.M. & Kamat, V.R. (2009) Evaluation of position tracking technologies for user localization in indoor construction environments. *Automation in Construction, 18*(4), 444–457.

Leser, R., Schleindlhuber, A., Lyons, K. & Baca, A. (2014) Accuracy of an UWB-based position tracking system used for time-motion analyses in game sports. *European Journal of Sport Science, 14*(7), 635–642.

MacLeod, H., Morris, J., Nevill, A. & Sunderland, C. (2009) The validity of a non-differential global positioning system for assessing player movement patterns in field hockey. *Journal of Sports Sciences, 27*(2), 121–128.

Murgia, M., Sors, F., Vono, R., Muroni, A. F., Delitalia, L., Di Corrado, D. & Agostini, T. (2012) Using auditory stimulation to enhance athletes' strength: An experimental study in weightlifting. *Review of Psychology, 19*(1), 13–16.

Neave, N., McCarty, K., Freynik, J., Caplan, N., Hönekopp, J. & Fink, B. (2011). Male dance moves that catch a woman's eye. *Biology Letters, 7*(2), 221–224.

Ogris, G., Leser, R., Horsak, B., Kornfeind, P., Heller, M. & Baca, A. (2012) Accuracy of the LPM tracking system considering dynamic position changes. *Journal of Sports Sciences, 30*(14), 1503–1511.

Pallavi, V., Mukherjee, J., Majumdar, A.K. & Sural, S. (2008) Graph-based multiplayer detection and tracking in broadcast soccer videos. *Multimedia, IEEE Transactions on, 10*(5), 794–805.

Pelz, J.B., Hayhoe, M.M., Ballard, D.H., Shrivastava, A., Bayliss, J.D. & von der Heyde, M. (1999) Development of a virtual laboratory for the study of complex human behaviour. *Electronic Imaging '99* (pp. 416–426). International Society for Optics and Photonics.

Polhemus (2014) http://polhemus.com/_assets/img/FASTRAK_User_Manual_OPM00PI002-G.pdf

ProZone (2014) http://www.prozonesports.com/

Rampinini, E., Coutts, A.J., Castagna, C., Sassi, R. & Impellizzeri, F.M. (2007) Variation in top level soccer match performance. *International Journal of Sports Medicine, 28*, 1018–24.

Resch, A., Pfeil, R., Wegener, M. & Stelzer, A. (2012, June) Review of the LPM local positioning measurement system. In *Localization and GNSS (ICL-GNSS), 2012 International Conference on* (pp. 1–5). IEEE.

Richards, J.G. (1999) The measurement of human motion: a comparison of commercially available systems. *Human Movement Science*, 18(5), 589–602.

Sensorize (2015) http://www.sensorize.it/fipcf-sensorize/

Stelzer, A., Pourvoyeur, K. & Fischer, A. (2004) Concept and application of LPM – a novel 3-D local position measurement system. *Microwave Theory and Techniques, IEEE Transactions on, 52*(12), 2664–2669.

Strava (2015) http://www.strava.com/

Thewlis, D., Bishop, C., Daniell, N. & Paul, G. (2013) Next generation low-cost motion capture systems can provide comparable spatial accuracy to high-end systems. *Journal of Applied Biomechanics, 29*(1), 112–117.

US Naval Observatory (2014) http://tycho.usno.navy.mil/gpscurr.html

Wisbey, B., Montgomery, P.G., Pyne, D.B. & Rattray, B. (2010) Quantifying movement demands of AFL football using GPS tracking. *Journal of Science and Medicine in Sport, 13*(5), 531–536.

Woo, S., Jeong, S., Mok, E., Xia, L., Choi, C., Pyeon, M. & Heo, J. (2011) Application of WiFi-based indoor positioning system for labor tracking at construction sites: A case study in Guangzhou MTR. *Automation in Construction, 20*(1), 3–3.

Wu, F., Zhang, K., Zhu, M., Mackintosh, C., Rice, T., Gore, C., Hahn, A. & Holthouse, S. (2007) An investigation of an integrated low-cost GPS, INS and magnetometer system for sport applications. In *the 20th International Technical Meeting of the Satellite Division of the Institute of Navigation ION GNSS.*

Xiang, Z., Song, S., Chen, J., Wang, H., Huang, J. & Gao, X. (2004) A wireless LAN-based indoor positioning technology. *IBM Journal of Research and Development, 48*(5.6), 617–626.

22 Interpersonal coordination tendencies in competitive sport performance

Issues and trends for future research

Keith Davids

The chapters of this book have illustrated how interpersonal coordination tendencies are a prominent feature of the functioning of collective systems in many different walks of life, including work, education, military and sport contexts. Different chapters have comprehensively shown how interactive phenomena emerge in everyday actions, sometimes when people are unaware of them, such as when negotiating a crowded street, bus or train, walking on a street while talking with a friend, or driving in a traffic jam. Other contributions to this book have revealed how fundamental behaviours in sport performance, whether in individual contexts like sprinting in athletics, in team sports and in creative physical activities like dance, ice skating and synchronized swimming, involve the co-adaptation of behaviours that are constrained, among other things, by the need to continually interact with other athletes.

Being intentional or unintentional, desired or undesired, due to competition or cooperation, emergent interpersonal coordination tendencies are strongly constrained by the locally created information that emerges from the interactive behaviours of individuals in a complex adaptive system. In competitive sport this information can be translated to simple rules that support dyadic (1v1) and collective system behaviours. Discovering these rules remains a major challenge for researchers in this field of study in the future. A key feature of understanding how interpersonal coordination tendencies emerge in social systems is that they are not just based on an individualized analysis, but emerge through studies of the continuous interactions between individuals performing in complex social collectives.

The chapters in this book have raised questions for future research on developing understanding of how competitive and cooperative contexts afford coordination for two or more independent individuals, and whether their continuous interactions may be grounded on variables that only emerge during these interactive behaviours. The reciprocal and mutual influences that emerge among performing individuals within these contexts creates a behavioural dependence. It is important to investigate whether nonlinearity is an outcome that is a general feature of emergent interactions between individuals in a performance context like sport. The characteristic of nonlinearity in complex social systems in sport signifies that it is not always possible to predict with complete accuracy what other individuals in the same complex social system will do next. This key property leads to dyadic and collective system behaviours that are emergent in performance contexts where adaptive variability is paramount for functional action. Due to its underlying nonlinearity, the term emergent refers to the variability of the outcomes of a complex system which contains two or more interacting individuals, even when the performers are not consciously engaged in direct interactions. Although this rationale has been quite prominent in studies of interpersonal coordination tendencies in complex social systems over the last two decades, it remains a key feature of research into sport performance that can enhance knowledge and develop understanding of emergent behaviours. An important effect of research into interpersonal interactions in sport is a stronger focus on the *dynamics* of these coordination tendencies, eschewing *categorical* thinking (with its associated binary decision making outlook) in favour of *complementary* thinking (exploring the relations between things), which James Gibson (1979) emphasized in his preferred person-environment scale of analysis (see Kelso, 2012). A commitment to adopt a preferential perspective on the complementary relations between people and between people and objects, events and properties of a performance environment requires investigative effort directed towards interpersonal coordination in complex adaptive systems. As Kelso (2012) noted, these properties include system degeneracy (for examples in team sports see Passos & Davids, 2015), inherent self-organization tendencies and the emergence of synergies (whether consciously or sub-consciously regulated; see Varlet & Richardson, 2015), multistability and metastability (for examples in sport see studies by Hristovski et al., 2006; Pinder et al., 2012).

An important point to note is that like other features of a complex adaptive system, interpersonal coordination tendencies that emerge are highly susceptible to the influence of task and environmental constraints. Some task constraints in sport involve competition whereas others require cooperation, with a varied mix between the two being typical. To perform some tasks people need to use tools (e.g. rackets, violins, a steering wheel), whereas in other tasks people just need to maintain a 'functional' interpersonal distance to an adjacent individual (e.g. avoid a defender and shoot at a basket in basketball, or maintain an angle to a teammate to facilitate a pass). Despite variations in task and environmental constraints, some general features characterize interpersonal coordination tendencies. Understanding how performance task constraints might lead an athlete to coordinate with an(other) individual(s) is a significant challenge for future research. The findings of such studies can facilitate applications of information from research on interpersonal coordination tendencies to aid sport scientists and sport pedagogists to design learning and practice environments to improve performance.

In the task of explaining and predicting sport performance and its outcomes, further research is also needed to continue to develop a theoretical basis for understanding emergent interpersonal coordination tendencies in different sport performance contexts at different scales of analysis (e.g. local and global, micro- and macro-structure, performance and learning). A key challenge of research programmes in the study of interpersonal coordination tendencies is to identify variables that bound the ways that individuals relate to each other during performance. A second challenge is to continue to develop our understanding of how interpersonal coordination tendencies can be measured, not only from an individual perspective, but through interactions between people. Many chapters have drawn attention to the need to continue to develop understanding of the nature of the tools to be used to measure and quantify coordinative behaviours among individuals under different task constraints. Some chapters have traced the move from simpler to more complex tools, utilizing intricate statistical and mathematical procedures to describe and explain continuously emerging interactive behaviours. A key feature of this research area has been the transition from reliance on kinematic variables of performance, such as relative angles and velocities of individuals, to relative phase analysis, principal component analysis (PCA), the uncontrolled manifold (UCM) hypothesis and neural networks. An important feature is the depth of mathematical modelling that might create the future grounding for behavioural predictions and the articulation of theoretical ideas.

Theoretical advances in interpersonal coordination can address challenges of the datafication of team sport performance

Regardless of the mathematical and statistical procedures used, the domain of sport is awash with data at the elite and developmental level of performance. This is because in order to understand successful performance in competitive sport environments, the collection of valid data is a key issue, forming the basis of sports analytics. Sports analytics is a term which defines the management of multi-dimensional data using predictive analytic models, with the aim of providing decision-makers with an advantage in performance contexts (Alamar & Mehrotra, 2011a). Currently, sports analytics greatly emphasize procedures and approaches like data cataloguing and mining to attempt to predict the future performance of an athlete or a team. During the past decades, the capacity to produce information that results in a substantive description of the performance of an individual athlete or a sports team to sustain the decision-making of coaches and managers has been vastly improved by rapid technological developments (Liebermann et al., 2002). Indeed, the sports sciences are entering a new era with the advent of new digital tools for the analysis of athletes' training, performance and health behaviours (on and off the performance arena). These (continuing) massive technological changes emerging over the past years have led to a new issue, captured in the term *Big Data*. A prevailing assumption has been that developing methods to collect data will complement analytic methods to improve sport performance in significant ways, such as fine-tuning training plans or identifying the specific strengths and weaknesses in the performance patterns of competitors. With recent technological developments (e.g. game analysis software or motion tracking systems), it has become possible to record performance data in real time, both in training and competition, exponentially increasing the amount of data available to practitioners. In turn, new statistical methods have been developed (i.e. from linear to stochastic methods) to model performance in sport (Alamar, 2011; Nevill, Atkinson & Hughes, 2008).

For these reasons, sport science and sport analytics is awash with data: waves and waves of statistics are available to be used to enhance athlete performance. How can we avoid being overwhelmed by this enormous volume of information? A clear challenge for future research in sport science, sport analytics and sport pedagogy is to understand which data matter and how to interpret them. Information is needed on performance, learning, training and recovery in sport programmes, but there are other overarching issues for future research. What to do with all these data, and how should we approach opportunities to collect more data? What specific types of data are actually needed by athletes and practitioners in team sports, for example? How can we harness the power of large data sets, and at the same time avoid the reductionism of athletes being treated identically, due to the thousands of measurements recorded during team practice and performance? These and other related questions are increasingly being posed in academic circles and development programmes, based on concerns about the impact of 'Big Data', 'Datafication or 'Dataveillance' (Williams & Manley, 2014; Couceiro, Dias, Araújo & Davids, under review) on athlete behaviours during practice, training and competition, as well as pedagogical practice.

Further work is needed to integrate theoretical knowledge, empirical data and research methodology of a quantitative and qualitative nature, as well as experiential knowledge from pedagogical practice and application, for sports pedagogists, practitioners and performance analysts in the future. More specifically, future research on interpersonal coordination in team sports is needed to: i) undertake experimental investigations by directly manipulating constraints on sport performers in high quality simulations of the contexts of action, ii) undertake computerized simulations, and iii), systematically record and analyse observations of behaviours in team sport performance and training settings. Measured variables in research need to be typically eco-kinematic, psychophysical, physiological, neuroscientific and notational (in terms of individual and collective behaviours), in a form which can be mathematically modelled. The equipment used needs to include Global Positioning Systems and other remote sensing technologies, video-based motion analysis, ball projection machines and neurophysiological sensors which support performance analysis in team sports environments, including competitive events and simulated training conditions such as small-sided and conditioned games. Additional methods that need to be used include interviews, questionnaires and document analysis, to access the experiential knowledge of expert performers and coaches in team sports. This range of different methods is needed to provide varied insights into how interpersonal coordination tendencies emerge during performance, and how practice tasks can be designed to facilitate performance advances.

Some of the chapters of this book have illustrated convincingly how scientists and practitioners in this field will be required to be conversant with relevant theoretical frameworks, research methods and measurements, and data analysis approaches which embrace the inherent nonlinearity in sport performance environments conceptualized as complex adaptive systems. This knowledge needs to be considered in applications to practice contexts to provide the type of support needed by elite, sub-elite and development programmes in sport. This research suggests that performance analysts will be required to work in teams with pedagogues, psychologists, skill acquisition specialists, physiologists, biomechanists and engineers to collect and verify, discard and retain, analyse and interpret data collected during different phases of competition, preparation and recovery. This integration of knowledge and skills will continue to be extremely important because it has been noted that each type of (sport) performance environment has a specific *form of life* (as Wittgenstein noted, predicated on unique patterns of behaviour, knowledge and skills as well as social customs and cultural constraints) (Rietveld & Kiverstein, 2014). According to these ideas, athletes need to become aware of the *form of life* surrounding different sport contexts where they need to interact with others. It is feasible that performance analysts of the future will need to have a good grasp of a theory explaining how these interactive behaviours during competition and practice emerge, so that an athlete's relationship with a particular performance environment can be made more functional as a result of the quality of their training, rather than the merely the hours spent in preparation. Two fundamental questions need to be addressed in future work: what is the nature of the coordinative tendencies in performance, and how should we conceptualize the interacting performer in sport? Answering these questions will provide support for performance analysts, because without these models performance analysis in team sports will become very reactive, rather than prospective.

Theoretical modelling to enhance understanding of the nature of performance under the task constraints of different sports is a key progression in future research, which might address criticisms that the approach to analytics in sport science is still somewhat limited (e.g. Alamar & Mehrotra, 2011b; Glazier, 2010; Vilar et al., 2012).

The theoretical framework proposed later in this chapter identifies (cooperating and competing) attackers and defenders in team games as intricate components (degrees of freedom) of a self-organizing system linked by informational fields which surround them, e.g. acoustic and visual. In team sports considered as *complex adaptive systems*, the individual performer is conceptualized as the base unit degree of freedom. Team sports are dynamic, complex performance environments because of the continuous changes in the location, positioning and movements of competing and cooperating players. Due to the highly interactive nature of the performance environment in team sports, opportunities for action are constrained in time and space. These *affordances* or invitations for action emerge from a process of co-adaptation between individuals. Co-adaptation can emerge in many different forms, including interpersonal coordination, which provides a platform for performance in all sports at elite and sub-elite levels, and even sub-consciously (Varlet & Richardson, 2015). For example, in a published observation an analysis of two elite sprinters in a world championship 100m final revealed spontaneous and intermittent (i.e. unintended) synchronized entrainment of stride frequencies when competing against each other. Despite obviously different structural constraints such as height and limb lengths of the athletes, leading to variations in stride lengths and the number of steps taken to complete the race, spontaneous and unintended co-adaptations were observed during parts of the competitive event. Varley and Richardson (2015) argued that this type of co-adaptation is only possible when natural behaviours of such systems (e.g. stride frequencies) are not too dissimilar to begin with, supporting a compensatory role of perceptual coupling. These data confirmed earlier findings showing synchronized entrainment between individuals reported in studies of team game performance such as in football, rugby union and basketball (e.g. Davids, Button, Araújo, Renshaw & Hristovski, 2006; Duarte, Araújo, Correia, Davids, Marques & Richardson, 2013). However, the interesting feature of the study of sprinters is the spontaneous and unplanned nature of the co-adaptive processes, which reveals an interesting line for future research in performance contexts like sport and work. Observations like these have implications for the representative design of practice task constraints due to the conscious and sub-conscious regulation of action.

In team games, co-adaptation has been shown to result in the continuous emergence and dissolution of spontaneous pattern-forming dynamics between system degrees of freedom (athletes) during performance (for a review see Passos & Davids, 2015). Like organisms in other complex adaptive systems, such as schools of fish and flocks of birds, individual athletes in sports teams use fairly simple local behavioural rules to (re)create rich structures and patterns at a collective level that are much more nuanced than the behaviour of any single individual playing in the team. This 'complexity from simplicity' model for understanding behaviours of complex adaptive system requires more research in sport performance contexts involving continuous interpersonal interactive tendencies between athletes. Research has shown how local rules for interactions between single system components can lead to the emergence of sophisticated 'macro-states' of organisation in a global system. What is needed in future work is the continued development of nonlinear methods for analysing patterns of emerging behaviours between individual athletes. A range of such methods can continue to develop our understanding of how sophisticated attacking and defending patterns of play in team sports emerge from continuous attacker-defender interactions. The aim of further work should be to clarify how the constraints of a competitive environment in team games force performers to continuously co-adapt to behaviours of teammates and opponents, typically in close proximity to specific locations of a playing area (such as the goal or sidelines). Important performance variables like relative angles between competing individuals, values of interpersonal distances between an attacker and defender, the relative velocity between two moving competitors and so on have been empirically verified as relevant variables for understanding interpersonal coordination tendencies of team sports players as agents in a complex adaptive system. Additionally, variables capturing the shape of attacking and defending sub-units or formations, such as team centroids (a variable showing the centre of performance gravity of a team) and surface areas, as well as width and length, provide information on coherent patterns in sports teams during performance. These behaviours emerge from processes of co-adaptation in team sport performance, providing valuable insights for coaches, teachers and practitioners. Despite this existing research, a major task for the future is for researchers in interpersonal coordination to work closely with practitioners to create a programme of small-sided and conditioned games that support informationally-constrained dynamical interactions in space and time, created by imposing changes in the relative positioning of attackers and defenders in many different team sports through careful manipulations of practice task constraints.

Despite some attempts to use predictive models of performance, most data captured in sports analytics currently remains focused on cataloguing and grouping discrete behaviours (McGarry, 2009), with little reference to the performance context in which behaviours emerge (Vilar, Araújo, Davids, & Button, 2012). Attempting to describe complex sport behaviours by aggregating an extensive range of discrete variables reduces performance analysis to a number of micro-components. This reductionist approach tends to neglect the performance conditions and the interactions between individuals that provide a background for the emergence of specific technical or tactical behaviours. It could be argued that, currently, sports analytics is driving performance analysis towards a probabilistic approach that coaches and managers are challenged to convert into useful information (Hughes & Franks, 2008; McGarry, 2009; Travassos, Araújo, Correia, & Esteves, 2010). As in other areas of research (e.g. Aerts et al., 2007), the prevalent approach to studying and predicting performance in sports provides opportunities for individuals to gain information about competitive performance (see Brillinger, 2007), but does little to help coaches and practitioners to understand the principles that regulate performance achievement. These issues are summarized in a quote from Analytics Magazine: "...despite the remarkable growth in the amount and variety of data available of examination and analysis, the world of sports analytics still faces the same ubiquitous challenge: How to get meaningful information into the hands – and minds – of the people who are in a position to make effective use of it" (Alamar & Mehrotra, 2011a).

In team sports, for example, the achievement of a performance goal emerges from complex non-linear behaviours that are constrained by intermittent intra- and inter-personal couplings between players in space and time (Duarte et al., 2013; Travassos et al., 2013). This perspective should not be based on an operational, atheoretical analysis of opportunities for successful performance that are based on calculated probabilities of a sequence of actions. Rather, analysis of team sports performance should be based on a theoretical rationale outlining the emergence of complex spatial-temporal interpersonal relations between performers, in which key events are considered as *perturbations* which facilitate transitions in the organizational state of the performance context (Vilar et al., 2013; Vilar et al., 2014). In this conceptualization, the production of performance data for coaches and managers can aid them to understand the mechanisms that regulate the interactive behaviours of individual players and teams, i.e. how players work individually and together in different sub-phases of a game to produce competitive behaviours. Later in this chapter data will be discussed that indicates how individuals and teams might harness system metastability in switching between the needs of individuals and teams to satisfy performance constraints in the sport of international cricket (Gauriot & Page, 2015). These data imply that performance in sport needs to be investigated at different levels of analysis (individual, dyadic (between players), collective (between teams)) in order to reveal the conditions under which individual athletes successfully perform and the contributions to collective performance made by each player, and how the overlap between individual and team needs can drive success. To enhance understanding of competitive behaviours in sport, future effort in research is needed to discover regions of metastability in the performance landscape that can provide an advantage for game management.

The introduction of the concept of metastability raises a key question for future work on interpersonal coordination tendencies concerning how sports analytics can be advanced beyond mere statistical description and prediction of performance outcomes based on operational models. Transiting beyond simple operational levels of performance analysis in future work will afford more holistic perspectives of athlete behaviours by focusing on their continuous interactions, rather than their individual behaviours in isolation, allowing more comprehensive analyses of the performance environments in team sports. Achieving this more challenging scientific goal will require a powerful theoretical rationale of performance in collective sports and activities. Ecological dynamics is one such theoretical rationale which can improve understanding of interpersonal coordination tendencies in athlete behaviour, predicated on dynamical systems theory and principles of ecological psychology (Araújo & Davids, 2009; Araújo, Davids, & Hristovski, 2006; Travassos et al., 2010; Passos & Davids, 2015; Duarte et al., 2012a).

Ecological dynamics

Ecological dynamics is a framework that integrates knowledge from ecological psychology and dynamical systems theory, creating the potential to enrich scientific understanding of interactions between individuals under different task constraints (Araújo et al., 2006; Warren, 2006; Vilar et al., 2012). In collective sport contexts, this theoretical framework is grounded on the notion that the goal-directed behaviours of players and teams are prospectively oriented on the mutual relationship between players/teams and the key constraints of a competitive performance environment (Araújo et al., 2006; Davids, Araújo & Shuttleworth, 2005; Travassos et al., 2012). Performer-environment relations are exploited over time and dyadic and team couplings are forged and broken according to changes in the informational constraints of the performance environment and the specific intentions of individual athletes (McGarry et al., 2002; Passos et al., 2008; Passos & Davids, 2015).

This theoretical framework has been applied in sport science and performance analysis to understand how spatial-temporal inter-relations constrain the actions of performers towards functional solutions. This viewpoint proposes that to understand team sports performance, it is important to identify and measure the changes in the informational constraints that regulate the interpersonal relations between individuals and teams (Passos et al., 2008; Travassos, Araújo, Vilar, & McGarry, 2011; Vilar et al., 2012). An important assumption in ecological dynamics is that game behaviours emerge through self-organization processes under constraints from the high number of action probabilities (Araújo et al., 2006; Davids et al., 2005; McGarry et al., 2002; Palut & Zanone, 2005). Performance behaviours emerge as players attempt to achieve more adaptive functional solutions, i.e. functional interpersonal patterns of coordination (Bourbousson, Sève & McGarry, 2010; Palut & Zanone, 2005; Passos et al., 2011; Travassos et al., 2011; Vilar et al., 2013). Understanding the nature of performer/team-environment relations that afford successful actions, as well as the maintenance of stable patterns of coordination or the transition to new states of organization, is needed in further work to explain performance outcomes in sport (Araújo et al., 2006; Araújo, Travassos & Vilar, 2010; Davids et al., 2005; McGarry et al., 2002; Vilar et al., 2014; Esteves et al., 2015). The decomposition of a competitive game into subsystems/functional units or sub-phases in future work will enable scientists to better identify the most reliable data to capture performance in sports (Travassos et al., 2010). This approach will continue to facilitate understanding of the interactive behaviours of players and teams based on spatial-temporal relations according to situated performance outcomes and goals.

In ecological dynamics one of the key issues is to explain how players optimize their performance of skills in competitive performance. This can be done by measuring spatial-temporal relations between players, ball and goal. Another important feature of performance concerns constraints on successful dribbling in team sports like association basketball, football and rugby union; this means understanding the way that attacking players in possession of the ball manage interactive values of key variables such as interpersonal distance and relative velocity with a marking defender (Duarte et al., 2010; Passos et al., 2008).

Additionally, more work is needed on understanding how performance area location can shape outcomes. For example, when interpersonal interactions emerge close to a scoring area, the intentionality of performers appears to become more conservative and less oriented towards risk taking (Headrick et al., 2012). Defending players do not tend to risk an expansive attempt at dispossessing the attacker because of the challenge of being the last player defending the goal area, and consequently attackers wait for an optimal moment to change the alignment with the defender and the goal target in order to create a shooting opportunity. More work is needed to observe the actions of an attacker to successfully move past a defender with the ball, in order to understand how the intentions of a performer shape how he/she manages spatial-temporal relations with a marking defender at different locations on the field. However, the number of passing options available for an attacker to perform a pass to an adjacent team-mate is also extremely important for understanding probabilities of a successful outcome in an attacking team (Vilar et al., 2012).

The capacity of defending players to coordinate behaviours in space and time in order to recover ball possession is also an important issue in predicting the quality of defensive performance in team sports. More than merely evaluating the behaviours of individual performers, there is a need to measure the interpersonal relations within a team, as well as between teams, in achieving specific performance goals. A relevant variable for predicting outcomes in defensive performance could be to investigate the coupling between the displacement of each defender relative to the positioning of other defenders and the movement of the ball.

To summarize so far, future research needs to continue to emphasize the analysis and prediction of future performance in sport contexts, using methods to capture the capacity of performers to manage informational constraints in different competitive performance contexts. An integration of spatial-temporal analysis in non-linear or stochastic models will help develop understanding of the principles that regulate performance achievement in collective sports.

How the specificity of task constraints shapes performance in team sports: the case of cricket batting in international test matches

The discussion in the previous section led to the conclusion that a major feature of future research in ecological dynamics concerns how the specificity of task constraints in interactive sports like team invasion games shapes performance outcomes, with an intricate and subtle interaction emerging between intentions, perception and action, providing an overarching informational constraint on performance behaviours. However, future research also needs to consider how the intertwined relations between perception, cognitions and action shape interactions in other sports which are not invasion games. Consider, for example, a study by Gauriot and Page (2015) on cricket batting in five-day international matches, known as test matches, which typically occur in a series of two to five matches. They documented the tensions between the performance needs of each individual batsman and the team captain in making decisions for the strategic benefit for the team, which they termed individual and collective performance incentives. Successful batting in cricket can be determined in different ways, but conventionally there are traditional landmark performances that are documented in record books, such as when an individual batsman scores 50, 100 or 200 runs. Terminating a successful batting performance of an individual in the team just prior to the achievement of one of these landmarks, to enhance the team's probability of winning the game by putting the opposing team in to bat in order to get them out, is a strategic decision that needs careful consideration by a team captain. This is because of the significance of these landmarks for individual professionals in existing statistical records. The morale of the individual or the team could be affected if the needs for achievement from both aspects are not balanced.

The significance of these performance milestones for individual cricket professionals was observed by Gauriot and Page (2015), using McClary tests to reveal clear evidence of effects on the rate of scoring immediately prior to and after the landmark of 100 runs (known as a century in the sport). Batsmen seemed to play more conservatively before the 100-run landmark than after, regardless of team needs. These data have some important implications for future research on the interpersonal relations between an individual performer and a team in a non-invasion game, through a lens focusing on the needs of the athlete and the collective.

To conclude this chapter, it may be instructive to interpret these data from an ecological dynamics perspective, to illustrate the complementary thinking that was discussed earlier (see Kelso, 2012). First, the data emphasize that the nature of the decision making tendencies of a captain in cricket is similar in many ways to decision making under other task constraints, such as invasion games. This is especially the case in regions of criticality, such as near landmark performance goals for an individual athlete. These similarities highlight the significance of conceptualizing sports teams as complex adaptive systems. An implication is that in both types of competitive performance environments there are no hard and fast rules that categorize decision making tendencies; instead. these emerge under task and environmental constraints. As in any complex adaptive collective system, they also function at different timescales. Players function as individual elements within a team, but they cannot operate completely independently of the team. Hence there is a switching between individual performance and team performance achievements at the timescale of perception and action. This might be conceptualized as an example of system metastability, which is an inherent part of sport performance that has been observed in a cricket batting study and in the task of hitting a boxing heavybag (see Pinder et al., 2012 Hristovski et al., 2006 respectively). Dwelling too long in one region of the landscape formed by components of a complex system might harm the performance achievements of a player and a team, with effects on the reputations of both over a longer timescale. At other timescales of performance – for example, over a series of test matches or one-day international (ODI) games (between two to five), or perhaps over the number of games in a season – these decisions at critical landmark points can influence player behaviours through effects on team motivation, morale, reputation and likely series outcomes. So, there are potential effects of these decisions, for players and teams, over two distinct but interrelated timescales. Future work on this issue might address a number of specific considerations such as the following:

1   Captains in professional teams now have sport scientist support and coaches and performance directors to discuss these decision making ideas with, and

so there is less emphasis on the relations between a captain's decision and a player's personal performance profile because this could be a group decision. The datafication of sport (which provided information for the analyses by Gauriot and Page (2015)) leaves this type of decision making less dependent on interpersonal relations, since there is likely to be computerized analysis of the effects of target scores at which an innings may be terminated by a captain.

2   Context is key, as argued in ecological dynamics, and it would be interesting to observe the cut-off points for decisions on terminating an innings near a landmark in a single ODI – for example, in a world cup knock-out round – compared to a series of matches. The longer timescale dynamics may have less of an influence in the former context, and it might be expected that a scoring discontinuity, as observed in the data of Gauriot and Page (2015), would be less evident in a data set of single games compared to the data sets of test series as examined in that study.

3   The implications of these data suggest that it would be useful for players to experience game simulations in training which set up scenarios for them to practice switching between satisfying personal and team performance goal needs, enhancing their capacity to assess the criticality of a performance context at the timescale of perception and action, with one eye on a series outcome.

4   Sometimes allowing opponents to observe a crowd celebration and the reaction to a player's landmark achievement can change the dynamics of the performance context, varying motivation and emotional responses in competing teams. This process can lead to a competitive advantage, as observed in the team game of association football when key performers in a winning team are substituted with one or two minutes of the match remaining. Is this a pointless decision by the coach? Not necessarily, since it could enhance the mood of a supporting crowd and boost the motivation of key players, although such a decision could go wrong if the scores are too close, giving an advantage to the opposition.

References

Aerts, D., Apostel, L., De Moor, B., Hellemans, S., Maex, E., Van Belle, H., et al. (2007) *Worldviews: from fragmentation to integration.* VUB Press: Brussels. Available online at http://pespmc1.vub.ac.be/CLEA/reports/WorldviewsBook.html

Alamar, B. (2011) The Next Step. *Journal of Quantitative Analysis in Sports, 7*(4).

Alamar, B. & Mehrotra, V. (2011a) Beyond 'Moneyball': The rapidly evolving world of sports analytics, Part I. *Journal, September/October.* Retrieved from http://www.analytics-magazine.org/septemberoctober-2011/405-beyond-moneyball-the-rapidly-evolving-world-of-sports-analytics-part-i.html

Alamar, B. & Mehrotra, V. (2011b). Sports Analytics, part 2. *Journal, November/December.* Retrieved from http://viewer.zmags.com/publication/7f1ab3ad#/7f1ab3ad/9

Araújo, D. & Davids, K. (2009) Ecological approaches to cognition and action in sport and exercise: Ask not only what you do, but where you do it. *International Journal of Sport Psychology, 40*(1), 5–37.

Araújo, D., Davids, K. & Hristovski, R. (2006) The ecological dynamics of decision making in sport. *Psychology of Sport and Exercise, 7*(6), 653–676.

Araújo, D., Travassos, B. & Vilar, L. (2010) Tactical skills are not verbal skills: A comment on Kannekens and Colleagues. *Perceptual and Motor Skills, 110*(3), 1086–1088.

Bourbousson, J., Sève, C. & McGarry, T. (2010) Space-time coordination patterns in basketball: Part 1 – Intra- and inter-couplings amongst player dyads. *Journal of Sport Sciences, 28*(3), 339–347.

Brillinger, D.R. (2007) A potential function approach to the flow of play in soccer. *Journal of Quantitative Analysis in Sports, 3*(1), 3.

Couceiro, M., Dias, G., Araújo, D. & Davids, K. (under review) Are practice, evaluation, training and pedagogies in sports being left behind by massive technolgical challenges?

Davids, K., Araújo, D. & Shuttleworth, R. (2005) Applications of dynamical systems theory to football. In T. Reilly, J. Cabri & D. Araújo (Eds.), *Science and Football V: The Proceedings of the Fifth World Congress on Sports Science and Football* (pp. 537–550). Routledge: London.

Davids, K., Button, C., Araújo, D., Renshaw, I. & Hristovski, R. (2006) Movement models from sports provide representative task constraints for studying adaptive behavior in human movement systems. *Adaptive Behavior, 14,* 73–95. Available online at http://dx.doi.org/10.1177/105971230601400103

Duarte, R., Araújo, D., Correia, V., Davids, K., Marques, P. & Richardson, M.J. (2013) Competing together: Assessing the dynamics of team–team and player–team synchrony in professional association football. *Human Movement Science, 32,* 555–566. http://dx.doi.org/10.1016/j.humov.2013.01.011

Duarte, R., Araújo, D., Correia, V. & Davids K. (2012a) Sport teams as superorganisms: Implications of sociobiological models of behaviour for research and practice in team sports performance analysis. *Sports Medicine, 42*(8), 633–642.

Duarte, R., Araújo, D., Davids, K., Travassos, B., Gazimba, V. & Sampaio, J. (2012) Interpersonal coordination tendencies shape 1v1 sub-phase performance outcomes in youth soccer. *Journal of Sports Sciences, 30,* 871–877.

Duarte, R., Araújo, D., Fernandes, O., Fonseca, C., Correia, V., Gazimba, V., et al. (2010) Capturing complex human behaviors in representative sports contexts with a single camera. *Medicina, 46*(6), 408–414.

Esteves, P., Araújo, D., Vilar, L., Travassos, B., Davids, K. & Esteves, C. (2015) Angular relationships regulate coordination tendencies of performers in attacker–defender dyads in team sports. *Human Movement Science, 40,* 264–272.

Gauriot, R. & Page, L. (2015) I Take Care of My Own: A Field Study on How Leadership Handles Conflict between Individual and Collective Incentives. *American Economic Review: Papers and Proceedings, 105,* 414–419.

Gibson, J.J. (1979) *The ecological approach to visual perception.* Boston: Houghton Mifflin.

Glazier, P.S. (2010) Game, set and match? Substantive issues and future directions in performance analysis. *Sports Medicine, 40,* 625–634.

Headrick, J., Davids, K., Renshaw, I., Araújo, D., Passos, P. & Fernandes, O. (2011) Proximity-to-goal as a constraint on patterns of behaviour in attacker-defender dyads in team games. *Journal of Sport Sciences, 30*(3), 247–253.

Hristovski, R., Davids, K. & Araújo, D. (2006) Affordance-controlled bifurcations of action patterns in martial arts. *Nonlinear Dynamics, Psychology, and Life Sciences, 10*(4), 409–44.

Hughes, M. & Franks, I. (2005) Analysis of passing sequences, shots and goals in soccer. *Journal of Sport Sciences, 23*(5), 509–514.

Hughes, M. & Franks, I. (2008). *The essentials of performance analysis*. Routledge: New York.

Kelso, J.A.S. (2012) Multistability and metastability: Understanding dynamic ocordination in the brain. *Philosophical transactions of the Royal Society of London, Series B: Biological Sciences, 376*, 906–918.

Liebermann, D.G., Katz, L., Hughes, M.D., Bartlett, R.M., McClements, J. & Franks, I.M. (2002) Advances in the application of information technology to sport performance. *Journal of Sports Sciences, 20*(10), 755–769.

McGarry, T. (2009). Applied and theoretical perspectives of performance analysis in sport: Scientific issues and challenges. *International Journal of Performance Analysis in Sport, 9*(1), 128–140.

McGarry, T., Anderson, D.I., Wallace, S.A., Hughes, M.D. & Franks, I.M. (2002) Sport competition as a dynamical self-organizing system. *Journal of Sports Sciences, 20*(10), 771–781.

Nevill, A., Atkinson, G. & Hughes, M. (2008) Twenty-five years of sport performance research in the Journal of Sports Sciences. *Journal of Sports Sciences, 26*(4), 413.

Palut, Y. & Zanone, P.G. (2005) A dynamical analysis of tennis: Concepts and data. *Journal of Sports Sciences, 23*(10), 1021–1032.

Passos, P., Araújo, D., Davids, K., Gouveia, L., Milho, J. & Serpa, S. (2008) Information-governing dynamics of attacker-defender interactions in youth rugby union. *Journal of Sports Sciences, 26*(13), 1421–1429.

Passos, P. & Davids, K. (2015) Learning design to facilitate interactive nehaviours in team sports. *RICYDE (Revista Internacional de Ciencias del Deporte), 39*(11), 18–32.

Passos, P., Milho, J., Fonseca, S., Borges, J., Araújo, D. & Davids, K. (2011) Interpersonal Distance Regulates Functional Grouping Tendencies of Agents in Team Sports. *Journal of Motor Behavior, 43*(2), 155–163.

Pinder, R.A., Davids, K. & Renshaw, I. (2012) Metastability and emergent performance of dynamic interceptive actions. *Journal of Science and Medicine in Sport, 15*(5), 437–443.

Reed, D. & Hughes, M. (2006) An Exploration of Team Sport as a Dynamical System. *International Journal of Performance Analysis in Sport, 6*(2), 114–125.

Rietveld, E. & Kiverstein, J. (2014) A rich landscape of affordances. *Ecological Psychology, 26*, 325352.

Travassos, B., Araújo, D., Correia, V., & Esteves, P. (2010). Eco-Dynamics Approach to the study of Team Sports Performance. *The Open Sports Sciences Journal, 3*, 56–57.

Travassos, B., Araújo, D., Davids, K., Vilar, L., Esteves, P. & Correia, V. (2012) Informational constraints shape emergent functional behaviors during performance of interceptive actions in team sports. *Psychology of Sport & Exercise, 13*(2), 216–223.

Travassos, B., Davids, K., Araújo, D. & Esteves, P. (2013) Performance analysis in team sports: Advances from an Ecological Dynamics approach. *International Journal of Performance Analysis in Sport, 13*, 83–95.

Travassos, B., Araújo, D., Vilar, L. & McGarry, T. (2011) Interpersonal coordination and ball dynamics in futsal (indoor football). *Human Movement Science, 30*, 1245–1259. http://dx.doi.org/10.1016/j.humov.2011.04.003

Travassos, B., Araújo, D., Vilar, L. & McGarry, T. (2011) Interpersonal coordination and ball dynamics in futsal (indoor football). *Human Movement Science, 30*, 1245–1259.

Varlet, M. & Richardson, M. (2015). What would be Usain Bolt's 100-meter sprint world record without Tyson Gay? Unintentional interpersonal synchronization between the two sprinters. *Journal of Experimental Psychology: Human Perception and Performance*, *41*(1), 36–41. doi: 10.1037/a0038640

Vilar, L., Araújo, D., Davids, K. & Button, C. (2012) The role of Ecological Dynamics in analysing performance in team sports. *Sports Medicine*, *42*, 1–10.

Vilar, L., Araújo, D., Davids, K. & Bar-Yam, Y. (2013) Science of winning soccer: emergent pattern-forming dynamics in Association Football. *Journal of Systems Science and Complexity*, *26*, 73–84.

Vilar, L., Araújo, D., Davids, K., Travassos, B., Duarte, R. & Parreira, J. (2014) Interpersonal coordination tendencies supporting the creation/prevention of goal scoring opportunities in futsal. *European Journal of Sport Science*, *14*, 28–35.

Warren, W.H. (2006) The dynamics of perception and action. *Psychological Review*, *113*(2), 358–389.

Williams, S. & Manley, A. (2014) Elite coaching and the technocratic engineer: Thanking the boys at Microsoft! *Sport, Education and Society*. doi: 10.1080/13573322. 2014.958816

Index

Taylor & Francis eBooks

Helping you to choose the right eBooks for your Library

Add Routledge titles to your library's digital collection today. Taylor and Francis ebooks contains over 50,000 titles in the Humanities, Social Sciences, Behavioural Sciences, Built Environment and Law.

Choose from a range of subject packages or create your own!

Benefits for you

» Free MARC records
» COUNTER-compliant usage statistics
» Flexible purchase and pricing options
» All titles DRM-free.

Benefits for your user

» Off-site, anytime access via Athens or referring URL
» Print or copy pages or chapters
» Full content search
» Bookmark, highlight and annotate text
» Access to thousands of pages of quality research at the click of a button.

REQUEST YOUR **FREE** INSTITUTIONAL TRIAL TODAY

Free Trials Available
We offer free trials to qualifying academic, corporate and government customers.

eCollections – Choose from over 30 subject eCollections, including:

Archaeology	Language Learning
Architecture	Law
Asian Studies	Literature
Business & Management	Media & Communication
Classical Studies	Middle East Studies
Construction	Music
Creative & Media Arts	Philosophy
Criminology & Criminal Justice	Planning
Economics	Politics
Education	Psychology & Mental Health
Energy	Religion
Engineering	Security
English Language & Linguistics	Social Work
Environment & Sustainability	Sociology
Geography	Sport
Health Studies	Theatre & Performance
History	Tourism, Hospitality & Events

For more information, pricing enquiries or to order a free trial, please contact your local sales team:
www.tandfebooks.com/page/sales

 **Routledge**
Taylor & Francis Group

The home of
Routledge books

www.tandfebooks.com